Computational Methods for General Sparse Matrices

T0332544

Mathematics and Its Applications

Volume 65

Computational Methods
for
General Sparse Matrices

by

Zahari Zlatev

Ministry of the Environment,
National Environmental Research Institute,
Roskilde, Denmark

KLUWER ACADEMIC PUBLISHERS
DORDRECHT / BOSTON / LONDON

Library of Congress Cataloging-in-Publication Data

```
Zlatev, Zahari, 1939-
   Computational methods for general sparse matrices / by Zahari
Zlatev.
     p.   cm. -- (Mathematics and its applications (Kluwer Academic
Publishers) ; v. 65)
   Includes index.
   ISBN 0-7923-1154-X (acid-free-paper)
   1. Sparse matrices--Data processing.   I. Title.  II. Series.
QA188.Z49  1991
512.9'434--dc20                                           91-23219
                                                              CIP
```

ISBN 0-7923-1154-X

Published by Kluwer Academic Publishers,
P.O. Box 17, 3300 AA Dordrecht, The Netherlands.

Kluwer Academic Publishers incorporates
the publishing programmes of
D. Reidel, Martinus Nijhoff, Dr W. Junk and MTP Press.

Sold and distributed in the U.S.A. and Canada
by Kluwer Academic Publishers,
101 Philip Drive, Norwell, MA 02061, U.S.A.

In all other countries, sold and distributed
by Kluwer Academic Publishers Group,
P.O. Box 322, 3300 AH Dordrecht, The Netherlands.

Printed in the Netherlands

SERIES EDITOR'S PREFACE

'Et moi, ..., si j'avait su comment en revenir,
je n'y serais point allé.'

Jules Verne

The series is divergent; therefore we may be
able to do something with it.

O. Heaviside

One service mathematics has rendered the
human race. It has put common sense back
where it belongs, on the topmost shelf next
to the dusty canister labelled 'discarded non-
sense'.

Eric T. Bell

Mathematics is a tool for thought. A highly necessary tool in a world where both feedback and non-linearities abound. Similarly, all kinds of parts of mathematics serve as tools for other parts and for other sciences.

Applying a simple rewriting rule to the quote on the right above one finds such statements as: 'One service topology has rendered mathematical physics ...'; 'One service logic has rendered computer science ...'; 'One service category theory has rendered mathematics ...'. All arguably true. And all statements obtainable this way form part of the raison d'être of this series.

This series, *Mathematics and Its Applications*, started in 1977. Now that over one hundred volumes have appeared it seems opportune to reexamine its scope. At the time I wrote

"Growing specialization and diversification have brought a host of monographs and textbooks on increasingly specialized topics. However, the 'tree' of knowledge of mathematics and related fields does not grow only by putting forth new branches. It also happens, quite often in fact, that branches which were thought to be completely disparate are suddenly seen to be related. Further, the kind and level of sophistication of mathematics applied in various sciences has changed drastically in recent years: measure theory is used (non-trivially) in regional and theoretical economics; algebraic geometry interacts with physics; the Minkowsky lemma, coding theory and the structure of water meet one another in packing and covering theory; quantum fields, crystal defects and mathematical programming profit from homotopy theory; Lie algebras are relevant to filtering; and prediction and electrical engineering can use Stein spaces. And in addition to this there are such new emerging subdisciplines as 'experimental mathematics', 'CFD', 'completely integrable systems', 'chaos, synergetics and large-scale order', which are almost impossible to fit into the existing classification schemes. They draw upon widely different sections of mathematics."

By and large, all this still applies today. It is still true that at first sight mathematics seems rather fragmented and that to find, see, and exploit the deeper underlying interrelations more effort is needed and so are books that can help mathematicians and scientists do so. Accordingly MIA will continue to try to make such books available.

If anything, the description I gave in 1977 is now an understatement. To the examples of interaction areas one should add string theory where Riemann surfaces, algebraic geometry, modular functions, knots, quantum field theory, Kac-Moody algebras, monstrous moonshine (and more) all come together. And to the examples of things which can be usefully applied let me add the topic 'finite geometry'; a combination of words which sounds like it might not even exist, let alone be applicable. And yet it is being applied: to statistics via designs, to radar/sonar detection arrays (via finite projective planes), and to bus connections of VLSI chips (via difference sets). There seems to be no part of (so-called pure) mathematics that is not in immediate danger of being applied. And, accordingly, the applied mathematician needs to be aware of much more. Besides analysis and numerics, the traditional workhorses, he may need all kinds of combinatorics, algebra, probability, and so on.

In addition, the applied scientist needs to cope increasingly with the nonlinear world and the extra mathematical sophistication that this requires. For that is where the rewards are. Linear models are honest and a bit sad and depressing: proportional efforts and results. It is in the non-linear world that infinitesimal inputs may result in macroscopic outputs (or vice versa). To appreciate what I am hinting at: if electronics were linear we would have no fun with transistors and computers; we would have no TV; in fact you would not be reading these lines.

There is also no safety in ignoring such outlandish things as nonstandard analysis, superspace and anticommuting integration, p-adic and ultrametric space. All three have applications in both electrical engineering and physics. Once, complex numbers were equally outlandish, but they frequently proved the shortest path between 'real' results. Similarly, the first two topics named have already provided a number of 'wormhole' paths. There is no telling where all this is leading - fortunately.

Thus the original scope of the series, which for various (sound) reasons now comprises five subseries: white (Japan), yellow (China), red (USSR), blue (Eastern Europe), and green (everything else), still applies. It has been enlarged a bit to include books treating of the tools from one subdiscipline which are used in others. Thus the series still aims at books dealing with:

- a central concept which plays an important role in several different mathematical and/or scientific specialization areas;
- new applications of the results and ideas from one area of scientific endeavour into another;
- influences which the results, problems and concepts of one field of enquiry have, and have had, on the development of another.

Mathematics is like Janus. It is about understanding structures, which helps to calculate them; it is also about calculating structures in order to understand them. By and large we understand mathematics quite well. But the structures one might meet in practice can get quite large (size $10^5 \times 10^5$ for instance) and calculating with these becomes both a computational and theoretical challenge. Fortunately, many of these large matrices arising from applications tend to have special properties. For instance they can be *sparse*, which, roughly, means 'lots of zeros'. And the theoretical and computational challenge becomes how to make effective use of that property.

There are several ways of exploiting sparsity. This book, by a world expert, surveys them and compares them and doing so it discusses quite a few topics which are not included (to date) in other books on sparse matrices; for instance complexity, parallelism, orthogonalization, and condition number aspects.

As almost always there is no best way of exploiting sparseness and no single algorithm is satisfactory in all cases. Thus one needs guidance in using computer packages and devising one's own algorithms. This book will provide that.

The shortest path between two truths in the real domain passes through the complex domain.

J. Hadamard

La physique ne nous donne pas seulement l'occasion de résoudre des problèmes ... elle nous fait pressentir la solution.

H. Poincaré

Never lend books, for no one ever returns them; the only books I have in my library are books that other folk have lent me.

Anatole France

The function of an expert is not to be more right than other people, but to be wrong for more sophisticated reasons.

David Butler

Amsterdam, August 1991   Michiel Hazewinkel

C O N T E N T S

P R E F A C E

Matrices that contain many zero elements are called **sparse**. If the matrix is sparse then it is often possible to exploit this property so as to reduce

(i) the storage

and/or

(ii) the computing time

needed when basic linear algebra operations are carried out. Various sparse matrix techniques can be applied in the effort to achieve these two aims. Sparse matrix techniques that are designed for **general** matrices are the main topic of this book.

The attempt to reduce the storage and/or the computing time in the computations involving sparse matrices is often carried out by avoiding both the storage of zero elements and the computations with zero elements. In many situations this approach, which will be called **the classical manner of exploiting sparsity**, is very efficient. However, in some cases many fill-ins are created during the performance of certain linear algebra operations when the classical manner of exploiting sparsity is used. A fill-in is a new non-zero element, created at a position where the original matrix contains a zero element. Experience indicates that many fill-ins are often small in some sense. Therefore it may sometimes be worthwhile to apply a second approach: check for small elements and remove these from the arrays where the non-zero elements are kept (or, in other words, to consider these elements as zero elements during the subsequent computation). Some of the small elements produced during the calculations may be caused by rounding errors. Assume that an element will be a zero element if the computations are performed by exact arithmetic. The computer will as a rule produce a non-zero element (instead of a zero element) in this situation because the computations are performed with rounding errors. Thus, some non-zeros appear only because the computer is in general not able to treat correctly the cases where zero elements will be produced by exact arithmetic. However, such non-zero elements are small elements and they will normally be removed when the second approach is applied. This is the reason for calling the second approach **the computer-oriented manner of exploiting sparsity**. Both approaches are studied in detail in this book and many numerical results achieved in comparing of the two approaches are presented.

Assume that the problem $x=A^{\dagger}b$ is solved by a direct method ($A \in R^{mxn}$, $b \in R^{mx1}$, $x \in R^{nx1}$, $m \geq n$, $rank(A)=n$). Then the computer-oriented manner can be used as an attempt to detect cancellations (i.e. to detect the situations where the modified elements would be zero elements if exact arithmetic were used). One should remove only very small elements in such an application. However, the computer-oriented manner of exploiting sparsity is normally more efficient with regard to storage and computing time when it is allowed to remove also elements that are not very small. In the latter situation both the storage and the computing time may be reduced considerably, but, on the other hand, the solution computed in this way may be inaccurate. Therefore it is necessary to try to **improve** the solution **iteratively**. The simplest way to do this is to apply straightforward iterative refinement, which is studied in detail in Chapters 3, 5, 8 and 15. However, other iterative methods can also be applied. Conjugate gradient-type methods are discussed in Chapters 11 and 16.

The computer-oriented manner of exploiting sparsity (together with some procedure for iterative improvement) is **not** always better than the first approach (where no fill-ins are discarded and the solution obtained by the direct method is accepted without any attempt to improve it iteratively). However, many experiments indicate that when the first approach performs better than the second one then the difference is not great. On the other hand, some very large scientific and engineering problems can be handled numerically only when the second approach is applied. An example, a large-scale problem arising in nuclear magnetic resonance spectroscopy, is given as an illustration of this statement in Chapter 8. Even on large modern computers it is extremely expensive, in terms of both storage and computing time, to treat this problem numerically when the first approach is used, while the second approach performs quite satisfactorily. **The conclusion is that in a good package for sparse matrices it is desirable to have options in which the first approach is applied, but there must also be options in which the second approach is applied and the solution is improved iteratively.**

The second approach, the computer-oriented manner of exploiting sparsity, may be preferred even in the case where the classical manner of exploiting sparsity performs better with regard to storage and computing time. This is so because the accuracy that can be achieved by the second approach is normally higher than that achieved by the first approach and, what is even more important, a reliable error estimate can be obtained when the second approach is used (no error estimate can be found in an easy way when the first approach is chosen).

Consider a system of linear algebraic equations Ax=b (which can be obtained from the more general problem mentioned above by setting m=n). Assume that Gaussian elimination (GE) is used. Then it is popular to use an **ILU** (incomplete LU) factorization as a preconditioner and apply an iterative method. The computer-oriented manner of exploiting sparsity can be considered as an **ILU** factorization. The name ILU factorization is not used in this book, however, in order to avoid misunderstanding. Indeed, all versions of the **ILU** factorization proposed in the literature are in fact applicable to only very special matrices. This is so because by definition any **ILU** factorization is performed under the assumption that at least two conditions is satisfied:

(i) the GE process is carried out without pivoting,

(ii) a sparsity pattern is prescribed before the beginning
     of the computations and all newly created non-zero
     elements that do not fit into this pattern are removed
     (i.e. non-zero elements are removed not because they
     are small, but because there is no place for them in
     the pattern prescribed in advance).

Consider the first condition. It is well known that the completion of the GE process without pivoting can be guaranteed when the matrix is either symmetric and positive definite or diagonally dominant. Assume that the matrix to be factorized belongs to the first of these two classes. Assume also that the Cholesky factorization is used. Then some extra requirements must be imposed in order to guarantee the completion of the factorization when the sparsity pattern is prescribed in advance and all elements that do not fit into it are removed (as, for example, the requirement that the matrix is an **M-matrix**, i.e. $a_{ij} \leq 0$ for $i \neq j$ and the entries of the inverse of A are positive). Thus, if the completion of the ILU factorization is to be guaranteed, then the second requirement is in some cases more restrictive than the first one. If a sufficiently fast convergence rate is desired, then some additional requirements may be needed.

Any pivoting can be applied when the computer-oriented manner is used. However, there is no guarantee that the GE process will be completed when small elements are dropped. Even if the GE process is completed, there is no guarantee that the iterative method chosen will converge. Thus, in this regard the situation is similar to the case where the ILU factorization is applied to general matrices, but it is not the same. The main difference is that if the computer-oriented manner is used, then an action can be taken both when the GE process is not completed and when the iterative method does not converge: the criterion by which one decides whether an element is small or not should be made more stringent (this remedy is discussed in detail in Chapter 8, Chapter 11 and Chapter 16). It is not clear what could be done when a similar situation occurs when an ILU factorization is used.

This book can be divided into three parts. In **the first part**, consisting of the first three chapters, some common principles concerning the exploitation of sparsity are discussed. In **the second part**, Chapter 4 - Chapter 11, exploiting sparsity in the solution of linear algebraic equations is studied. It is assumed the GE process is used. Some related problems (solving linear least squares problems by augmentation, calculating approximate solutions of systems of ordinary differential equations, and estimating condition numbers of sparse matrices) are also studied in the second part. Finally, some parallel methods for sparse matrices are described and tested in Chapter 10 and Chapter 11 of the second part. **The third part**, from Chapter 12 to Chapter 16, is devoted to the orthogonalization methods for sparse matrices and their use in the numerical solution of large linear least squares problems.

This book discusses several topics that are not discussed in any of the other books on sparse matrices to date:

(i) **Parallel methods (Chapter 10 and Chapter 11),**

(ii) **Orthogonalization methods (Chapter 12-Chapter 16),**

(iii) **A detailed description of an application (Chapter 8) to illustrate how much the complexity of the problem solved is increased when operations with the sparse matrices are only a part of the computational process,**

(iv) **Condition number estimators for sparse matrices (Chapter 9),**

(v) **A general scheme for the direct methods that allows one to study the common properties of the direct methods and to define a common approach for treatment of problems that involve computations with sparse matrices (Chapter 3).**

The other books on sparse matrices discuss either the direct methods or the iterative methods (with special preconditioners as, for example, preconditioners based on the use of different versions of ILU factorization). This book attempts to build a bridge between the direct methods and the preconditioned iterative methods, with preconditioners found by using the natural assumption that it is justified to remove elements only if these are small in some sense (and not because these do not fit into a sparsity pattern prescribed before the calculation of the elements that are candidates for removing). No method is best in all cases. There are classes of matrices for which the direct methods perform better (typically matrices that produce only a few fill-ins). There are other classes of matrices for which the preconditioned iterative methods are the better choice (typically matrices that produce many of fill-ins).

There are many discussions concerning the choice of the best iterative method. In many cases the choice of an iterative method is not the most important issue. More important is to try adaptively to improve the quality of the preconditioner so that **the preconditioned matrix becomes suitable for the iterative method used** (Chapter 11 and Chapter 16). The adaptive improvement of the preconditioner during the computational process (by the code itself) is especially efficient when a long sequence of matrices is to be handled (Chapter 8).

It is perhaps in place to mention here that I was encouraged to write a book on sparse matrices, with emphasis on preconditioning by removing small non-zero elements, by the late **Richard Bellman**. He read my first paper on this subject ("Use of iterative refinement in the solution of sparse linear systems", SIAM. J. Numer. Anal., 19(1982), 381-399) and wrote a letter to me in 1983 in which he suggested that I should extend these results and describe them in book form. Various matters delayed the book, but I never forgot the kind letter from Richard Bellman and finally finished the project. The first title of the book was: "Iterative improvement of direct solutions of large and sparse problems", the choice of which was influenced by the suggestion made by Richard Bellman. However, one of the referees of this book noted that in fact the book contains much more than what the title suggests. Looking once again at the contents of the different chapters, I realized that the remark made by the referee was correct and decided to use the present title: "Computer methods for general sparse matrices".

It is assumed that the reader is familiar with only a few fundamental concepts from numerical linear algebra. He should know, for example, the definition of a matrix, the different types of matrices (symmetric, positive definite, orthogonal, triangular, etc.) and something about operations involving matrices (together with vector norms and subordinate matrix norms). This material is normally contained in the first courses of higher mathematics. Therefore the book might be used by students from both natural sciences departments and engineering departments.

The different chapters of this book, and even some sections, can be studied independently. Moreover, short cross-references are in general avoided in the book. That is if a few formulae or several sentences from an earlier chapter are needed then they **are** repeated. Further, the tables are supplied with as many explanations as possible so as to allow the reader to understand their contents without searching for the page in the text where the contents of the table under consideration are explained. Therefore the book might be used as a reference book.

A C K N O W L E D G E M E N T S

Some of the results in this book were obtained in cooperation with colleagues: **V. A. Barker, H. B. Nielsen** and **P. G. Thomsen** from the Institute for Numerical Analysis at the Technical University of Denmark (Lyngby, Denmark), **K. Gallivan** and **A. Sameh** from the Center for Supercomputing Research and Development at the University of Illinois at Urbana-Champaign (Ilinois, USA), **J. Moth** and **J. Wasniewski** from the Computing Centre for Research and Education (UNI-C, Lyngby, Denmark), **K. Schaumburg** from the Department of Physical Chemistry at the University of Copenhagen, **Ph. Vu** from the Cray Research, Inc. (Houston, USA), **O. Østerby** from the Department of Computer Science at Aarhus University (Århus, Denmark). I should like to thank all of them very much for the stimulating discussions.

V. A. Barker (Technical University of Denmark, Lyngby, Denmark), **T. Davis** (University of Florida, Gainesville, USA) and **G. Fairweather** (University of Kentucky, Lexington, USA) read chapters of this book. They corrected my language and made many suggestions for improvements. I should like to thank them very much.

I should like to thank **all members** of the Institute for Numerical Analysis at the Technical University of Denmark (where I worked in the period 1974-1978) for helping me when I came to Denmark as well as when I returned to my work in numerical analysis after a pause of five years. I should like to thank very much **Prof. Emer. Th. Busk** (the Head of the Institute for Numerical Analysis, when I was there) for giving me the possibility of working at the Institute.

I should like to thank very much **L. P. Prahm** (now Director of the Danish Meteorological Institute), who introduced me to the field of large air pollution models. Also thanks to **H. Flyger** and **Chr. Lohse** for giving me excellent working conditions at the Division of Emissions and Air Pollution at the Danish Environmental Research Institute. Many thanks also to **all my colleagues** from the Division of Emissions and Air Pollution, especially to the scientists from the former Section of Meteorology.

During the work on this project I received several grants from the **Danish Natural Science Research Council** and the **Danish Technical Science Research Council**. I should like to thank both Councils very much for supporting my research.

Finally, I should like to thank very much my wife **Ida Zlateva** and our children **Tanja** and **Anna Irina** for supporting me the whole time and for accepting my time away from them (also at home) during the many years I worked on this project.

NOTATION AND ABBREVIATIONS

Single capital Latin letters are normally used for **the matrices**. If, say, A, B and C are matrices, then their elements in the i'th row and the j'th column are denoted by a_{ij}, b_{ij} and c_{ij}, respectively.

Vectors are normally denoted by single lower case Latin letters. If, say, a, b and c are vectors, then their i'th componets are denoted by a_i, b_i and c_i, repectively. However, it should be noted that indices are also used to denote the i'th vector in a sequence of vectors (as, for example, the successive approximations obtained during the application of an iterative method). It should be clear from the context if the index refers to the i'th component of a vector or to the i'th vector in a sequence of vectors.

The following notation is used in this book for **the arrays** used to store non-zero elements of sparse matrices or information about the non-zero elements (row numbers, column numbers, pointers, etc.):

AORIG -	the non-zero elements of the original matrix are stored in this array.
RNORIG -	the row numbers of the non-zero elements of the original matrix are stored in this array.
CNORIG -	the column numbers of the non-zero elements of the original matrix are stored in this array.
ALU -	the non-zero elements in the factors L and U obtained during Gaussian elimination are stored in this array.
RNLU -	the row numbers of the non-zero elements in the factors L and U obtained during Gaussian elimination are stored in this array.
CNLU -	the column numbers of the non-zero elements in the factors L and U during Gaussian elimination are stored in this array.
AQDR -	the non-zero elements in the upper triangular matrix R obtained during the orthogonal decomposition QDR (Q is a matrix with orthonormal columns, while D is a diagonal matrix) are stored in this array.
RNQDR -	the row numbers of the non-zero elements in matrix R (see above) are stored in this array.
CNQDR -	the column numbers of the non-zero elements in matrix R (see above) are stored in this array.
HA -	is an auxiliary array in which useful information about row starts and ends, column starts and ends, pivotal interchanges, etc. is stored.

Some important quantities used in connection with all sparse matrix techniques discussed in this book are:

T -	the drop-tolerance; a parameter used in the decision whether a non-zero element is small or not. All small elements together with their row and column numbers are removed from the arrays where these quantities are kept. No element is removed if $T \leq 0$.
NZ -	the number of all non-zero elements in the original matrix (before the beginning of the computations).
NZ1 -	the number of non-zero elements in the original matrix (before the beginning of the computations) that are not "small" according to the drop-tolerance chosen.
COUNT -	the maximal number of non-zero elements kept at any stage of the computational process in array ALU when Gaussian elimination is used or in array AQDR when an orthogonal decomposition is applied. At the end of the decomposition the number of non-zero elements in any of these two arrays may be less than COUNT when a positive value of the drop-tolerance T is specified.

The abbreviations used in this book are listed below:

BLAS - basic linear algebra subprograms.

BDF - backward differentiation methods.

CG - conjugate gradients.

CGS - conjugate gradients squared (a conjugate gradient-type method).

DIRK - diagonally implicit Runge-Kutta methods.

DMT - dense matrix technique.

DS - direct solution.

GE - Gaussian elimination.

GMFS - generalized minimun fill-in pivotal strategy.

GMRES - generalized minimum residual (a conjugate gradient-type method).

GMS - generalized Markowitz pivotal strategy.

IGMFS - improved generalized minimum fill-in pivotal strategy.

IGMS - improved generalized Markowitz pivotal strategy.

ILU   -   incomplete LU factorization.

IR   -   iterative refinement

MFLOPS - million floating point operations per second.

MDIRK - modified diagonally implicit Runge-Kutta methods.

ODE   -   ordinary differential equation.

PDE   -   partial differential equation.

SMT   -   sparse matrix technique.

SVD   -   singular value decomposition.

CHAPTER 1

EXPLOITING SPARSITY

Let $A \in \mathbb{R}^{m \times n}$ ($m \geq n$, $m \in \mathbb{N}$, $n \in \mathbb{N}$, $rank(A)=n$) be a **general** matrix, which means that A has neither a special property (such as symmetry or positive definiteness) nor a special structure (such as bandedness). Assume that many of the elements $a_{ij} \in A$ ($i=1,2,\ldots,m$, $j=1,2,\ldots,n$) are equal to zero. Then matrix A is called **sparse**. General sparse matrices occur in the numerical treatment of many engineering and scientific models. Both computing time and storage can be saved when sparsity is exploited. Moreover, many large problems can successfully be solved only when sparsity is exploited. Hence the relevance of the following question:

How can sparsity be exploited?

In the general case, where no particular property of the matrix considered can be exploited in an efficient way, there are two different (in principle) manners by which the sparsity can be treated:

(i) the classical manner

and

(ii) the computer-oriented manner.

> In the classical manner of exploiting sparsity one attempts (1) to store only non-zero elements in the computer memory and (2) to work only with non-zero elements during the computations.

It must be emphasized immediately that this definition is not very precise. For example, one can avoid the storage of non-zero elements that are equal to one, and this is commonly used when the unit triangilar factor L obtained by Gaussian elimination is stored. Even in dense codes the ones on the main diagonal are not stored (only the elements of L under the main diagonal are stored). However, the definition reflects the main principle used when the classical manner is specified and, therefore, it is quite sufficient in the case where the two manners of exploiting sparsity are compared.

It must also be emphasized that some non-zero elements may become zero elements during the computations. Such elements are normally kept when the classical manner of exploiting sparsity is in use. This means that one also works with some zero elements during the computations. Moreover, there are sparse codes where some zero elements, which will gradually become non-zero elements, are stored from the very beginning (and, of course, one works with these zero elements during the computations). Thus, the requirement to store only non-zero elements and to work only with non-zero elements is not always satisfied. In fact one merely attempts, as stated in the definition above, to store only non-zero elements and to work only with non-zero elements when the classical manner of exploiting sparsity is used.

> In the computer-oriented manner of exploiting
> sparsity one attempts, at some stages of the
> computational process at least, to avoid both
> the storage of "small" non-zero elements and
> the computations with such non-zero elements.

It is obvious that one must define an appropriate criterion to determine whether an element is small. In general it is not easy to define such a criterion. On the other side, some classes of very large problems can be handled on the computers available at present **only** when the computer-oriented manner is applied, even if the criterion for deciding whether a non-zero element is small is rather crude.

The influence of pure mathematics on the classical manner of exploiting the sparsity is fairly obvious: one assumes that the elements of the matrix (especially the zero elements) can be calculated by **exact arithmetic** and, moreover, that all transformations during the solution process that lead to zero elements can be performed **exactly**. On computers this assumption does not hold and, normally, non-zero elements will be produced where the exact computation produces zeros. Therefore, it is illustrative to demonstrate, by examples, the influence of pure mathematics on the classical manner of exploiting sparsity.

Example 1.1. Consider the solution of a system of linear algebraic equations:

$$(1.1) \quad Ax=b \quad (A \in R^{n \times n}, \quad b \in R^{n \times 1}, \quad x \in R^{n \times 1}, \quad rank(A)=n)$$

by Gaussian elimination (GE), where the coefficient matrix A is decomposed into two triangular matrices $L \in R^{n \times n}$ and $U \in R^{n \times n}$. The transformations carried out during the GE process are based on the formula:

$$(1.2) \quad a_{ij}^{(s+1)} = a_{ij}^{(s)} - a_{is}^{(s)} (a_{ss}^{(s)})^{-1} a_{sj}^{(s)}, \quad s=1,2,\ldots n-1, \quad i,j=s+1,s+2,\ldots,n,$$

$$a_{ss}^{(s)} \neq 0, \quad a_{ij}^{(1)} = a_{ij} \in A, \quad i,j=1,2,\ldots n.$$

The elements of the unit lower triangular matrix L are given by

$$(1.3) \quad l_{ij} = -a_{ij}^{(j)}/a_{jj}^{(j)}, \quad j=1,2,\ldots,n-1, \quad i=j+1,j+2,\ldots,n,$$

$$(1.4) \quad l_{ii} = 1, \quad i=1,2,\ldots n,$$

$$(1.5) \quad l_{ij} = 0 \quad j=2,3,\ldots,n, \quad i=1,2,\ldots,j-1.$$

The elements of the upper triangular matrix U are given by

$$(1.6) \quad u_{ij} = a_{ij}^{(i)}, \quad i=1,2,\ldots,n, \quad j=i,i+1,\ldots,n,$$

$$(1.7) \quad u_{ij} = 0, \quad i=2,3,\ldots,n, \quad j=1,2,\ldots,i-1.$$

It is clear from (1.2) that $a_{ij}^{(s+1)}$ may be a non-zero element even if $a_{ij}^{(s)}$ is equal to zero. This is so when both $a_{is}^{(s)}$ and $a_{sj}^{(s)}$ are non-zeros. If $a_{ij}^{(s+1)} \neq 0$ while $a_{ij}^{(s)} = 0$, then a new non-zero element, called **a fill-in**, is created in position (i,j) of matrix A.

If the calculations are performed in exact arithmetic, then the product LU will be equal to the original matrix A. However, the calculations are, in general, performed with rounding errors, and therefore

$$(1.8) \quad LU = A + E, \quad E \in \mathbf{R}^{n \times n},$$

is satisfied, where E is called a perturbation matrix. It is well-known, [99,100,133,231,236,259,270-277], that the straightforward application of the above formulae **may** lead to perturbation matrices E with large elements e_{ij} ($i=1,2,\ldots,n$, $j=1,2,\ldots,n$). Pivoting is normally used in an attempt to keep the elements of E small. If matrix A is sparse and if sparsity is exploited, then the following fact is (or at least may be for some matrices) as important as the attempt to keep the elements of E small. Consider (1.2). It is apparent that, when a computer arithmetic is in use, the calculated value of $a_{ij}^{(s+1)}$ may be a non-zero element even if the exact value of $a_{ij}^{(s+1)}$ is equal to zero (or, in other words, fill-ins may be produced only because the calculations are carried out with rounding errors). This means that **if the classical manner of exploiting sparsity is applied, then one could obtain factors L and U which are denser than the corresponding factors that would be obtained if the calculations are carried out in exact arithmetic.** The reason for this peculiar behaviour is the attempt to apply a principle from pure mathematics (the attempt to avoid the storage of and the work with **exact** zero elements) in circumstances that are not entirely appropriate for such a principle (in the presence of rounding errors in the calculated results when computer arithmetic is used). ∎

The above example shows that elements, which should not be considered if the sparsity is to be exploited, have sometimes both to be stored in the computer memory and to be used in the calculations when the classical manner of exploiting sparsity is applied. The following examples show how one could attempt to avoid such a situation.

Example 1.2. Consider the previous example. It is obvious that if the exact value of $a_{ij}^{(s+1)}$ is zero, while its calculated value is different from zero, then the calculated value will be small in some sense. Therefore, it may be worthwhile to set $a_{ij}^{(s+1)} = 0$ when it becomes small according to some well-defined criterion. This means that small elements are effectively removed from the computer memory and, moreover, not involved in the subsequent computations. Such elements are said to be **dropped** during the GE. The simplest way to implement the idea of dropping in practice can be described as follows. Let T be a non-negative number. Then an element obtained during the GE is considered as small and dropped if

$$(1.9) \quad |a_{ij}^{(s+1)}| \leq T \quad (s=1,2,\ldots,n-1, \ i=s+1,s+2,\ldots,n, \ j=s+1,s+2,\ldots,n).$$

The parameter T used in the above criterion is called the **drop-tolerance**. It is clear that the application of a positive drop-tolerance in the computational process may lead to calculating inaccurate factors L and U. If this happens, then the approximation

(1.10) $x_1 = (LU)^{-1}b$

of the exact solution $x = A^{-1}b$ of   (1.1)   will normally also be inaccurate.
Therefore it is necessary to attempt to regain the accuracy lost by dropping
small elements. This could be done by applying **iterative refinement**:

(1.11) $r_i = b - Ax_i$, $d_i = (LU)^{-1}r_i$, $x_{i+1} = x_i + d_i$, $i = 1, 2, \ldots, p-1$,

starting with some approximate solution x_1 that is either calculated by
(1.10)   or in some other way and using appropriate stopping criteria.

If the drop-tolerance   T   is sufficiently large, then the calculated
factors   L   and   U   will in general be sparser, and sometimes much
sparser, than the corresponding factors that can be obtained by the use of the
classical manner (i.e. without dropping small elements). Many numerical
examples that illustrate this statement will be presented in this book. ∎

A demonstration of the practical use of the computer-oriented manner of
exploiting sparsity is given by **Example 1.2.** It is clear that the classical
manner of exploiting sparsity can be considered as a special case of the
computer-oriented manner (obtained by setting   T = 0   and by accepting the
first approximation x_1 as a final solution). If a sufficiently large drop-
tolerance T is used, then one **expects** to save both **storage** (because many
elements are dropped during the computations) and **computing time** (because the
reduction of the computing time for performing GE is normally greater than
the extra computing time spent to carry out the iterative refinement).
However, it is quite clear that savings will be achieved only when the two
important conditions are satisfied:

> (i) **many small elements are dropped during the Gaussian**
>   **elimination process,**

> (ii) **the iterative refinement (1.11) is convergent and**
>   **the rate of convergence is sufficiently fast.**

If these two conditions are not satisfied, then the use of the classical
manner is more profitable. This is demonstrated by the example given below.

Example 1.3. Let NZ be the number of non-zero elements in matrix A.
Let COUNT be the maximal number of non-zero elements apearing at any stage
s (s = 1, 2, \ldots, n-1) of the GE.

If the classical manner of exploiting the sparsity is used, then COUNT
is the number of non-zero elements in the factors L and U (without the
diagonal elements of L), and the total number of fill-ins, NFILL, is equal
to COUNT - NZ. If some zero elements are produced during the computations and
if such elements are kept in the array where the non-zero elements are stored,
then the number of these elements is added to COUNT.   Thus, strictly
speaking, COUNT may be greater than the number of non-zeros in the factors
L and U. However, the important fact is that some zeros are added to COUNT
only when the code treats them in the same way as it treats the non-zero
elements.

If the computer-oriented manner of exploiting sparsity is used, then the total number of non-zero elements in exact factors L and U from A=LU, again without the diagonal elements in L, is in general greater than COUNT (because of the dropping of small elements).

Both when the classical manner is used and when the computer-oriented manner is applied COUNT is a good measure for the storage used; this will be discussed in more detail in the following chapters.

The first matrix used in this experiment, see **Table 1.1**, is taken from the Harwell set of test-matrices, [76] (now a much larger set, the set of the Harwell-Boeing sparse test-matrices, [71,72], is available). It is clearly seen that even when T = 0 (i.e. when the classical manner of exploiting sparsity is applied) **no fill-in appears**. This is not a surprise, because a careful examination shows that the matrix   A = SHL 0   from the Harwell set is a permutation of a triangular matrix. Nevertheless, this example is very illustrative for our purposes, because it shows that the use of a positive drop-tolerance is not efficient when the classical manner of exploiting the sparsity is performed with no fill-ins (or with a few fill-ins only).

Exploiting the sparsity	COUNT	COMPUTING TIME
Classical manner	1687	0.61
Computer-oriented manner	1687	1.11(3)

Table 1.1
Results obtained in the solution of the system Ax=b with A = SHL 0 from the Harwell set of test-matrices; SHL 0 is a matrix of order 663 with 1687 non-zero elements. The number of iterations, when the computer-oriented manner of exploiting the sparsity is chosen, is given in brackets. The drop-tolerance used in the latter case is T = 0.001.

Exploiting the sparsity	COUNT	COMPUTING TIME
Classical manner	105074	15.90
Computer-oriented manner	74462	30.23(114)

Table 1.2
Results obtained in the solution of the system Ax=b with A = E(3600,20) from [291]. The number of non-zero elements is NZ=17958. The number of iterations, when the computer-oriented manner of exploiting the sparsity is chosen, is given in brackets. The drop-tolerance used in the latter case is T = 0.001.

The second matrix used in this experiment, see **Table 1.2**, belongs to the matrices of class E(n,c). These matrices were introduced in [324] and used in   [291,295,301,311,331,341].   The number of non-zero elements in a

matrix $A = E(n,c)$ is $NZ = 5n-2c-2$, where n is the order of the matrix
and c is a parameter by which the location of certain non-zero elements can
be varied. More precisely, the non-zero elements of a matrix of this class are
given by

(1.12) $a_{ii} = 4$, $i=1,2,\ldots,n$,

(1.13) $a_{i,i+1} = a_{i+1,i} = -1$, $i=1,2,\ldots,n-1$,

(1.14) $a_{i,i+c} = a_{i+c,i} = -1$, $i=1,2,\ldots,n-c$,

where the following two conditions are satisfied:

(1.15) $n \geq 3$ $\wedge$ $2 \leq c \leq n-1$.

While the condition **(i)** is not satisfied for the matrix used to obtain
the results in **Table 1.1** (because no small elements are dropped during the
computations), condition **(ii)** is not satisfied for the matrix of class
$E(n,c)$ with n=3600 and c=20 used in **Table 1.2** (because the speed of
convergence is too slow). It is seen from **Table 1.2** that COUNT is reduced
considerably when the computer-oriented manner is applied, but this does not
lead to a reduction of the computing time, because 114 iterations are
performed. ∎

The results given in **Table 1.1** and in **Table 1.2** show that **for some
classes of matrices the classical manner of exploiting sparsity is more
efficient than the computer-oriented manner.** However, there are classes of
matrices for which the computer-oriented manner performs much better than the
classical manner. This is demonstrated by the example given below.

Example 1.4.   The matrix considered in this example appeared in the
numerical treatment of a model simulating different biological patterns,
(**Hunding, [152]**). The order of the matrix is n = 2000 and the number of its
non-zero elements is NZ = 14800.   The results, obtained by using the two
manners of exploiting sparsity are given in **Table 1.3**. It is seen that the
classical manner leads to rather dense factors L and U (COUNT ≈ 20*NZ ≈
141*n). This is the reason for the very large computing time spent when the
classical manner is applied. If the computer-oriented manner with T = 1.0
is applied, then both storage and computing time are reduced dramatically. The
storage needed is reduced by a factor of approximately 10. The computing time
is reduced by a factor of about 41. It should be mentioned here that the
biological model described in **[152]** consists of a system of time-dependent
partial differential equations involving first-order derivatives. After a
space discretization a system of non-linear ordinary differential equations
is obtained. The use of a quasi-Newton iterative procedure leads to the
solution of a long sequence of systems of linear algebraic equations. Thus,
the solution of systems Ax=b is a very large part of the total computational
work needed in the treatment of this model.

Exploiting the sparsity	COUNT	COMPUTING TIME
Classical manner	283189	470.73
Computer-oriented manner	28003	11.44(23)

Table 1.3
Results obtained in the solution of the system   Ax=b
which appears in the numerical treatment of a model
simulating biological patterns, [152]. The order of the
matrix is   n = 2000 and the number of its non-zero
elements is   NZ=14800.   The number of iterations used
when the computer-oriented manner is used is given in
brackets. The drop-tolerance applied used is T=1.0.

The results given in Table 1.3 indicate that the biological model can
easily be treated numerically if the computer-oriented manner of exploiting
sparsity is applied, while it is practically untractable numerically when the
classical manner is chosen. A similar situation appears when some very large
systems of linear ordinary differential equations arising in nuclear magnetic
resonance spectroscopy are to be solved ([333-337]). The spectroscopic models
will be discussed in detail in Chapter 8) ∎

Example 1.5. The previous example shows that the use of a large drop-
tolerance may lead to a very efficient solution process. Therefore, it is
advocated to use such a large drop-tolerance together with some kind of
iterative improvement (not necessarily the iterative refinement procedure used
in this chapter). However, for some problems even a very small drop-tolerance
can lead to a great reduction of the storage needed. This is illustrated in
Table 1.4, where the matrix steam2   from the Harwell-Boeing set of sparse
test-matrices, [71,72], is factorized with several small values of the drop-
tolerance T.

T	COUNT	ERROR
0.0	21145	6.7E-16
1.0E-12	8543	1.1E-11
1.0E-10	7614	2.0E-10
1.0E-08	5819	5.7E-08

Table 1.4
Results obtained for matrix   steam2 from the Harwell-
Boeing set of test-matrices with several small values of
the drop-tolerance   T.   The matrix is of order   600
with   5660   non-zero elements. The max-norm of the
absolute error of the solution vector is given under
"ERROR".

It is interesting to note that for this particular example the accuracy achieved was of the same order of magnitude as the drop-tolerance used (when the drop-tolerance is positive but sufficiently small). Of course, in general this will not be the case.

It is necessary to explain how the accuracy of the computed solution is evaluated. Vector b is calculated so that $x_i = 1$ for $i=1,2,\ldots,n$. Let x^* be the approximate solution calculated by the code. The max-norm of $x-x^*$, $\|x-x^*\|_\infty$, is used to evaluate the error in **Table 1.4** (i.e. an absolute accuracy check is applied there). Relative accuracy checks, $\|x-x^*\|_\infty/\|x\|_\infty$, will normally be used in the other chapters of the book. Other vector norms will be applied when appropriate.

This example shows that cancellations or, at least, situations where many very small elements are created do occur sometimes. Therefore it is not always justified to ignore the cases where cancellations take place and/or many very small non-zero elements are produced. ■

Two important conclusions can be drawn from the experiments described in this chapter:

(A)	**The classical manner of exploiting sparsity is useful for some classes of problems.**
(B)	**Some very large problems become tractable on the computers available at present only when the computer-oriented manner of exploiting sparsity is applied with an apprprioate criterion for dropping small non-zero elements.**

This means that both an option where the classical manner is implemented and an option where the computer-oriented manner of exploiting the sparsity is implemented should be available in a general purpose package of sparse subroutines. Therefore both manners will be studied in the following chapters. However, before the beginning of the detailed description of the implementation of these two manners in a sparse matrix software, it is necessary to explain how general sparse matrices can be stored in the computer memory. This will be done in **Chapter 2**.

Solving large systems of linear algebraic equations as well as solving large linear least squares problems by using the two manners of exploiting sparsity will be the main topic in this book. However, many of the ideas may also be applied when other problems involving calculations with general sparse matrices are to be treated. For example, many of the devices used in this book may be applied when large eigenvalue problems are to be handled by the methods described in [47,215,216,252,262,272,279].

CHAPTER 2

STORAGE SCHEMES

Consider a square or rectangular sparse matrix A. The first task that has to be solved by a developer of a sparse matrix code is to determine the way in which the **non-zero elements** of matrix A are to be stored in the computer memory. There are two main requirements to the storage scheme that is to be applied:

> **(a)** the storage scheme should be (from the user's point of view) as simple as possible,

> **(b)** it should be possible to perform efficiently different basic linear algebra operations (as, for example, matrix-vector multiplications, products of two matrices, Gaussian elimination, orthogonal transformations, etc.).

These two requirements work in opposite directions and, therefore, a compromise has to be obtained. Such a compromise can be achieved by using the following strategy:

> **(i)** *Find an input storage scheme which is as simple as possible from the user's point of view.*

> **(ii)** *If a basic linear algebra operation can be performed efficiently by the storage scheme chosen in (i), then apply the scheme to the basic linear algebra operation under consideration.*

> **(iii)** *If a basic linear algebra operation cannot be performed efficiently by the storage scheme chosen in (i), then develop a subroutine which reorders the elements of the matrix so that the basic linear algebra operation considered can efficiently be carried out using the new storage scheme.*

Any storage scheme based on the above strategy is a good compromise. It does not ensure that the problem solved is treated in the most efficient way. However, the work required from the user is minimized. This is so both because the input storage scheme is very simple and because the reordering process (when such a process is necessary) is automatically performed by the software and the user is not forced to think how to order in an optimal way the **non-zero elements** in connection with different basic linear algebra operations.

2.1. An input storage scheme

Let matrix A be sparse and let NZ be the number of elements in A. Assume that the elements of A are to be stored in a **REAL** array **AORIG** of length at least equal to NZ. Assume that two **INTEGER** arrays **RNORIG** and **CNORIG** of length at least equal to NZ are also available. The notation used in connection with the names of these arrays can be explained as follows:

(a) The letter A refers to the fact that the elements of matrix A are to be stored in array AORIG.

(b) RN is an abbreviation for "row numbers"; the row numbers of the elements of matrix A are to be stored in array RNORIG.

(c) CN is an abbreviation for "column numbers"; the column numbers of the elements of matrix A are to be stored in array CNORIG.

(d) ORIG stands for "original". The fact that the non-zero elements of the original matrix A, their row numbers and their column numbers are to be stored in arrays AORIG, RNORIG and CNORIG, respectively, is emphasized by the use of this abbreviation. If the computer-oriented manner of exploiting sparsity is used, then the contents of arrays AORIG, RNORIG and CNORIG may remain unchanged during the whole computational process (these arrays can be used in the iterative process). The original matrix is often not needed when the classical manner of exploiting sparsity is used. If this is so, then the contents of arrays AORIG, RNORIG and CNORIG may be modified during the computations. For example, if matrix A is square and if GE is used, then these arrays may be overwritten by the non-zero elements of the factors L and U together with information about their positions. Storage can be saved by such a procedure.

A simple input storage scheme based on the three arrays AORIG, RNORIG and CNORIG can be described as follows. The elements of matrix A are stored in the first NZ locations of array AORIG. **The order in which the non-zero elements are stored in array AORIG is arbitrary.** However, if an element a_{ij} of matrix A is stored in position K ($1 \leq K \leq NZ$) in array AORIG (or, in other words, if AORIG(K) = a_{ij}), then RNORIG(K) = i and CNORIG(K) = j must also be assigned. This is probably the simplest input scheme. **Only the non-zero elements of matrix A together with their row and column numbers are stored in arrays AORIG, RNORIG and CNORIG respectively. It is not necessary to order the non-zero elements in any special way; any order of the non-zero elements is acceptable.** Any of the other input storage schemes used in sparse matrix codes imposes some extra requirements which have to be satisfied by the user (as, for example, ordering the non-zero elements by rows and/or columns, counting the number of non-zero elements in each row and/or column, etc.). In many cases it is rather easy to satisfy such extra requirements. However, sometimes this may be a very difficult task. For example, if the elements of matrix A are generated by a complicated numerical algorithm, then it may be desirable to store element a_{ij} in array AORIG (while its row number i and its column number j are stored in arrays RNORIG and CNORIG) immediately after its calculation.

The first well-documented code where the simple input storage scheme was implemented is probably **ST**, [324], a code for solving sparse systems of linear algebraic equations by GE. Now it is used in two well-known sparse matrix packages: **MA28** ([63,77]), and **Y12M** ([329,331,341]).

2.2. Performance of static linear algebra operations

If the elements of matrix A remain unchanged during a basic linear algebra operation, then the operation is called **static**. Some examples of static linear algebra operations are given below:

(i)   Calculation of matrix-vector products (y=Ax or z=y+Ax).
(ii)   Calculation of residual vectors (r=b-Ax).
(iii) Calculation of the one-norm of a matrix (ANORM1).
(iv) Calculation of the max-norm of a matrix (ANORMM).

The input storage scheme described in the previous section can be used to perform any static linear algebra operation. Let us consider, for example, the expression z=y+Ax in order to illustrate this statement. It is necessary to emphasize here that the expression z=y+Ax is not chosen arbitrarily. This is an important linear algebra operation that appears in the numerical treatment of many large scientific and engineering models handled by sparse matrix techniques (see, for example, [213,336,337]). If the models are time-dependent, then several expressions of the type z=y+Ax must be calculated at each time-step (see the above references again). Assume that N is the length of the vectors x, y and z. Let the components of these three vectors be stored in arrays X, Y and Z, respectively (each of them of length at least equal to N). Let N be the order of matrix A and let NZ be the number of its elements. Assume that the input storage scheme described in the previous paragraph is in use. This means that the elements of matrix A, their row numbers and their column numbers are stored in arrays AORIG, RNORIG and CNORIG, respectively (the order of the elements being arbitrary). Then the **static** linear algebra operation z=y+Ax can be performed using the piece of code given in **Fig. 2.1**.

```
        DO 10 I=1,N
           Z(I)=Y(I)
     10 CONTINUE
  C
        DO 20 I=1,NZ
           Z(RNORIG(I))=Z(RNORIG(I)+AORIG(I)*X(CNORIG(I))
     20 CONTINUE
```

Figure 2.1
A piece of code for calculating the expression z=y+Ax.

It is clear that the code can easily be modified for the calculation of z=Ax (in the second line Y(I) should be replaced by 0.0) or for the calculation of a residual vector z=y-Ax (in the fifth line + should be replaced by -).

2.3. Dynamic linear algebra operations

If the elements of matrix A vary during the performance of a linear algebra operation, then the operation is called **dynamic**.

Some **dynamic** linear algebra operations, namely these in which no new non-zero elements (**fill-ins**) are created, can efficiently be performed by the use of the input storage scheme described in **Section 2.1**. Two examples of **dynamic** linear algebra operations of this type are given below.

(i) Multiplication of a vector by a constant.
(ii) Scaling a matrix.

Some important **dynamic** linear algebra operations cannot be performed by the use of the input storage scheme from **Section 2.1**. Examples of such **dynamic** linear algebra operations are:

(i) Gaussian elimination.
(ii) Orthogonal transformations.

Typical for the dynamic operations of the latter type is the fact that some zero elements may be transformed into non-zero elements during the computational process (i.e. fill-ins may appear). Such dynamic linear algebra operations will be called **essentially dynamic**. The sparsity pattern of a matrix is normally changed when an **essentially dynamic** linear algebra operation has been performed (due to new elements, **fill-ins**). The dynamic linear algebra operations that do not change the sparsity pattern of the matrix treated will be called **simply dynamic** operations.

The input storage scheme from **Section 2.1** is a **static** storage scheme. The position of each non-zero element is fixed when a static storage scheme is used. That is the elements remain on their place from the beginning to the end of the solution process, although the values of the non-zero elements may be varied during the computations. It is easy to apply a **static** storage scheme when the elements of matrix A remain at the same locations during the whole computational process. This means that a **static** storage scheme is easily applicable for **static and simply dynamic** linear algebra operations. Algorithms for advanced use of **static** storage schemes, in the case where **fill-ins** appear, will shortly be discussed in **Section 2.11**. However, for **essentially dynamic** linear algebra operations it is more natural to apply a **dynamic** storage scheme, where it is possible to insert new elements immediately after such elements are produced. A fairly general **dynamic** storage scheme, applicable both to GE and to the orthogonal transformations, will be described in the next section.

2.4. Dynamic storage scheme

Assume that an **essentially dynamic** linear algebra operation is to be performed. Two examples, namely GE and orthogonal transformations, were mentioned in the previous section. However, it must be emphasized that fill-ins can appear in **any** particular algorithm within the general scheme for solving the linear algebraic problem

$$(2.1) \quad x = A^\dagger b, \quad x \in R^{n \times 1}, \quad A \in R^{m \times n}, \quad b \in R^{m \times 1}, \quad m \geq n, \quad rank(A) = n,$$

which will be studied in **Chapter 3**. It should also be emphasized that the **dynamic** storage scheme, which will be presented in this section, could be applied, possibly after some slight modifications, to many particular algorithms within the general scheme for solving (2.1). The case where GE is to be performed is considered in this section, but this is done only in order to facilitate the exposition of the results.

Assume that a system of linear algebraic equations Ax=b is to be solved by the GE process. Let the elements of matrix A be stored in the first NZ locations of the **REAL** array **AORIG**, while their row and collumn numbers are stored in the corresponding locations of two **INTEGER** arrays **RNORIG** and **CNORIG** (or, in other words, assume that the input storage scheme from **Section 2.1** is applied to matrix A). Assume also that a **REAL** array **ALU(NN)** and three **INTEGER** arrays **RNLU(NN1)**, **CNLU(NN)** and **HA(NHA,11)** are available. The parameters **NN**, **NN1** and **NHA** are such that **NN ≥ NZ**, **NN1 ≥ NZ** and **NHA ≥ N**, where **NZ** is the number of elements in matrix A, and where **N = n** is the order of matrix A. A special subroutine, which will be called **ORDER** here, can be used to reorder the elements of matrix A from the static input scheme described in **Section 2.1** to a dynamic storage scheme that can be used to carry out efficiently the calculations needed in the GE process. After a call of this subroutine the elements that are in an arbitrary order in array **AORIG** are stored by rows in array **ALU**. This means that the following structure is prepared in array **ALU** by subroutine **ORDER**: first the elements of the first row of matrix A are stored, then the elements of the second row of matrix A are stored, and so on. Moreover, the non-zero elements of matrix A are stored in the first **NZ** locations of array **ALU**. This implies that the structure prepared by subroutine **ORDER** in array **ALU** is **compact**. That is there are no free locations: (a) between the last non-zero element of row i and the first non-zero element of row i+1 for any $i \in \{1,2,\dots,n-1\}$ and (b) between any two non-zero elements in a row. The order of the non-zero elements within any row i, i=1,2,...,N, is arbitrary. Subroutine **ORDER** does not attempt to order the non-zero elements in row i by increasing column numbers (if a_{ij} is stored, in array **ALU**, before a_{ik}, then j may be larger than k). This is so because no special order of the non-zero elements within a row is needed in the computations, and the storage scheme described above is very efficient for the factorization of matrix A by the GE process.

Assume that the non-zero element $a_{ij} \in A$ is stored by subroutine **ORDER** in ALU(K), where $1 \leq K \leq NZ$. Then CNLU(K) = j is also assigned by subroutine **ORDER**. This means that the column numbers of the elements are stored at the same positions in array **CNLU** as the non-zero elements in **ALU**.

The structure applied in arrays **ALU** and **CNLU** cannot be used without some additional information concerning the first and last positions of the non-zero elements of row i (i=1,2,...,N) in array **ALU**. Subroutine **ORDER** prepares such information. Let the elements of row i be located between position K_i and position K_i^* in array **ALU** (their column numbers being located between the same positions in array **CNLU**). Then subroutine **ORDER** assigns HA(i,1)=K_i and HA(i,3)=K_i^*. HA(i,2) is equal to K_i at the beginning of the GE. At stage s (s=1,2,...,n-1) of GE, the column numbers of the elements of row i (i=s,s+1,...,N) that are stored between HA(i,1) and HA(i,2)-1 are smaller than s, while the column numbers of the elements that are stored between HA(i,2) and HA(i,3) are larger than or equal to s. This means that, while the order of the elements within a row is arbitrary at the beginning of GE, some ordering is performed in the course of GE. The reason for this action will become clear later, when the performance of the GE transformations at stage s is described. In this chapter the order of the elements within a row will only be used to explain how a new element, **fill-in**, can be stored by the use of such a structure (this

will be done in the next section). However, it should be mentioned already
here that this ordering results in a separation of the elements of the
matrices L and U at the end of GE. The elements in row i of L are
located between the positions HA(i,1) and HA(i,2)-1. The diagonal elements
of L are never stored (since $l_{ii} = 1$ for i=1,2,...,N). If HA(i,1) >
HA(i,2)-1, then there are no elements in row i of L excepting l_{ii}. The
elements of row i of matrix U are located between the positions HA(i,2)
and HA(i,3), excepting the diagonal element u_{ii}, which is stored in the
i'th location of a **REAL** array **PIVOT** of length at least equal to N. If
HA(i,2) > HA(i,3), then there are no elements in row i of U excepting
u_{ii}. This structure is useful for the forward and back substitutions.

The structure in arrays **ALU** and **CNLU** as well as in the first three
columns of **HA** (prepared by subroutine **ORDER**) is sufficient to carry out
the computations needed to solve Ax=b. However, the solution process will
be very inefficient if only this structure is used. This is so because it is
sometimes necessary to scan the non-zero elements in a column (as, for
example, during the search for pivotal elements; this will be discussed in
detail in the following chapters). Another structure is needed in order to
facilitate the scan of the non-zero elements in a column and subroutine **ORDER**
prepares such a structure by storing in array **RNLU** the row numbers of the
non-zero elements in matrix A ordered by columns: first the row numbers of
the non-zero elements of the first column, then the row numbers of the non-
zero elements of the second column, and so on. The row numbers of the non-
zero elements ordered by columns are stored in the first NZ locations in
RNLU. The structure is **compact**. This means that there are no free locations:
(a) between the row number of the last non-zero element of column j and the
row number of the first non-zero element of column   j+1   for any   j ∈
{1,2,...,n-1} and (b) between the row numbers of any two non-zero elements
in a column. The order of the elements within any column j,   j=1,2,...,N,
is arbitrary. Subroutine **ORDER** does not attempt to order the row numbers of
the non-zero elements in column j by increasing column numbers (if i is
stored, in array **RNLU**, before k, then i may be larger than k). This
is so because no special order of the row numbers of the non-zero elements
within a column is needed in the computations, and the storage scheme
described above is very efficient for the factorization of matrix A by the
GE process.

The structure in array **RNLU** cannot be used without some additional
information concerning the first and the last positions in the part of array
RNLU where the row numbers of the non-zero elements in column j are stored
(j=1,2,...,N). Such information is also prepared by subroutine **ORDER**. Let
the row numbers of the elements in column j be located, in array **RNLU**, from
position L_j to position L_j^*. Then subroutine **ORDER** assigns HA(j,4)=L_j
and HA(j,6)=L_j^*. HA(j,5) is equal to L_j at the beginning of the GE. The
order, in array **RNLU**, of the row numbers of the elements in any column j
is **arbitrary** before the beginning of the GE process. However, at the end
of stage s   (s=1,2,...,N-1)   the row numbers stored between   position
HA(j,4) and HA(j,5)-1 in array **RNLU** are smaller than s, while the row
numbers stored between positions HA(j,5) and HA(j,6) in **RNLU** are larger
than or equal to s (j=s+1,s+2,...,N). Thus, some ordering is obtained in
the course of the GE. The reason for this action will become clear later
(when the performance of the GE transformations is described). In this
chapter the information about this ordering will only be used to explain how
a row number of a fill-in can be stored in array **RNLU**. This will be done in
the next section.

The description of the dynamic storage scheme is based on several assumptions which are not really needed and which were made only in order to facilitate the exposition of the results. The following remarks explain how these assumptions can be avoided.

Remark 2.1. The dynamic storage scheme described in this section is appropriate when the computer-oriented manner of exploiting sparsity is in use. It is assumed that two groups of arrays are needed. The first group contains **AORIG, RNORIG** and **CNORIG** (the arrays needed in the input storage scheme). The second group contains **ALU, RNLU, CNLU** and **HA** (the arrays needed in the dynamic storage scheme). If the classical manner of exploiting sparsity is used and if the data stored in the arrays **AORIG, RNORIG and CNORIG** are not needed for other purposes, then the contents of these three array can be overwritten by the contents of **ALU, RNLU and CNLU** respectively. Thus, **AORIG, RNORIG and CNORIG** are in fact not needed in the latter situation. This is exploited in package **Y12M**, [329,331,341]. In the particular algorithm used in **Y12M** the requirement $NN \geq 2*NZ$ is imposed when **AORIG, RNORIG and CNORIG** are to be overwritten by **ALU, RNLU and CNLU**. Since the length of **ALU** and **CNLU** should be large enough in order to accumulate all **fill-ins** (see the next section), **NN** should often be greater than $2*NZ$. Therefore in many cases the requirement $NN \geq 2*NZ$ is not a restriction. Nevertheless, it is better to remove this requirement. This could be done by an algorithm described in [341]. If the algorithm from [341] is used, then $NN \geq NZ$ is required (as in the beginning of this section), but let us repeat that in practice **NN** as a rule should be considerably larger than $2*NZ$ (this will be demonstrated by many numerical examples in the end of this chapter). ∎

Remark 2.2. The dynamic scheme described above is good, as mentioned in **Remark 2.1**, for the case when the computer-oriented manner of exploiting sparsity is in use. When this is the case, the contents of arrays **AORIG, RNORIG and CNORIG** are needed for the iterative refinement process (1.11), or for some other iterative process. Therefore these arrays cannot be overwritten by **ALU, RNLU and CNLU** as in the case where the classical manner of exploiting sparsity is in use. If one decides to store **all** non-zero elements of matrix A during the initialization of the dynamic scheme, then the procedure described in this section can be used without any changes when the computer-oriented manner is in use. However, for some classes of matrices this is not the best decision. It may be much more efficient to remove all small elements (according to some appropriate criterion) during the preparation of the dynamic storage scheme. Assume that the number of elements of matrix A that are not small is $NZ1$. It is clear that $NZ1 \leq NZ$. The length of arrays **ALU** and **CNLU** should be $NN \geq 2*NZ1$ and the length of array **RNLU** should be $NN1 \geq NZ1$ when the computer-oriented manner is applied, while the corresponding lengths should satisfy $NN \geq 2*NZ$ and $NN1 \geq NZ$ when the classical manner of exploiting the sparsity is used. In some cases $NZ1$ is considerably smaller than NZ. This is so for the problems arising in nuclear magnetic resonance spectroscopy ([336,337]). Some other examples that demonstrate this statement will be given at the end of this chapter. ∎

Remark 2.3. The dynamic storage scheme described in this section consists of two groups of arrays. In the first group of arrays the elements of matrix A together with their column numbers are ordered by rows in **ALU** and **CNLU**; some pointers concerning this structure are kept in the first three columns of **HA**. It is said that **ALU, CNLU** and the first three columns of **HA** form

the row ordered list   or   **the row-oriented structure**.   In the second group
of arrays the row numbers of the elements in matrix A are ordered by columns
in array **RNLU**; some pointers concerning this structure are given in the next
three columns of array **HA**. It is said that **RNLU** together with the fourth,
the fifth and the sixth columns of **HA** form   **the column ordered list**   or
the column-oriented structure.   ∎

        Remark 2.4. The dynamic storage scheme discussed here is designed for
GE with transformations carried out by rows. If the Gaussian transformations
are carried by columns, then the elements of matrix A should be ordered by
columns and moved to the column ordered list. Such a dynamic scheme, in which
the column ordered list is more important than the row ordered list, can be
considered instead of the dynamic scheme described in this section when
appropriate.   ∎

        Remark 2.5. It is assumed that a system of linear algebraic equations
$Ax=b$ is solved by the GE process in the discussion of the dynamic scheme
here. However, the same ideas can also be applied for other methods for
solving systems of linear algebraic equations as well as in the case where the
more general problem of computing $x=A^\dagger b$ (with $x \in R^{nx1}$, $A \in R^{mxn}$, $b \in R^{mx1}$,
$m \geq n$, $rank(A) = n$) is solved. In the latter case it is convenient to split
the array **HA** into two arrays **HA1** and **HA2**. Array **HA1** is used in
connection with the row ordered list. Therefore the length of the columns in
this array should be at least equal to m. Array **HA2** is used in connection
with the column ordered list. The length of the columns in this array should
be at least equal to n. A dynamic storage scheme modified as described in
this remark has been applied in code **LLSS01** (see [320-322]. The problem
$x=A^\dagger b$ ($x \in R^{nx1}$, $A \in R^{mxn}$, $b \in R^{mx1}$, $m \geq n$, $rank(A) = n$) is solved by
carrying out orthogonal transformations to decompose matrix A when **LLSS01**
is used. **LLSS01** will be discussed in detail in **Chapter 12 - Chapter 16**. ∎

2.5. Storage of fill-ins

        The dynamic storage scheme is much more complicated than the input
storage scheme. The only reason for developing the former scheme is the fact
that fill-ins arise during GE as well as in many other linear algebra
processes and have to be stored in a convenient way. This is impossible when
the input storage scheme is used. In this section it will be explained how a
fill-in is stored when the dynamic storage scheme is implemented in sparse
matrix software. It will be assumed again that GE is used, but the same
ideas can be applied with obvious modifications to many other linear algebra
processes.

        If a new fill-in, say $a_{ij}^{(s)}$, is created at stage s of the GE, then
a copy of row i is made after the position where the last element is kept
in **ALU**. A corresponding copy of the column numbers of the elements in row
i is made at the end of **CNLU** and the contents of $HA(i,\mu)$, $\mu = 1,2,3$, are
modified. The new element is stored at the end of the copy of row i in array
ALU. Its column number is stored at the same position in array **CNLU**. The
introduction of the new element at the end of the row ordered list is taken
into account in the modification of $HA(i,3)$. The locations occupied before
these operations by the elements of row i and by their column numbers (in
arrays **ALU** and **CNLU** respectively) are freed. The free locations are
indicated by setting zeros in the positions in array **CNLU** where the column

numbers of the elements of row i were stored before the new element was created. There are no such free locations before the beginning of the GE (the structure is initially compact). Thus the row ordered list is in general not compact after the start of the GE. However, the structure within each row is still compact: there are no free locations between two elements of any row i at any stage s and the same is true for their column numbers.

The fact that there could be free locations between two rows after the start of the GE indicates that, before making a copy of row i at the end of the row ordered list, it is worthwhile to check about free locations either before the position where the first element of row i is stored or after the position where the last element of row i is stored. A position K is a free position when $CNLU(K)=0$. Assume that the contents of array **HA** that are of interest when a fill-in $a_{ij}^{(s)}$ is created are: $HA(i,1)=K_i$, $HA(i,2)=K_i^{**}$ and $HA(i,3)=K_i^*$. This means that: (i) the position of the first element of row i in array **ALU** is K_i, (ii) the position of the first element of row i with column number greater than or equal to s in array **ALU** is K_i^{**} (iii) the position of the last element of row i in array **ALU** is K_i^* and (iv) the same statements as (i)-(iii) are also true for the column numbers of the elements of row i stored in **CNLU**. Assume also that there are free locations before position K_i (or, in other words, at least $CNLU(K_i-1)=0$). Then the element stored in position $K_i^{**}-1$ is moved to position K_i-1, the fill-in $a_{ij}^{(s)}$ is stored in position $K_i^{**}-1$ and the contents of the i'th components of the first two columns of array **HA** are updated by setting $HA(i,1) = K_i-1$ and $HA(i,2) = K_i^{**}-1$. Similar operations are performed in array **CNLU**: the column number stored in position $K_i^{**}-1$ is moved to position K_i-1 and the column number of the fill-in $a_{ij}^{(s)}$ is stored in position $K_i^{**}-1$.

Assume now that there are free locations after position K_i^* in array **ALU**; or, in other words, at least $CNLU(K_i^*+1)=0$. Then the fill-in $a_{ij}^{(s)}$ is stored in position K_i^*+1 in array **ALU**, its column number j is stored at the same position in **CNLU** and $HA(i,3)$ is set equal to K_i^*+1.

The number of copies made in the course of the computations is normally reduced when an attempt to exploit the free locations at the beginning and end of a row is carried out as described above. Nevertheless, as a rule many new copies have to be made. This means that the capacity of arrays **ALU** and **CNLU** may be exceeded. If this happens, then it is necessary to compress the structure in these two arrays by performing so-called **garbage collections**. After such an action there are no free locations between the rows in **ALU** and between the column numbers of the non-zero elements of any two two rows in **CNLU**. The performance of a garbage collection will be described in the following section. It is important to emphasize here that not only is the number of copies of rows reduced when fill-ins are stored at the beginning or end of a row (when there are free locations there), but also the number of garbage collections may be reduced. This second effect is often the more important.

A similar procedure is used in connection with the column ordered list (the array **RNLU** as well as the fourth, fifth and sixth columns of array **HA**). Consider again a fill-in $a_{ij}^{(s)}$. Assume that there are no free locations at the beginning or end of column j (or, more precisely, at the beginning or end of the part of array **RNLU** where the row numbers of the non-zero elements of column j are stored). In this case a copy of column j (i.e. a copy of the part of the array **RNLU** where the row numbers of the non-zero elements

of column   j   are stored) is made after the last occupied position in array
RNLU,   and the row number   i   of the fill-in $a_{ij}^{(s)}$ is stored at the end of
the copy. The contents of   HA(j,4),   HA(j,5)   and   HA(j,6)   are updated and
the locations occupied by the row numbers of the elements of column   j   before
these operations are freed. As in the row ordered list free locations are
marked by zeros: RNLU(K)=0 means that the K'th location of array **RNLU**
is free. From this description it follows that there can be free locations
between two columns in array **RNLU**. In the beginning of   GE   the structure
is compact, but this is in general not true after the start of the computations. The same was true for the structure in the row ordered list. The
structure within each column is compact during the whole GE process: there
are no free locations between the row numbers (stored in **RNLU**) of any two
elements of column   j,   j=1,2,...,N.

The fact that there could be free locations between two columns after the
start of GE means that one could sometimes avoid copying a column at the end
of the column ordered list. Assume that the contents of array **HA** that are
of interest when a free location for the row number of the fill-in $a_{ij}^{(s)}$ has
to be found are: HA(j,4)=L_j,   HA(j,5)=L_j^{**} and   HA(j,6)=L_j^*. This means
that: (i) the position of the row number of the first non-zero element of
column j in array **RNLU** is L_j, (ii) the position of the row number of the
first non-zero element of column   j   with row number greater than or equal to
s   in array **RNLU** is L_j^{**},   and   (iii) the position of the row number of
the last non-zero element of column   j   in array **RNLU** is L_j^*. Assume also
that there are free locations before position L_j (or, in other words, at
least RNLU(L_j-1)=0). Then the row number stored in position L_j^{**}-1 is moved
to position L_j-1,   the row number   i   of the fill-in $a_{ij}^{(s)}$ is stored in
position L_j^{**}-1 and the contents of the   j'th   components of HA(j,4)   and
HA(j,5)   are updated by setting   HA(j,4) = L_j-1   and   HA(j,6) = L_j^{**}-1.

Assume now that there are free locations after position L_j^* in array
RNLU; or, in other words, that at least   RNLU(L_j^*+1)=0. Then the row number
i   of the fill-in $a_{ij}^{(s)}$ is stored in   position L_j^*+1 in array **RNLU** and
HA(j,6)   is set equal to L_j^*+1.

As for the row ordered list, the number of copies made in the course of
the computation is normally reduced considerably when free locations at the
beginning and end of the columns are exploited. Nevertheless many copies may
still have to be made. Therefore the capacity of array **RNLU** may be exceeded.
If this happens, then it is necessary to compress the structure in this array
by performing a **garbage collection**. The performance of garbage collections
will be described in the next section. As for the row ordered list, it should
be emphasized that not only is the number of copies of columns reduced when
the above operations (storing row numbers of fill-ins either at the beginning
or end of the column under consideration when there are free locations there)
are implemented in the sparse matrix software, but also the number of garbage
collections may be reduced. This second effect is often the more important.

The above discussion shows that the dynamic modifications in the column
ordered list (arrays **RNLU, HA(*,4), HA(*,5) and HA(*,6)**) are very
similar to those in the row ordered list (arrays **ALU, CNLU, HA(*,1),
HA(*,2) and HA(*,3)**). However, there are two significant differences.

(i) **When the column ordered list is to be modified one has to work with
the row numbers of the elements only, while both the elements and their column
numbers are used in the modification of the row ordered list.**

(ii) The row numbers of the non-zero elements in column s are not needed after stage s (s=1,2,...N-1) of the GE. Therefore these row numbers can be removed at the end of stage s.

A final remark is appropriate in connection with the dynamic storage scheme described in this section. The pivotal elements $a_{ss}^{(s)}$ (s=1,2,...,N-1) are stored in a special **REAL** array **PIVOT** of length at least equal to N. The locations occupied by the pivotal element $a_{ss}^{(s)}$ and by its row and column numbers in arrays **ALU**, **RNLU** and **CNLU**, respectively, are freed at the beginning of stage s (s=1,2,...,N-1) of the GE and could be used to store fill-ins.

2.6. Performance of garbage collections

When a fill-in, $a_{ij}^{(s)}$ is created in the course of the computation, it may be necessary to make copies of the contents of certain parts of **ALU** and **CNLU** at the ends of these arrays. However, if many copies have been made, there may be not enough space for another copy. On the other side, there may be free locations between the rows (after a row is copied the locations originally occupied by the row are freed). This means that a compression of the structure in the row ordered list may result in freeing many locations at the end of the list so that necessary copying can be performed. Such a compression has occasionally to be carried out. It is known under the name of **"garbage collection"**.

The situation is similar when the column ordered list is considered. Garbage collections may also be needed during the dynamic modifications of this list, but not necessarily simultaneously with the garbage collections in the row ordered list. The only difference is that, while the contents of both **ALU** and **CNLU** have to be compressed when a garbage collection in the row ordered list is performed, the contents of one array only, **RNLU**, have to be compressed during a garbage collection in the column ordered list. The performance of garbage collections in the two ordered lists will now be discussed in detail.

Assume that a garbage collection in the row ordered list has to be carried out. **This action is performed in two steps.** In the first step some information to be used in the second step is prepared. The second step is the actual compression of the structure.

During the first step of the garbage collection the positions where the first non-zero elements of the rows are located in array **ALU** are marked by setting HA(i,1)=CNLU(HA(i,1)) and CNLU(HA(i,1))=-i for i=1,2,...,N. In this way a negative value of CNLU(K) will indicate the beginning of a new row at position K in array **ALU**. Moreover, the value of -CNLU(K) will show which row starts at this position. Finally, no information is lost because the column number of the element stored in ALU(K) will be found in HA(-CNLU(K),1). The first step of the garbage collection in the row ordered list can be performed in O(N) operations.

Consider now the second step of a garbage collection in the row ordered list. Let **NCNLU** be the last occupied location in arrays **ALU** and **CNLU** (**NCNLU** ≤ **NN**, where **NN** is the length of these two arrays). It is clear that **NN-NCNLU** is small when a garbage collection is to be performed (because there is no place for a copy of a row at the end of **ALU** if a garbage collection is needed). Consider a variable **NEWPOS** and assume that NEWPOS=1 before the beginning of the garbage collection. Then the garbage collection can be performed as follows. Scan the components CNLU(K) for K=1,2,...,NCNLU. If CNLU(K)=0, then location K is free and nothing is done. If CNLU(K)<0, then a new row starts at position K of array **ALU**. In such a case set:

```
(i)    ALU(NEWPOS) = ALU(K)
(ii)   CNLU(NEWPOS) = HA(-CNLU(K),1)
(iii)  HA(-CNLU(K),1) = NEWPOS
(iv)   HA(-CNLU(K),2) = NEWPOS + HA(-CNLU(K),2) - K + 1
(v)    HA(-CNLU(K),3) = NEWPOS + HA(-CNLU(K),3) - K + 1
(vi)   NEWPOS = NEWPOS + 1
```

If CNLU(K)>0, then the element stored in position K of array **ALU** has to be moved to position **NEWPOS**. The same action is performed for its column number. Finally, the value of **NEWPOS** has to be updated. These operations are performed following the rules:

```
(i)    ALU(NEWPOS) = ALU(K)
(ii)   CNLU(NEWPOS) = CNLU(K)
(iii)  NEWPOS = NEWPOS+1
```

At the end of this process the structure is fully compressed. The last occupied position in arrays **ALU** and **CNLU** is NEWPOS-1. It is assumed that there are sufficiently many locations at the end of the row ordered list so that a copy of the row, which could not be made before the garbage collection (and was the reason for the garbage collection), can be made now. If this is not the case, then the software returns with an error message ("**the length NN of the arrays ALU and CNLU is too small**").

The operational cost of the second step is O(NN) taking into account the fact that, as mentioned above, **NN-NCNLU** is small and, thus, the parameter **NCNLU**, which is in general not known before the actual performance of a garbage collection, is approximately equal to **NN**.

A garbage collection in the column ordered list is performed in a quite similar way. Only one significant modification is needed during the first step. Since all row numbers of the non-zero elements in the columns j < s have been removed from array **RNLU** before the performance of stage s of the GE (see the previous section), the positions of the column starts in array **RNLU** are marked by setting RNLU(HA(j,4))=-j and HA(j,4) = RNLU(HA(j,4)) for j=s,s+1,...N (it is assumed here that the garbage collection is to be performed at stage s of the GE).

The operational costs of the two steps of a garbage collection in the column ordered list are O(N-s+1) for the first step and O(NN1) for the second step, where **NN1** is the length of array **RNLU**.

The garbage collection in the column ordered list is performed only if the row numbers of the non-zero elements in some column are to be copied at the end of array **RNLU** and there are not sufficiently many locations there. It is expected that there will be sufficiently many locations after the

garbage collection, so that a copy of the row numbers of the non-zero elements
in the column can be made. If this is not the case (i.e. if there is still not
enough room at the end of RNLU), then the software returns with an error
message ("the length NN1 of array RNLU is too small").

2.7. Need for elbow room

Consider the three main arrays, ALU, RNLU and CNLU, used in the
dynamic storage scheme described in the previous sections. The lengths of
these arrays are NN, NN1 and NN, respectively. The minimal length of
these arrays, which is sufficient to initialize the dynamic storage scheme,
is NN=NN1=NZ1. NZ1 is equal to the number NZ of non-zero elements in
matrix A when the classical manner of exploiting sparsity is in use. NZ1
is the number of non-zero elements that are not small according to the
criterion chosen when the computer-oriented manner of exploiting sparsity is
applied; NZ1 ≤ NZ.

Since fill-ins arise during GE (and in many other methods for solving
algebraic problems), it is clear that the minimal lengths of the main arrays
in the dynamic storage scheme are not sufficient for performing the
computations after the initialization of the dynamic scheme. The lengths of
the arrays should be considerably larger than NZ1 in order to ensure
locations for the fill-ins and for making copies of rows and columns at the
end of the main arrays (see Section 2.5). This means that some extra
locations, which are free at the initialization of the dynamic storage scheme
(i.e. before the beginning of the GE process), are needed. These extra
locations form the so-called elbow room in the main arrays of the dynamic
storage scheme.

It is very important to determine how many locations are needed for elbow
room. Unfortunately, the number of these locations depends on the number of
fill-ins created during the GE (or, more generally, on the particular method
used to decompose matrix A) and therefore is not known before the start of
the computations. Experiments carried out with several thousands of matrices
indicated that as a rule the choice NN ∈ [5NZ,10NZ] is a good one when the
classical manner of exploiting sparsity is in use. However, some matrices
produce many fill-ins (20NZ or even more). NN > 20NZ should be specified
when this is so. The choice of NN1 is much easier (if a good choice of NN
has already been made). The experiments show that as a rule NN1 = 0.6NN is
a very good choice when the classical manner is applied. If the computer-
oriented manner of exploiting sparsity is applied, then NN ∈ [2NN,3NN] is
often a good choice. If NN has been determined, then again NN1 = 0.6NN is
probably the best choice.

The statements made in this section are based on numerical results
obtained from extensive computations with sparse matrices. Some experiments
are given below to demonstrate the need of elbow room when the dynamic
storage scheme is used.

Example 2.1. The matrix JPWH 991 from the Harwell set of sparse test-
matrices, [76] (see also [71,72]), is used in this experiment. The order of
this matrix is N=991 and the number of its elements is NZ=6027. The
maximal number of non-zero elements kept in the row ordered list during the
GE is denoted, as in Chapter 1, by COUNT. The maximal number of row
numbers kept in the column ordered list is denoted by COUNT1. The results

obtained by the two manners of exploiting sparsity are given in **Table 2.1**.
It is seen that the elbow room needed is very large when the classical manner
of exploiting the sparsity is used: NN > 8NZ must be specified. If a good
NN is selected, then, as stated above, NN1 = 0.6NN is a good choice.

The computer-oriented manner of exploiting the sparsity leads to great
savings both in computing time and storage. It should be mentioned that the
absolute drop-tolerance, T, as in the experiments in **Chapter 1**, is
applied. The value $T = 2^{-5}$ is not the best choice when the computing time
is taken into account. The global computing time can be reduced by choosing
$T = 2^{-6}$. However, more storage is needed when the latter drop-tolerance is
applied.

Compared characteristics	Classical manner	Computer-oriented manner
NZ1	6027	6027
COUNT	49023	7976
COUNT1	27437	6027
Factorization time	15.82	0.54
Solution time	0.09	2.21(46)
Total computing time	15.91	2.75
Accuracy achieved	5.03E-4	3.95E-5

Table 2.1
Comparisons of some characteristics obtained in the
solution of a system Ax=b, where A = JPWH 991 is a
matrix from the Harwell set with N=991 and NZ=6027.
The computer-oriented manner is used with $T = 2^{-5}$, the
number of iterations performed is given in brackets.

The runs performed in this experiment were carried out on an IBM 3081
computer under the FORTH compiler, and the computing time is measured in
seconds. ∎

Example 2.2. An attempt to find out how large **NN** and **NN1** must be if
one wants to avoid the performance of **garbage collections** is carried out in
this experiment. Again the matrix A = JPWH 991 is used. In **Example 2.1**
both **NN** and **NN1** were very large; NN = NN1 = 199999, so that the
factorization has been performed without gargage collections. In this
experiment NN = COUNT + kN and NN1 = COUNT1 + kN with k=2(2)32 is used
(the values of **COUNT** and **COUNT1** were found in the previous experiment:
COUNT=49023 and COUNT1=27427). For k=2,4,6 the number of garbage
collections is very large. For k>10 the number of garbage collections is
quite reasonable. However, k must be large if one wants to avoid garbage
collections. Garbage collections in the column ordered list are avoided for
NN1 = COUNT1 + 32N = 59149. The smallest number **NN** for which garbage
collections in the row ordered list can be avoided has not been found (for
NN = 99999 the code performed one garbage collection.

The experiment shows that a good compromise is to allow a small number
of garbage collections. It must be emphasized here that no problems with
garbage collections arise when this particular system is solved using the
computer-oriented manner of exploiting sparsity. ∎

k	NN	NN1	GC in ROL	GC in COL	Computing time
2	51005	29419	29	127	27.68
4	52987	31401	17	47	20.98
6	54969	33383	12	30	19.41
8	56951	35365	9	15	17.98
10	58933	37347	6	10	17.37
12	60915	39329	5	7	17.07
14	62897	41311	4	5	16.75
16	64879	43293	3	4	16.63
18	66861	45275	3	3	16.44
20	68843	47257	2	2	16.29
22	70825	49239	2	2	16.27
24	72807	51221	2	1	16.16
26	74789	53203	1	1	16.10
28	76771	55185	1	1	16.11
30	78753	57167	1	1	16.09
32	80735	59149	1	0	16.00

Table 2.2
Comparison of the choice of the lengths of the main
arrays (NN and NN1) and the number of garbage collections ("GC in ROL" and "GC in COL" are abbreviations
for garbage collections in the row ordered list and
garbage collections in the column ordered list respectively). The classical manner of exploiting sparsity is
applied. There are no problems with garbage collections
during the factorization of JPWH 991 when the computer-oriented manner of exploiting the sparsity is used.

For the matrix used in **Example 2.1** the number **NZ1** of elements in the
row ordered list of the dynamic storage scheme before the beginning of the
GE is equal to the number **NZ** of non-zero elements. This means that no non-
zero element in matrix A = JPWH 991 is small according to the criterion
(1.9) with $T = 2^{-5}$ (or, in other words, there is no element $a_{ij} \neq 0$ for
which $|a_{ij}| < T$). The next example shows that this is not always the case
and that for some matrices **NZ1** can be considerably smaller than **NZ**.

Example 2.3. Sparse matrix technique has been used in the numerical
treatment of some chemical models described by systems of linear ordinary
differential equations (**ODE's**) $y' = A(t)y + b(t)$, $y(0) = \eta$, $y \in R^{nx1}$,
$A(t) \in R^{nxn}$, $b(t) \in R^{nx1}$, $\eta \in R^{nx1}$, $t \in [0,T]$; $A(t)$, $b(t)$ and η being
given. The problem is discretized on a set of non-equidistant grid-points
$t_0=0, t_1, t_2, \ldots, t_K=T$. At each grid-point $t_k \in \{t_1, t_2, \ldots, t_K\}$ an approximation y_k to the solution $y(t_k)$ is calculated by using the modified
diagonally implicit Runge-Kutta method from **[294]**. It is said that an
integration step, **step k**, is performed when such an approximation is
calculated. Two systems of linear algebraic equations of the type $A_k x_k^i = c_k^i$
($i=1,2$)   are to be solved at each **step k** ($k \in \{1,2,\ldots,K\}$).   The
coefficient matrix A_k is not factorized at each step. An old factorization
$L_m U_m$ of A_m ($m<k$) is also used at **step k** with iterative refinement and
a new factorization $L_k U_k$ is calculated only if the iterative refinement does
not converge. It is said that a **substitution** is performed when the
subroutine calculating the solution of a system of linear algebraic equations
by a forward and back substitution is called. It is clear that the number of

substitutions is equal to the total number of iterations carried out in the
solution of the system of **ODE's**. This information is sufficient for the
purposes of this section. However, this problem will be discussed in more
detail in **Chapter 8**.

N	Compared characteristics	Classical manner	Computer-oriented manner
63	NZ	839	839
	NZ1	839	315
	COUNT	1972	475
	Number of steps	774	774
	Number of factorizations	82	92
	Number of substitutions	5038	5984
	Total computing time	63.09	30.12
255	NZ	3323	3323
	NZ1	3323	2283
	COUNT	20170	6634
	Number of steps	473	484
	Number of factorizations	67	81
	Number of substitutions	3015	3852
	Total computing time	606.52	148.60
1023	NZ	64517	64517
	NZ1	64517	13571
	COUNT	352470	75013
	Number of steps	23(*)	788
	Number of factorizations	4(*)	76
	Number of substitutions	127(*)	4701
	Total computing time	3864.14(*)	2124.00

Table 2.3
Comparison of some results obtained in the solution of
systems of linear differential equations arising in
nuclear magnetic resonance spectroscopy. It is extremely
expensive to solve the large spectroscopic problem using
the classical manner of exploiting the sparsity; only
1.55% of the time-interval has been integrated in this
case (asterisks are used to indicate that the integration of the system of **ODE's** has not been completed).

Three problems, a small problem with 63 equations, a medium problem
with 255 equations, and a large problem with 1023 equations, are
considered. It is seen from **Table 2.3** that for all three problems **NZ1** is
considerably smaller than **NZ** when the computer-oriented manner of exploiting
sparsity is used. This leads to great savings in both **storage** (compare the
values of **COUNT** for the two manners of exploiting sparsity and take into
account that **COUNT1** is approximately equal to 0.6*COUNT) and **computing
time**. It must be emphasized that it was not possible to solve the large
problem with 1023 equations by using the classical manner, because it is
extremely expensive. This problem has been integrated over only 1.55% of the
time-interval.

Two additional remarks are needed in connection with this experiment.

The **first** remark is connected with the determination of the value of the drop-tolerance T. The code attempts automatically to determine an optimal (or nearly optimal) value of T starting with a rather large value and decreasing T after each failure. This device will be discussed in **Chapter 8**. Here it is sufficient to point out that the automatic determination of the drop-tolerance is the reason for the larger numbers of factorizations when the computer-oriented manner of exploiting the sparsity is in use. The code performs several extra factorizations with large values of T until a good value is found.

The **second** remark is connected with the iterative refinement process. When systems of ODE's are solved, iterative refinement has also to be used with the classical manner of exploiting sparsity, because the factors $L_m U_m$ are normally inaccurate when these are used at step k with k>m and, therefore, the solution found by using $L_m U_m$ should be improved by iterative refinement (or by some other iterative method). Nevertheless, the total numbers of substitutions are considerably smaller when the classical manner of exploiting the sparsity is in use. This is so because when the computer-oriented manner is applied the factors L_m and U_m are inaccurate not only because these are calculated at a previous step (m<k), but also because some "small" elements are removed during the GE process.

The fact that both the number of factorizations and the number of substitutions are increased when the computer-oriented manner of exploiting sparsity is applied in the numerical treatment of this class of chemical problems (and this is also true for some other classes of problems) does not lead to an increase of the total computing computing time (in comparison with the classical manner). In fact, the computing time is reduced considerably when the computer-oriented manner is used. When any sparse matrix technique is in use it is most important to preserve the sparsity of the original matrix as much as possible, and this is achieved by reducing NZ1 and by keeping COUNT as small as possible during the GE process. The success of the computer-oriented manner of exploiting sparsity for this class of problems is due to the fact that the sparsity is preserved much better than in the case where the classical manner of exploiting sparsity is applied. ∎

2.8. Application of the storage schemes on a small matrix

A small matrix with N=5 and NZ=11 is used in this section to demonstrate the implementation of the input storage scheme and the dynamic storage scheme. The matrix is given by (2.2).

The input storage scheme is given in **Fig. 2.2**. The names of the arrays are as in **Section 2.1**. The order of the elements is arbitrary. If an element is stored in position K of array **AORIG**, then its row and column numbers are stored at the same position in arrays **RNORIG** and **CNORIG**. For example, the non-zero element $a_{35}=3.0$ is stored in AORIG(5). At the same position, K=5, in arrays **RNORIG** and **CNORIG** its row and column numbers are stored.

The contents of the dynamic storage scheme before the beginning of the GE are shown in **Fig. 2.3** (for the row ordered list) and in **Fig 2.4** (for the column ordered list).

$$(2.2) \qquad A \; = \; \begin{array}{|ccccc|} \hline 6.0 & 0.0 & 0.0 & 3.0 & 0.0 \\ 2.0 & 1.0 & 0.0 & 0.0 & 0.0 \\ 0.0 & 1.0 & 3.0 & 0.0 & 3.0 \\ 0.0 & 0.0 & 1.0 & 2.0 & 0.0 \\ 0.0 & 0.0 & 0.0 & 2.0 & 9.0 \\ \hline \end{array}$$

Position	1	2	3	4	5	6	7	8	9	10	11
REAL array AORIG	9.0	2.0	2.0	1.0	3.0	3.0	1.0	1.0	2.0	3.0	6.0
INTEGER array RNORIG	5	5	4	4	3	3	3	2	2	1	1
INTEGER array CNORIG	5	4	4	3	5	3	2	2	1	4	1

Figure 2.2
The input storage scheme for the small test-matrix from (2.2); N=5, NZ=11.

Position	1	2	3	4	5	6	7	8	9	10	11	12	13	14	15
ALU	6.0	3.0	2.0	1.0	1.0	3.0	3.0	1.0	2.0	2.0	9.0				
CNLU	1	4	1	2	2	3	5	3	4	4	5				
Row No.	Row 1		Row 2		Row 3			Row 4		Row 5		Free space			
HA(*,1)	1	3	5	8	10	positions of the row starts in ALU									
HA(*,2)	1	3	5	8	10	positions of the separators in ALU									
HA(*,3)	2	4	7	9	11	positions of the row ends in ALU									

Figure 2.3
The row ordered list before the beginning of the Gaussian elimination.

Position	1	2	3	4	5	6	7	8	9	10	11	12	13	14	15
RNLU	1	2	2	3	3	4	1	4	5	3	5				
Column No.	Col. 1		Col. 2		Col. 3		Col. 4			Col. 5		Free space			
HA(*,4)	1	3	5	7	10	positions of the column starts in RNLU									
HA(*,5)	1	3	5	7	10	positions of the separators in RNLU									
HA(*,6)	2	4	6	9	11	positions of the column ends in RNLU									

Figure 2.4
The column ordered list before the beginning of the Gaussian elimination.

The length **NN** of arrays **ALU** and **CNLU** is NN=15 (see **Fig. 2.3**),
but only the first NZ=11 locations are occupied before the beginning of the
GE. A special variable, **NCNLU**, is used as a pointer to the last occupied
location in the row ordered list. At the beginning of the GE NCNLU=NZ
(NCNLU=NZ1 when the computer-oriented manner is applied). If an non-zero
element is stored in position **K** ($1 \le K \le 11$) in array **ALU**, then its
column number is stored in the same position in array **CNLU**. For example,
element a_{43} = 1.0 is stored in position K=8 in array **ALU** (i.e.
ALU(8)=1.0). Its column number is stored at the same position, K=8, in array
CNLU (CNLU(8)=3). The row starts and ends are stored in the first and the
third columns of array **HA**. For example, HA(3,1)=5 and HA(3,3)=7 shows
that the elements of the third row are located between ALU(5) and ALU(7).
The column numbers of the elements of the third row are located between the
same positions, 5 and 7, in array **CNLU**. The second column of **HA**
contains the same information as the first one at the beginning, but later on
it will contain the positions that separate the non-zero elements in the
active submatrix from the other non-zero elements.

The length NN1 of array **RNLU** is NN1=15 (see **Fig. 2.4**). In this case
NN=NN1, but this is not typical: for large matrices NN1<NN may successfully
be specified (see, for example, **Table 2.2** in the previous section). Only the
first NZ=11 locations of the **RNLU** are occupied in the beginning. As for
the row ordered list, a special variable, **NRNLU**, is used as a pointer to the
last occupied position in array **RNLU**. At the beginning of the GE process
NRNLU=NZ (NRNLU=NZ1 when the computer-oriented manner is applied). The
column starts and ends are stored in the fourth and the sixth columns of array
HA. For example, HA(4,4)=7 and HA(4,6)=9 shows that row numbers of
elements of the fourth column are located between RNLU(7) and RNLU(9). The
fifth column of **HA** contains the same information as the fourth one at the
beginning, but later on it will contain the positions that separate the row
numbers of the non-zero elements in the active submatrix from the row numbers
of the other non-zero elements.

Consider now the first stage, s=1, of the GE process, assuming that
a_{11} is chosen as a pivot. For the small example considered in this section
the only operation during the first stage concerns the element a_{24}. This
element is originally equal to zero (see the matrix given in the beginning of
this section). However, since both a_{21}=2.0 and a_{14}=3.0 are non-zero
elements, a fill-in, $a_{24}^{(2)}$=-1, is created in position (2,4). This new
element should be incorporated both in the row ordered list (the element
itself and its column number) and in the column ordered list (the row number,
2, of the new element). The operations needed to insert the new element,
together with its column number, in the row ordered list can be described as
follows. There is no free location at the beginning of row 2. There is no
free location at the end of row 2 either. Therefore a copy of the second row
at the end of the row ordered list is made and the fill-in is stored at the
end of the copy (see **Fig. 2.5**). The locations in which the second row was
stored in the row ordered list are freed. This is done by setting zeros in the
locations in array **CNLU** where the column numbers of the elements of the
second row were stored before the copy. The second components of the first
three columns of array **HA** are updated (compare **Fig. 2.3** and **Fig. 2.5**).
Note that now we have HA(2,2) $\neq$ HA(2,1), because an element of matrix L
is created in row 2 (this element is denoted by ∎ in **Fig. 2.5**). The
structure of the row ordered list is no more compact: there are free locations
between the first and the third row. The updated row ordered list (during
stage 1 of the GE process) is given in **Fig. 2.5**.

Position	1	2	3	4	5	6	7	8	9	10	11	12	13	14	15
ALU	6.0	3.0	2.0	1.0	1.0	3.0	3.0	1.0	2.0	2.0	9.0	■	1.0	-1.0	
CNLU	1	4	0	0	2	3	5	3	4	4	5	1	2	4	
Row No.	Row 1		Free		Row 3			Row 4		Row 5		Row 2			

HA(*,1)	1	12	5	8	10	positions of the row starts in ALU
HA(*,2)	1	13	5	8	10	positions of the separators in ALU
HA(*,3)	2	14	7	9	11	positions of the row ends in ALU

Figure 2.5

The row ordered list after the introduction of the new
non-zero element $a_{24}^{(2)}=-1$. A copy of the second row is
made at the end of the row ordered list. By ■ the
element l_{21} is denoted (see (1.3)).

Position	1	2	3	4	5	6	7	8	9	10	11	12	13	14	15
RNLU	1	2	2	3	3	4	0	0	0	3	5	1	4	5	2
Column No.	Col. 1		Col. 2		Col. 3		Free space			Col. 5		Col. 4			

HA(*,4)	1	3	5	12	10	positions of the column starts in RNLU
HA(*,5)	1	3	5	13	10	positions of the separators in RNLU
HA(*,6)	2	4	6	15	11	positions of the column ends in RNLU

Figure 2.6

The column ordered list after the introduction of the
row number of the new non-zero element $a_{24}^{(2)}=-1$. A copy
of the fourth column is made at the end of the column
ordered list.

Similar transformations are to be performed in the column ordered list
in order to insert the row number of the new non-zero element $a_{24}^{(2)}=-1$. There
are neither free locations at the end of column 4 nor free locations at the
end of column 4. Therefore a copy of the row numbers of the elements in the
fourth column is made at the end of the column ordered list, and the row
number, 2, of the fill-in is stored at the end of the copy. The locations,
in which the row numbers of the non-zero elements of the fourth column were
stored before stage 1 are now free. This is shown be setting zeros in the
locations in array RNLU where the row numbers of the elements of the fourth
column were before the copy (from position 7 to position 9). The fourth
components of the fourth, fifth and sixth columns of array HA are updated.
HA(4,4) is different from HA(i,5) after these operations because a row
number of an element of matrix U is stored now in position 12 of array
RNLU. In fact, a row number of an element of U is stored also in position
1 of array RNLU (after removing the row number of the pivot). This is not

indicated in the updated column ordered list because all row numbers of the non-zero elements in the first column will nevertheless be removed after the first stage of GE (see also the discussion concerning the next fill-in). The structure is no longer compact. There are free locations between the third and the fifth columns. The structure of the column ordered list after the insertion of the fill-in $a_{24}^{(2)}=-1$ is given in **Fig. 2.6**.

Position	1	2	3	4	5	6	7	8	9	10	11	12	13	14	15
ALU	3.0	3.0	2.0	■	1.0	3.0	3.0	1.0	2.0	2.0	9.0	■	1.0	-1.0	
CNLU	4	0	0	2	4	3	5	3	4	4	5	1	2	4	
Row No.	R.1	Free		Row 3				Row 4		Row 5		Row 2			
HA(*,1)	2	12	4	8	10	positions of the row starts in ALU									
HA(*,2)	1	13	5	8	10	positions of the separators in ALU									
HA(*,3)	1	14	7	9	11	positions of the row ends in ALU									

Figure 2.7
The row ordered list after the introduction of the new non-zero element $a_{34}^{(3)}=1$. A copy of the second row is made at the end of the row ordered list. By ■ the elements of matrix L that are already calculated are denoted. HA(i,2) < HA(i,1) means that there are no non-zero elements in row i of matrix L (excepting the diagonal element $l_{ii} = 1$ which is never stored). R.1 is an abbriviation of Row 1.

Position	1	2	3	4	5	6	7	8	9	10	11	12	13	14	15
RNLU	2	3	3	4	3	5	1	2	5	4	3	1	4	5	2
Column No.	Col. 2		Col. 3		Col. 5		Col. 4					Free space			
HA(*,4)	1	1	3	7	5	positions of the column starts in RNLU									
HA(*,5)	1	1	3	9	5	positions of the separators in RNLU									
HA(*,6)	2	2	4	11	6	positions of the column ends in RNLU									

Figure 2.8
The column ordered list after the introduction of the row number of the new non-zero element $a_{34}^{(3)}=1.0$. The information stored in the last four locations of array RNLU as well as the information stored in the first components of the fourth, the fifth and the sixth columns of array HA will not be needed in the further computation.

Consider now **stage** 2 of the GE. Assume that the element $a_{22}^{(2)}=1.0$ is the pivotal element at this stage. As in the previous stage, only one operation has to be carried out and a fill-in, $a_{34}^{(3)}=1.0$, is created when this operation is performed. Information about this new non-zero element has to be inserted in both lists of the dynamic storage scheme.

It is not very difficult to insert $a_{34}^{(3)}=1.0$ and its column number in the row ordered list, because there are free locations before the fifth position in arrays **ALU** and **CNLU** (where the first element of row 3 is stored). Therefore there is no need to make a new copy of row 3. However, it is necessary to explain several intermediate operations in the row ordered list. The first pivotal element $a_{11}^{(1)}=6.0$ is already stored in array **PIVOT** (see the end of **Section 2.5**) and its location in the row ordered list is freed (the necessary modification being performing in HA(1,k), k=1,2,3). Some ordering is carried out; see **Section 2.4**. The elements are ordered so that in each row first the elements with column numbers smaller than or equal to s=2 are given and then the other elements. The positions in which elements with column numbers greater than 2 starts are given in the separators HA(i,2), i=1,2,...,5. In fact this ordering is finished only at the end of stage 2, but in this simple case, where only one operation is performed at stage 2, these changes could be represented immediately after the insertion of $a_{34}^{(3)}=1.0$. The row ordered list of the dynamic storage scheme at the end of stage 2 of the GE is given in **Fig. 2.7**.

The insertion of the row number, 4, of the new element $a_{34}^{(3)}$ in the column ordered list of the dynamic storage scheme is not very simple in this particular situation. There are no free location either before 4 or after column 4. Moreover, there is no elbow room left at the end of the column ordered list (i.e. there is no place for a copy of column 4 at the end of the column ordered list). Therefore it is necessary to perform a **garbage collection**. The situation after the performance of the garbage collection is shown in **Fig. 2.8**. Note that it is not necessary to set the contents of the last unoccupied locations of **RNLU** equal to zero (to indicate free locations), because only the position of last occupied location is sufficient.

Illustrations of all interesting situations were presented following the computations connected with the first two stages of the GE for the simple example (2.2): (1) new copies were made, (2) free locations adjacent to the row modified were exploited to store fill-ins and (3) the performance of a garbage collection was shown. No new situation appears in the further computations: (i) one more fill-in is produced, (ii) a garbage collection in the row ordered list is needed, and (iii) a copy of column 3 at the end of the column ordered list is made. Therefore the discussion of the small example is stopped here. However, several remarks are needed in connection with the illustrations given in this section.

Remark 2.6. In the example studied here it has been assumed that simple GE is applied. However, a careful examination of the discussion shows that the algorithm actually used is not very important for the modifications of the structure of the dynamic storage scheme. The dynamic storage scheme can be applied with many other algorithms. This is so because it does not matter how the fill-in is produced. The only important action in the dynamic modifications of the scheme is to find a free location for the fill-in. ∎

Remark 2.7. It has been assumed during this discussion and, partly, in that of the previous sections that the classical manner of exploiting sparsity is used. The dynamic storage scheme is not restricted to the classical manner. If the computer-oriented manner of exploiting sparsity is applied, then the same ideas can be used. The only significant change (for some classes of matrices at least) is the following. Since small non-zero elements are removed when the computer-oriented manner is applied, free locations between rows and/or columns in the dynamic scheme are created not only when copies of rows and/or columns are made at the end of the lists, but also when some elements $a_{ij}^{(s+1)}$ become small (even if $a_{ij}^{(s)} \neq 0$). This means that some copies and/or some garbage collections may be avoided when the computer-oriented manner is used and when many small elements are created. This is true not only for GE but also for many other algorithms. ∎

Remark 2.8. In this section it has been assumed that the pivot element is moved to array **PIVOT** at the end of the stage. Of course, this can be done, and is done in package **Y12M**, immediately after the determination of the pivot. The reason for the above assumption is the desire to describe as many different situations as possible, considering as small an amount of computations as possible and as small an example as possible. Indeed, all important cases in the treatment of the dynamic scheme are illustrated by using two operations only. If the pivot is moved to array **PIVOT** immediately after its selection and if the small example (2.2) is considered, then the whole GE process can be performed **without any copies and garbage collections** in the row ordered list of the dynamic storage scheme. ∎

2.9. Initialization of the dynamic storage scheme

It has been explained what type of structure is required for the two lists of the dynamic storage scheme before the start of the GE. Moreover, it has been explained how one can exploit this structure during the computation. The explanations have been illustrated by performing the basic operations on a small test-matrix. The only question that is still open is the following: **how can the ordered lists be initialized?** The answer to this question will be given in this section. Small pieces of code will also be presented in order to demonstrate the implementation of the main ideas.

The first step in the initialization of the dynamic storage scheme is to set the contents of the third and the sixth column of array **HA** equal to zero. This can be done by the following piece of code:

```
        DO 10 I=1,N
          HA(I,3)=0
          HA(I,6)=0
    10 CONTINUE
```

Figure 2.9
The first step of the initialization of the dynamic storage scheme. ∎

The second step in the initialization of the dynamic storage scheme is to find the numbers of non-zero elements in the rows and columns of matrix A. The code by which this step can be performed is given in **Fig. 2.10**.

```
      DO 20 I=1,NZ
          HA(RNORIG(I),3)=HA(RNORIG(I),3)+1
          HA(CNORIG(I),6)=HA(CNORIG(I),3)+1
      20 CONTINUE
```

Figure 2.10
The second step in the initialization of the dynamic storage scheme.

From the code given in **Fig. 2.10** it is clear that an assumption that the input storage scheme is available is made here, i.e. the non-zero elements of matrix A, together with their row and column numbers, are stored (in an arbitrary order) in arrays **AORIG, RNORIG** and **CNORIG** respectively.

After the second step the number of non-zeros in row i is stored in HA(i,3), while the number of non-zeros in column j is stored in HA(j,6). ■

During **the third step of the initialization of the dynamic storage scheme** the positions of the row starts in the row ordered list and the column starts in the column ordered list are determined. Information about these positions is stored in the first and the fourth columns of array **HA**. All these operations are carried out by the code given in **Fig. 2.11**.

```
      HA(1,1)=1
      HA(1,4)=1
      DO 30 I=1,N-1
          HA(I+1,1)=HA(I,1)+HA(I,3)
          HA(I+1,4)=HA(I,4)+HA(I,6)
          HA(I,3)=0
          HA(I,6)=0
      30 CONTINUE
      HA(N,3)=0
      HA(N,6)=0
```

Figure 2.11
The third step in the initialization of the dynamic storage scheme.

The contents of the third and sixth column of array **HA** are set equal to zero during the third step. ■

The row ordered list of the dynamic storage scheme is in fact prepared during the next step, **the fourth step**. The code by which this step is performed is shown in **Fig. 2.12**. ■

In the next step, **the fifth step**, the contents of the second and the third columns of array **HA** as well as the contents of array **RNLU** are prepared. The code for these operations is given in **Fig. 2.13**. ■

After the fifth step the row ordered list of the dynamic storage scheme is completely prepared. The contents of array **RNLU** and of the fourth column of array **HA** are also ready. It is only necessary to prepare the contents of the fifth and the sixth columns of array **HA**. This last part of the initialization of the dynamic storage scheme is performed during **the sixth step**, illustrated by the code in **Fig. 2.14**. ■

```
      DO 40 I=1,NZ
        K=RNORIG(I)
        L=HA(K,1)+HA(K,3)
        CNLU(L)=CNORIG(I)
        ALU(L)=AORIG(I)
        HA(K,3)=HA(K,3)+1
   40 CONTINUE
```

Figure 2.12
The fourth step in the initialization of the dynamic storage scheme.

```
      DO 60 I=1,N
        HA(I,2)=HA(I,1)
        HA(I,3)=HA(I,1)+HA(I,3)
        DO 50 J=HA(I,1),HA(I,3)
          L=CNLU(J)
          RNLU(HA(L,4)+HA(L,6))=I
          HA(L,6)=HA(L,6)+1
   50   CONTINUE
   60 CONTINUE
```

Figure 2.13
The fifth step in the initialization of the dynamic storage scheme.

```
      DO 70 I=1,N
        HA(I,5)=HA(I,4)
        HA(I,6)=HA(I,4)+HA(I,6)
   70 CONTINUE
```

Figure 2.14
The sixth step of the initialization of the dynamic storage scheme.

The pieces of codes given in **Fig. 2.9 - Fig. 2.14** can easily be connected in a subroutine for initialization of the dynamic storage scheme. Several additional operations are also needed during the initialization of the dynamic storage scheme: one should check:

(i) whether there are rows without non-zero elements,

(ii) whether there are columns without non-zero elements,

(iii) whether there are dublicated non-zero elements.

These checks, together with some other checks, are performed in the subroutine that initializes the dynamic storage scheme in package **Y12M**. If the subroutine finds any error during the initialization, then it stops the calculations and returns with an appropriate error message.

The codes given in **Fig. 2.9 - Fig. 2.14** are written for the case where the classical manner of exploiting sparsity is in use. However, the same ideas can also be applied to the computer-oriented manner, and are applied for this case in package **Y12M**.

It is assumed here that the arrays of the input storage scheme, **AORIG**, **CNORIG** and **RNORIG**, are different from the three main arrays, **ALU**, **CNLU** and **RNLU**, of the dynamic storage scheme. This will be the case when the computer-oriented manner is applied, and sometimes also when the classical manner is used. However, in the latter case it may be more profitable to use the same arrays in the input storage scheme and in the dynamic storage scheme (see **Remark 2.1** in **Section 2.4**). The initialization of the dynamic storage scheme in this situation can be performed by using similar ideas as those applied to obtain the codes given in this section; more details can be found in [341].

The case where matrix A is square is handled in this section. However, the ideas applied here can be used, with some obvious modifications, also when rectangular matrices are to be treated. As already mentioned in **Section 2.4** (see **Remark 2.5**), array **HA** should be split into two arrays, **HA1** and **HA2**. If this is done, then all of the ideas discussed here for square matrices can easily be applied also to rectangular matrices. This is done in package **LLSS01** (see [295,320-322]).

2.10. Use of linked list in the dynamic storage scheme

Ordered lists are used in the dynamic storage scheme described in the previous sections of this chapter. Of course, this is not the only scheme that can be used and probably not the best one for all situations. Another storage scheme, which was popular during the sixties, is based on so-called **linked lists**. The use of such a storage scheme together with its advantages and drawbacks will be outlined in this section.

The basic idea used when linked lists are created is a very simple one. If an element a_{ij} is stored in position K ($1 < K < NZ$) of array **ALU**, then the position where the next element of row i is located in array **ALU** can be found in RNLU(K), while the position where the next element in column j is located in **ALU** can be found in CNLU(K). Negative integers in **RNLU** and **CNLU** indicate the end of a row and a column respectively. The three large arrays **ALU**, **RNLU** and **CNLU** together with two **INTEGER** arrays of length N (HA(*,1) and HA(*,4) will be used here) are quite sufficient to carry out the computations. However, several extra **INTEGER** arrays may be added to the scheme in order to increase efficiency. In code **MA18** ([48]) 13 **INTEGER** arrays of length N are used.

The small matrix (2.2) is used to illustrate the linked lists in **Fig. 2.15**. Consider the element $a_{44} = 2.0$. This element is stored in position K=3 in array **ALU**. RNLU(3)=4 gives the position (in **ALU**) where the next element in row 4, $a_{45} = 1.0$, is stored. RNLU(4)=-4 indicates that a_{45} is the last element in the 4'th row. -RNLU(K) gives the number of the row under consideration when the value of RNLU(K) is negative. Consider again $a_{44} = 2.0$, which is stored in ALU(3). Since CNLU(3)=10, the next element in column j is stored in ALU(10). This is the last element in column 4, because CNLU(10)=-4.

It is very easy to insert fill-in when linked lists are in use. It has been seen in the previous section that a fill-in $a_{24}^{(2)} = 1.0$ is created during the first stage of the GE. Let us see how this new non-zero element can be inserted in the storage scheme based on linked lists. The new element is stored in ALU(12) (the first free position in the linked list). The first

element of row 2 is stored in ALU(3), because HA(2,1)=3. It is necessary
to set RNLU(12)=RNLU(3) and RNLU(3)=12 in order to update the row linked
list. The first element in column 4 is stored in ALU(2), because
HA(4,4)=2. Therefore the column ordered list is updated by setting
CNLU(12)=CNLU(2) and CNLU(2)=12.

Position	1	2	3	4	5	6	7	8	9	10	11	12	13	14	15
ALU	9.0	2.0	2.0	1.0	3.0	3.0	1.0	1.0	2.0	3.0	6.0				
RNLU	2	-5	4	-4	6	7	-3	9	-2	11	-1				
CNLU	5	3	10	6	-5	-3	8	-2	11	-4	-1				
HA(*,1)	10	8	5	3	1	positions of the row starts in ALU
HA(*,4)	9	7	4	2	1	positions of the column starts in ALU

Figure 2.15
The linked list for the matrix from (2.2). The non-zero
elemets are ordered as in Fig 2.1. The locations after
position 11 are free and can be used to store fill-ins.

The storage schemes based on linked lists have at least three advantages
(compared with the dynamic storage scheme discussed in the previous sections):

> **(i)**   No copies of rows and/or columns are needed.
> **(ii)** No garbage collections are made.
> **(iii)** It is very easy to insert fill-ins.

However, there are also several disadvantages of the codes based on
linked lists (see [66,341]). The **main disadvantage** of the linked list can be
described as follows. If the row or column number of an non-zero element is
needed (this will be the case if, for example, a row multiplied by a constant
is to be subtracted from another row, which is typical for the Gaussian
transformations), then one has to search through the row or column list until
the row or column number searched for is found. This is clearly a cumbersome
operation unless the matrix is very sparse and stays that way. Of course, the
row and column numbers can be stored along with the linked list, but two extra
arrays of length **NN** are to be used when this is done (this means that four
integers must be stored per each element).

The main disadvantage of the linked list is so serious that linked lists
have hardly been used in the last 20 years (an exception is [280]).
However, linked lists may be useful for some additional short arrays (of
length **N**). The following two examples are typical:

> **(a)**   a list linking the columns of the first
>   non-zero element in each row.
> **(b)**   a linked list of all rows and/or columns
>   with the same number of non-zero elements.

    Short linked lists are employed in code **MA28** (see [63-69,76-77]). A
very well-known code in which long linked lists are used is **MA18**, [48],
which has been developed in 1971.

2.11. Application of static storage schemes

    All storage schemes discussed until now are dynamic storage schemes
(excepting the input storage scheme from **Section 2.1**). The use of dynamic
storage schemes is rather expensive. The transformations in such a storage
scheme require both extra computing time and extra storage. Therefore it is
desirable to use a static storage scheme instead of a dynamic one if this is
possible and, moreover, if this is not connected with extra computations. The
application of static storage schemes will be discussed in this section.

    Assume first that an iterative method is used. Many such methods are
advocated by different authors, [3-4,10-17,29,45-46,80-81,87-89,93-96,139,
159-160,170,204-210,281-287]. It should be noted here that convergence is not
guaranteed for most of them when general sparse matrices are considered. One
can, for example, apply an iterative method to the normal system $A^TAx=A^Tb$
and, thus, convergence may theoretically be ensured, but in practice such a
method will either converge very slowly or not converge at all when matrix
A is not well-conditioned.

    A static storage scheme is very suitable for **purely** iterative methods,
because **no fill-in appears during the computations**. The input storage scheme
from **Section 2.1** can successfully be used. However, if one is prepared to
perform some extra work (as, for example, to order the elements by rows or by
columns), then a more economical storage scheme can be designed. It contains
only two long arrays (say, **ALU** and **CNLU**) of order **NZ** and one **INTEGER**
array of length **N** (say, the first column of the array **HA** used in the
previous sections). Thus, the storage needed is only **2NZ + N** (while **3NZ**
locations are needed for the input storage scheme). In **ALU** the non-zero
elements are stored. They are ordered by rows (or by columns); the order
within a row (a column) can be arbitrary. The column numbers (row numbers) of
the non-zero elements are stored in **CNLU** so that if a_{ij} is stored in
ALU(K), then j (i) is stored in CNLU(K). Information concerning row
starts (column starts) is stored in HA(*,1). Several extra **INTEGER** arrays
of length N may be needed in some particular implementations.

    The second case in which a static storage scheme can easily be applied is
the case where a direct method (the discussion of the direct methods for
sparse matrices will be started in the next chapter) is used under the
assumption that **no pivoting for numerical stability is needed**. In such a case
it is possible to perform the so-called **symbolic factorization** in order to
determine the pattern of decomposed matrix (the factors **L** and **U** when the
GE is in use). Some pivotal interchanges are normally to be used in order to
keep the number of fill-ins small (or, in other words, in order to preserve
sparsity better). At the end of the symbolic factorization (where one works
with the indices of the elements only and no arithmetic operations are
performed) the positions of **all** fill-ins that will appear during the actual
decomposition are determined (but their numerical values are not calculated).
With this information the code can prepare a static storage scheme in which
there are free locations for the fill-ins at appropriate positions in the
arrays used. Using this static storage scheme, the actual computations can be

carried out **(i)** without making copies of row/columns, **(ii)** without garbage collections and **(iii)** without any search for free locations for fill-ins during the part where the arithmetic operations are performed. However, it should be emphasized that the advantages **(i)-(iii)** are not achieved for nothing. The price is **(a)** the performance of a symbolic factorization and **(b)** the assumption that no pivotal search for preserving stability is needed. The most important restriction (in the context of this book) is the assumption of the numerical stability in absence of pivoting. For some classes of matrices (as, for example, diagonally dominant or positive definite matrices) no pivotal interchanges for stability are needed, and the method is applicable for such matrices. For matrices of this type one can apply the methods used in **SPARSPAK-A** (**[115-117]**) and in the symmetric part of the **YALE** package (**[82]**, see also **[83-85]**). For general sparse matrices, the main topic of this book, this method is not applicable.

George and Heath (**[112]**, see also **[113-114,118-121]**) have shown that a static storage scheme as that outlined in the previous paragraph can also be applied in the solution of linear least-squares problems. The method of **George and Heath** is implemented in **SPARSPAK-B**; this method will be discussed in **Chapter 11**.

In general one should carry out pivotal interchanges not only in order to preserve sparsity as well as possible, but also in order to preserve the numerical stability. This is certainly true when, for example, GE is carried out for general square matrices (**[99-100,133,155,236,249-250,259,270-277]**). It was believed until 1985 that the use of a static storage scheme is impossible when pivoting for preserving the numerical stability has to be performed. It is remarkable that **George and Ng [122-123]** developed an algorithm in which a static storage scheme can be applied for GE with partial pivoting. The algorithm of **George and Ng** can be outlined as follows. Assume that all diagonal elements of matrix A are non-zeros. This is not a restriction because when $rank(A)=n$ there exists a permutation matrix Q such that all diagonal elements of QA are non-zeros. Consider $B=A^TA$ and let P be a permutation matrix chosen so that the Cholesky factor R of B ($R^TR=PBP^T$) is sparse. Then it can be proved, **[122]**, that the use of GE to decompose matrix QA with any partial pivotal strategy will lead to triangular matrices L and U whose sparsity patterns are contained in the sparsity patterns of the factors R^T and R. This fact can be used to construct a static storage scheme that can be applied with the actual GE process. Five preliminary steps are necessary.

   (i)   A permutation matrix Q such that all diagonal elements of matrix QA are non-zeros must be found.

   (ii)   The sparsity pattern of the normal matrix $B = (QA)^TQA = A^TA$ must be determined.

   (iii) A permutation matrix P such that PBP^T has a sparse Cholesky factor R (with $R^TR = PBP^T$) must be found.

   (iv)   A symbolic factorization of matrix B must be performed (in order to determine the sparsity pattern of the Cholesky factor R).

(v)   The matrix $C = PQAP^T$ must be formed;   C   will
be used in the actual computations during   GE.

After these preliminary steps the actual (numerical) LU factorization
can be obtained by using partial pivoting with some permutation matrix P_{GE}
such that $LU = P_{GE}PQAP^T$.   The triangular factors   L   and   U   are stored in
the storage for the sparsity patterns of R^T and   R,   which is prepared
during the fourth step of the above algorithm.

The method based on the use of a static storage scheme has several
advantages (compared with the use of the dynamic storage scheme described in
the previous sections):

(a)   No copies of rows and/or columns are needed.
(b)   No garbage collections are needed.
(c)   No search for free locations for inserting fill-ins is necessary.

However, advantages   (a)-(c)   are not obtained for nothing: the five
preliminary steps   (i)-(v)   have to be performed before the beginning of the
actual computations. It is not quite clear whether the operations in the
preliminary steps   (i)-(v)   are cheaper than   making copies and possibly
garbage collections together with a search for free locations for fill-ins.
The comparisons made in   [122]   do not answer the question, because the
pivotal strategy in the version of   **MA28**   used in the experiments is very
expensive.

The method based on using a static storage scheme has several disadvantages (compared again with the dynamic storage scheme):

(A) The storage needed for the non-zero elements of the factors L and U may be overestimated.   In some extreme cases, as for example in presence of one or more dense rows, the normal matrix and, thus also R may be dense even if A is sparse.
(B) This method imposes a restriction on pivoting. The method works **only** if **partial pivoting** is used. **Partial pivoting is not very popular for sparse matrices** (see also **Chapter 4**).
(C) It is not clear how the   **computer-oriented manner of exploiting the sparsity**   can be applied.

An attempt to avoid disadvantage (A) has been made by **George and Ng**;
by treating dense rows in a special way. However, it is difficult to define
when a row should be considered as dense.

The idea of using a static storage scheme is very interesting and promising. It will become even more attractive if **(1)** the restriction to partial pivoting is removed, and **(2)** it is proved that it can also be applied together with the computer-oriented manner of exploiting the sparsity.

2.12. The storage scheme of Gilbert–Peierls

Assume that matrix A is stored in a static storage scheme by columns (i.e. first the non-zero elements of the first column, then the non-zero elements of the second column and so on). Assume also that some storage for the non-zero elements of L and U is available. Then **Gilbert and Peierls** ([125]) proposed to carry out GE column-wise. When a new column of L and U is calculated it is stored in the additional storage available. Thus, at each stage one new column is added to the scheme, and when the GE is completed the non-zero elements of L and U are ordered by columns in the appropriate arrays. This short description shows that the storage scheme of Gilbert-Peierls is in fact a dynamic storage scheme: it is updated at the end of each stage. However, it is a very special dynamic storage scheme and, therefore, deserves to be considered independently of the dynamic storage schemes described previously.

While the above description is by no means complete, it shows clearly that the **Gilbert and Peierls** storage scheme has the advantages of the static storage scheme proposed by **George and Ng** (see the previous section). However, it should be added immediately that it has also disadvantages **(B)** and **(C)** above. Therefore the final remark from the previous section is also applicable for this scheme.

It should be pointed out here that not only is the storage scheme proposed by **Gilbert and Peierls** interesting, but also the algorithm used for the decomposition of general sparse matrices by GE applied by these authors seems to be very efficient. Therefore this algorithm should be studied carefully when an algorithm for general sparse matrices is to be chosen.

2.13. The vectorizable storage scheme in ITPACK

The **ITPACK**, [159], vectorizable storage scheme is based on keeping the non-zero elements in a **two-dimensional REAL** array, say A. The first dimension of array A is equal to the order N of the matrix. Let r_i be the number of non-zeros in row i (i=1,2,...,N) of matrix A. The second dimension is equal to $r_{max}=\max\{r_i\}$, i=1,2,...N. In each row of array A the non-zero elements of row i of matrix A are stored; if $r_i<r_{max}$ then the last $r_{max}-r_i$ locations are filled with zeros. An INTEGER array of the same size as array A is also needed; the column numbers of the non-zero elements are stored in this array. The advantage of the scheme is that it is efficient when vector machines are used. The disadvantage is that if there are a few rows with many non-zero elements, then a lot of unnecessary computations will be needed and, moreover, the storage used will be unnecessarily large. This is a static scheme that is efficient for matrices arising after discretizing partial differential equations (then nearly all rows contain the same number of non-zeros). It is used with success for purely iterative methods where matrix-vector multiplications are the most time-consuming operations. There exist modifications of this scheme for some preconditioned systems in which

the preconditioners are obtained by calculating the so-called incomplete LU
(**ILU**) factorization or some of its modifications (the **ILU** factorizations
will be discussed in **Chapter 11**).

The vectorizable **ITPACK** storage scheme has been improved by **Paulini
and Radicati**, [190]. The rows of the matrix are first reordered by decreasing
number of non-zero elements. Then the same idea as in the original **ITPACK**
scheme is used but the rows of array A are not filled with zeros now. This
implies that the number of non-zero elements (occupied locations) in each
column must be given. The number of non-zeros in the first column is N, but
the last columns will normally contain less non-zeros. Thus, the unnecessary
operations with zeros are avoided (but the storage needed is not decreased)
when this scheme is in use.

2.14. Other storage schemes

Other storage schemes for sparse matrices have been proposed in the
literature (see, for example, [66]). The following storage schemes have been
used (but note that some of them are suitable for special problems only):

(a)	BIT-maps, [66].	
(b)	Hash coding, [66].	
(c)	Storage schemes for band matrices, [55,155,178,204].	
(d)	Frontal and multifrontal techniques, [78,201].	
(e)	Multigrid techniques, [268].	

The first two techniques are not used very much for general sparse
matrices (see again [66]). Matrices that are handled by the last three
techniques appear mainly in the numerical solution of partial differential
equations. There are many other techniques which can be used in this case (due
to the very regular structure of the data produced after the space discretization process. This is especially true when the **Laplace**, the **Poisson**
and the **Helmholtz** operators are discretized, because the so-called **"fast
solvers"** can often be applied after the discretization of these operators.
The special methods applicable to partial differential equations are beyond
the scope of this book (many methods commonly used in this situation as well
as the appropiate storage schemes are treated in [204]). It is interesting
to note here that the dynamic storage scheme together with the computer-
oriented manner is sometimes competitive with special methods in which special
properties of the matrices handled are exploited. An example will be given in
Chapter 5 (see **Table 5.5**), where package **Y12M** is compared with subroutines
written for symmetric, positive definite and banded matrices.

2.15. Using principles of graph theory

Concepts from graph theory are commonly used in sparse matrix algorithms.
This is especially true for symmetric and positive definite matrices (see, for
example, **George and Liu**, [115]).

An undirected graph $G = (X,U)$ consists of a set X of N vertices and a set U of NZ edges. The vertices are x_i $(i=1,2,...,N)$. An edge $\{x_i,x_j\}$ is a pair of two distinct vertices. If edge $\{x_i,x_j\} \in U$, then it is said that the vertices x_i and x_j are adjacent. Let vertex x_i be fixed. Then the adjacency set $\text{Adj}(x_i)$ contains **all** vertices of G that are adjacent to x_i. Let $Y \subseteq X$. Then the adjacency set $\text{Adj}(Y)$ is a set of vertices of X, each of which is an adjacent vertex to at least one vertex in set Y. The number of elements in set $\text{Adj}(x_i)$ is called the degree of vertex x_i. A graph $G^* = (X^*,U^*)$ is a subgraph of graph $G = (X,U)$ if both $X^* \subseteq X$ and $U^* \subseteq U$.

Assume that A is a symmetric matrix. Assume also that $a_{ii} \neq 0$ for all $i \in \{1,2,...,n\}$. Then graph $G = (X,U)$ represents the structure of matrix A when X contains N vertices $\{x_1,x_2,...,x_N\}$ and when $\{x_i,x_j\} \in U$ if and only if a_{ij} is a non-zero of matrix A.

By the above definition an equivalence between the structure of a symmetric matrix and the corresponding graph is established. This is a useful equivalence because:

(i) the graph is invariant in a certain sense under symmetric permutations of its associated matrix,

---

(ii) if the problem arises after a discretization of a partial differential equation, there could be a correspondence between the underlying grid-structure and the corresponding graph (these can be equivalent), so that one is in a sense closer to the underlying problem when one is working with the graph.

There are some other advantages of using graphs instead of the matrix structure. However, the two advantages given above are quite sufficient to explain why graphs are very popular in the case where the sparsity of symmetric matrices is exploited.

Undirected graphs cannot be applied to non-symmetric sparse matrices. It is necessary to introduce **directed or bipartite graphs** in order to establish a corespondence between the structure of a non-symmetric sparse matrix and its graph.

A bipartite graph $G = (X,Y,U)$ consists of three sets X, Y and U where

(a) set X contains M vertices $\{x_1,x_2,...,x_M\}$,

(b) set Y contains N vertices $\{y_1,y_2,...,y_N\}$,

(c) set U contains NZ edges, and each edge links one vertex of set X with one vertex of set Y.

Consider $A \in \mathbf{R}^{m \times n}$. A bipartite graph $G = (X,Y,U)$ describes the sparsity pattern of A if and only if the following conditions are satisfied:

    (A)   the number of vertices in X is equal to the
           number of rows in A,

    (B)   the number of vertices in Y is equal to the
           number of columns in A,

    (C) $\{x_i, y_j\} \in U$ is equivalent to $a_{ij} \neq 0$.

The equivalence between the sparsity pattern of a general matrix A and
its bipartite graph has been exploited in several studies; [115-121,166,167].
The exploitation of this equivalence has certain advantages and may be very
useful in some cases. On the other hand, the use of the graph representation
requires two sets of definitions. In the first set the definitions from graph
theory are used. In the second set it is necessary to determine the same
concepts in terms that are familiar in the theory of matrices. This means that
graph representations should be applied only when these are really needed. All
concepts used in the following part of this book can in a natural way be
defined in terms that are traditionally used in matrix theory. Therefore no
graph representation of the sparsity pattern of matrices will be used.
However, it must be reiterated that graphs are useful in some cases.

2.16. Reduction of the integer locations used in the lists

The number of integer locations used can sometimes be reduced by
exploiting the fact that some of the non-zero elements at the beginning of a
row may have the same structure as the tail of the previous row. This is
especially efficient when a symbolic factorization is used to prepare a static
storage scheme for symmetric and positive definite matrices. The algorithm has
been proposed by **Sherman** in [218,219]. Similar ideas can also be applied
when linear least-squares problems are solved; **Manneback** [167]. A good
description of the algorithm can be found in the book of **George and Liu**
[115], pp.138-141. The dynamic storage scheme will be used in the following
chapters and Sherman's scheme is not.very efficient when dynamic storage
schemes are used in the treatment of general sparse matrices. Therefore it
will not be discussed further. However, this scheme is very efficient for some
special problems; including problems for which the sparsity patterns of
adjacent rows are similar. Such a requirement is often satisfied for problems
that arise after the discretization of partial differential equations defined
by some simple differential operators on simple space domains.

2.17. Why are the methods studied here based on the dynamic
storage scheme?

In the following chapters it will be assumed that the dynamic storage
scheme (described in **Section 2.3 - Section 2.9**) is used. This is not because
the dynamic storage scheme is the best one in all situations. Such a statement
is certainly not true for all classes of matrices. However, for general sparse
matrices the dynamic storage scheme seems to be the best choice. If the
dynamic storage scheme is used, then some work has to be done when a fill-in
is created (see also the previous sections):

> **(i)** one has to search for **free locations** in the lists   (where the fill-in together with its row and column number could be stored),

> **(ii)** one has   occasionally   to **copy** rows and/or columns at the end of the lists

> **(iii)** in the worst case,   one has to compress   the structure by performing **garbage collections**.

These operations can be avoided by using some of the other schemes discussed in this chapter. However, it must be emphasized that while the other schemes have the advantage that the operations (i)-(iii) are avoided, they also have some disadvantages:

> **(a)** some price has to be paid in order to avoid the operations (i)-(iii) (one has to perform some extra steps),

> **(b)** all other storage schemes introduce some restrictions (either they are applicable only to special matrices, or there is a restriction on the pivoting strategy),

> **(c)** it is not clear whether these schemes can be applied together with the computer-oriented manner of exploiting sparsity or not.

The conclusion is that if the classical manner of exploiting sparsity is in use and if partial pivoting preserves well the sparsity, then the other schemes (the **George-Ng** static storage scheme and the storage scheme proposed by **Gilbert-Peierls**) are competitive with the dynamic storage scheme and sometimes could be even better. However, if the problems are such that the computer-oriented manner performs well, then the dynamic storage scheme with dropping small elements and with iterative improvement is often much more efficient. Some experiments that confirm this statement will be described in **Chapter 5** and in **Chapter 11**.

2.18. Concluding remarks and references for the storage schemes

The simple input storage scheme from **Section 2.1** was first applied in the code ST in 1976 ([324]). Subsequently this storage scheme was also used in all codes developed in Copenhagen: SSLEST ([311,312]), SIRSM ([318,319]), LLSS01 ([320-322]) and Y12M ([329,331]). This input storage scheme is also used in the   HARWELL   code   MA28   ([63-69])   as well as in the subroutines F01BRE and F04AXE from the NAG Library ([178]).

The dynamic storage scheme from **Section 2.4** is based on ideas proposed by **Gustavson** ([136,137]), see also [33]. This scheme is implemented in all codes mentioned in the previous paragraph. It should be mentioned that the code **MA28** and the codes from the **NAG Library** have an interesting option by which an attempt to reorder the matrix to a block triangular form with

square diagonal blocks is carried out. The use of this option is profitable
if matrix can be reordered as a block triangular matrix with several large
diagonal blocks. This idea can be generalized by removing the requirement for
square diagonal blocks. An algorithm in which **rectangular diagonal blocks** are
allowed will be discussed in **Chapter 10**.

The deletion of the row numbers of column s after stage s of the GE
(see the end of **Section 2.5**) was first implemented in **ST** and then in all
codes developed in Copenhagen. The implementation of a device of this type in
MA28 and the **NAG** subroutines seems to be more efficient.

The device for performing **garbage collections** was originally proposed
by **Curtis and Reid [48]** and then improved in [63,77].

Only the dynamic storage scheme is fully described in this chapter,
because only this storage scheme is applicable (until now at least) when the
computer-oriented manner of exploiting the sparsity is in use. Other storage
schemes, some of which were briefly discussed in this chapter, are studied in
[3-5,62,82-85,93-96,122-125,131,166-170,175-178,192,194,198-202,204,218-
219,234-235,237-248,256-257,286-287]. Rather detailed descriptions of storage
schemes are given in the five books on sparse matrices, [5,69,115,194,341].

CHAPTER 3

GENERAL SCHEME FOR LINEAR ALGEBRAIC PROBLEMS

Let $A \in R^{m \times n}$ and $b \in R^{m \times 1}$ be given. Assume that $m \geq n$ and $rank(A) = n$. Denote $A^{\dagger} = (A^T A)^{-1} A^T$. $A^{\dagger}$ is called a **pseudo-inverse** or a **generalized inverse** of matrix A, [174,191]. Consider the problem:

$$\text{find the vector } x = A^{\dagger} b .$$

Obviously, $x \in R^{n \times 1}$ can be considered:

(i) as the least squares solution, i.e. the vector that minimizes the norm $\| r \|_2$ of $r = b - Ax$,

(ii) as the solution of the system $Ax = b - r$ when $A^T r = 0$ is also satisfied.

The solution of the above problem is trivial when A is a product of certain special matrices as, for example, diagonal, tri-diagonal, orthogonal and triangular matrices. Such special matrices are called **easily invertible**. A method for calculating x belongs to the class of **direct methods** when matrix A (or a matrix B obtained in some way from A) is expressed as a finite product of easily invertible matrices.

There exist numerous direct methods for solving $x = A^{\dagger} b$. This is especially true for the case $m = n$ ($m = n \Rightarrow A^{\dagger} = A^{-1}$ when $rank(A) = n$). In this chapter it will be shown that **most of the direct methods for solving $x = A^{\dagger} b$ can be found as special cases of a quite general scheme.** It must immediately be emphasized however, that the general scheme is not a new method, but rather an attempt to develop **a tool for studying the common properties of a class of direct methods.** In the case where the matrix is **sparse** this common treatment of the direct methods can be used to construct a common approach which often leads to an improvement of the efficiency of the solution process (see **Section 3.3**). The usefulness of this approach, which can be applied in connection with any particular direct method within the general scheme, will be demonstrated by many numerical experiments.

3.1. Introduction of the general scheme

Definition 3.1. The computational process consisting of the following three steps is called **a general k-stage scheme for solving $x = A^{\dagger} b$.**

Step 1 - Transformation. The problem $x = A^{\dagger} b$ is replaced by another problem $y = B_1^{\dagger} c$ ($B_1 \in R^{p \times q}$, $c \in R^{q \times 1}$, $p \geq q$, $rank(B_1) = q$), where

(i) B_1 and c are calculated using A and b,

(ii) there is a simple relationship between x and y .

Step 2 - Generalized decomposition.   Calculate

$$(3.1) \quad \overline{B}_i = P_i B_i Q_i + E_i, \quad P_i \in R^{p \times p}, \quad Q_i \in R^{q \times q}, \quad E_i \in R^{p \times q}, \quad i=1,2,\ldots,k,$$

where

  (i) P_i and Q_i are permutation matrices,

  (ii) E_i are perturbation matrices,

  (iii) $\overline{B}_k$ is a product of easily invertible matrices,

  (iv) if $k > 1$,   then the matrices $\overline{B}_i$ are given by

$$(3.2) \quad \overline{B}_i = C_i \overline{C}_i D_i, \quad B_{i+1} = C_i^T C_i \overline{C}_i, \quad i=1,2,\ldots,k-1;$$

where C_i and $\overline{C}_i$ are matrices which depend on the particular method used (see the examples given in **Section 3.2**), while the matrices D_i are easily invertible.

Step 3 - Generalized substitution.   Compute

$$(3.3) \quad y_1 = Hc \quad with \quad H \overset{def}{=} (\prod_{i=1}^{k-1} Q_i D_i^{\uparrow}) Q_k \overline{B}_k^{\uparrow} P_k (\prod_{i=1}^{k-1} P_i^T C_i)^T$$

(here and below it is assumed that a product of matrices in which the upper index is smaller than the lower one, is equal to the identity matrix of an appropriate order; for example, the products in   (3.3)   are equal to the identity matrix when $k < 2$).   ■

The above definition is meaningful only if it allows us to calculate the **exact** solution x of the problem $x = A^{\uparrow}b$ (by using the relationship between x and y assumed in   Step 1) when all computations are carried out without rounding errors. It is obvious that this is the case if $y_1 = y$ in the absence of rounding errors. Therefore it is necessary to show that this condition is satisfied. The following lemma, in which the global perturbation matrix is expressed as a sum of matrix products H_j containing the factors E_j ($j=1,2,\ldots,k$),   is needed for the main result.

Lemma 3.1.   Consider

$$(3.4) \quad F = I - HB_1, \quad I \in R^{q \times q}, \quad F \in R^{q \times q}, \quad \text{where   I   is the identity matrix.}$$

Assume that the matrices $\overline{B}_k$ and D_i ($i=1,2,\ldots,k-1$)   are of full column rank. Then

$$(3.5) \quad F = \sum_{j=1}^{k} H_j,$$

where

(3.6) $H_j \overset{def}{=} (\prod_{i=1}^{k-1} Q_i D_i^\dagger) Q_k \overline{B}_k^\dagger P_k (\prod_{i=j}^{k-1} P_i^T C_i)^T P_j^T E_j Q_j^T (\prod_{i=1}^{j-1} Q_i D_i^T)^T$.

Proof. Use (3.4), (3.1), (3.2), (3.3) and (3.6) to get

(3.7) $F = I - (\prod_{i=1}^{k-1} Q_i D_i^\dagger) Q_k \overline{B}_k^\dagger P_k (\prod_{i=2}^{k-1} P_i^T C_i)^T B_2 (\prod_{i=1}^{1} Q_i D_i^\dagger)^T + H_1$.

Use successively (k-2 times) (3.1), (3.2) and (3.6) to transform (3.7) to

(3.8) $F = I - (\prod_{i=1}^{k-1} Q_i D_i^\dagger) Q_k \overline{B}_k^\dagger P_k B_k (\prod_{i=1}^{k-1} Q_i D_i^\dagger)^T + \sum_{i=1}^{k-1} H_j$.

Apply (3.1) to rewrite the last equality as

(3.9) $F = I - (\prod_{i=1}^{k-1} Q_i D_i^\dagger) Q_k \overline{B}_k^\dagger \overline{B}_k Q_k^T (\prod_{i=1}^{k-1} Q_i D_i^\dagger)^T + \sum_{i=1}^{k} H_j$.

The second term on the right-hand side of (3.9) is equal to the identity matrix (the assumption that the matrices involved in this term have full column rank should be used in the proof of this statement) and, thus, the lemma is proved. ■

Now the main result of this chapter can be formulated as follows.

Theorem 3.1. If $E_i = 0$ $(i=1,2,\ldots,k)$, then $H = B_1^\dagger$. If, moreover, the computations during Step 3 of the general k-stage scheme are carried out without rounding errors, then $y_1 = y$.

Proof. The proof of Theorem 3.1 follows immediately from the result proved in Lemma 3.1. ■

3.2.   Application to some particular methods

Six examples are given in this section to illustrate that some particular algorithms for the treatment of linear algebraic problems can be obtained from the general scheme defined in the previous section. These examples demonstrate also some common properties of the general k-stage scheme. Example 3.1, Example 3.4 and Example 3.5 show that Step 1 is not always needed. The first four examples are one-stage schemes, while the last two are two-stage schemes. To the author's knowledge no k-stage direct method with $k \geq 3$ is used in practice at present. The theory given in **Section 3.1** indicates that it is possible, in principle at least, to construct such a method.

Example 3.1. Assume that $m=n$ and consider Gaussian elimination (GE). This algorithm can be obtained from the general k-stage scheme by setting $k=1$ and

(3.10) $B_1 = A$, $c = b$, $p = q = n$, $y = x$,

(3.11) $\overline{B}_1 = LU$, $\overline{B}_1^\dagger = U^{-1}L^{-1}$ ($C_1 = L$, $\overline{C}_1 = I$, $D_1 = U$),

where L and U are the triangular matrices obtained by GE (see the previous chapters or [99,100,231,272]). For $k = 1$ the main formulae of the general k-stage scheme, (3.1), (3.2) and (3,3), are reduced to:

(3.12) $\overline{B}_1 = P_1 B_1 Q_1 + E_1$,

(3.13) $\overline{B}_1 = C_1 \overline{C}_1 D_1$,

(3.14) $y_1 = Hc = Q_1 \overline{B}_1^\dagger P_1 c$.   ∎

Example 3.2. Consider the method based on forming the normal equations (see, for example, [133,153,164,231]). This algorithm can be found from the general k-stage scheme with k=1 and

(3.15) $B_1 = A^T A$, $c = A^T b$, $p = q = n$, $y = x$,

(3.16) $\overline{B}_1 = L_1 D L_1^T$, $\overline{B}_1^\dagger = (L_1^T)^{-1} D^{-1} L_1^{-1}$ ($C_1 = L_1$, $\overline{C} = D$, $D_1 = L_1$),

where $L_1 \in \mathbf{R}^{n \times n}$ is a unit lower triangular matrix and $D \in \mathbf{R}^{n \times n}$ is a diagonal matrix. It is assumed that the symmetry of $A^T A$ is exploited by the use of some symmetric form of GE. If the Cholesky factorization is used, then $\overline{C} = I$.   ∎

Example 3.3. Consider the method of solving $x = A^\dagger b$ by forming augmented matrices ([23,25,164]). Assume that GE is used to factorize the augmented matrix. The algorithm based on this idea can be found from the general k-stage scheme by setting k=1 and

(3.17) $B_1 = \begin{vmatrix} \alpha I & A \\ A^T & 0 \end{vmatrix}$, $c = \begin{vmatrix} b \\ 0 \end{vmatrix}$, $p=q=m+n$, $y = \begin{vmatrix} \alpha^{-1} r \\ x \end{vmatrix}$, $\alpha \neq 0$,

(3.18) $\overline{B}_1 = LU$, $\overline{B}_1^\dagger = U^{-1}L^{-1}$ ($C_1 = L$, $\overline{C}_1 = I$, $D_1 = U$),

where $L \in \mathbf{R}^{q \times q}$ and $U \in \mathbf{R}^{q \times q}$ are the triangular factors obtained in the factorization of the augmented matrix B_1 by GE. The symmetry of B_1 is not exploited here, but (3.18) can easily be modified for some symmetric form of GE. However, the Cholesky factorization cannot be applied, because B_1 is not positive definite. Orthogonalization methods could be applied (instead of the GE process); see [21,22,30].   ∎

Example 3.4. Assume that an orthogonalization method is used in the solution of $x = A^\dagger b$. Such an algorithm can be obtained from the k-stage scheme with k=1 and

(3.19) $B_1 = A$, $c = b$, $p = m$, $q = n$, $y = x$,

(3.20) $\overline{B}_1 = QRD$, $\overline{B}_1^\dagger = R^{-1}D^{-1}Q^T$,   ($C_1 = Q$, $\overline{C} = D$, $D_1 = R$),

where the columns of matrix $Q \in \mathbf{R}^{m \times n}$ are orthonormal vectors (this means that $Q^T Q = I \in \mathbf{R}^{n \times n}$), $D \in \mathbf{R}^{n \times n}$ is a diagonal matrix and $R \in \mathbf{R}^{n \times n}$ is an upper triangular matrix. QRD is an orthogonal decomposition of matrix A. The orthogonalization methods will be studied in detail in **Chapter 12 - Chapter 16**. It is sufficient to mention now that if the Householder method or the Givens method ([127,128,150]) is applied to obtain this decomposition, then $D=I$. $D \neq I$ when the Gentleman-Givens method ([109,110]) is used. ∎

All of the algorithms discussed in the above examples are one-stage schemes. The first formula in each example describes the transformation step, while the second formula gives the result of the generalized decomposition for the algorithm under consideration. The next two examples are two-stage schemes. The first formula in each of these examples describes again the transformation step, while the second and the third formulae give the results obtained after the first and the second stage of the general scheme.

Example 3.5. Consider the **Peters-Wilkinson [193]** method. This algorithm can be obtained from the general k-stage scheme by setting k=2 and

$$(3.21) \quad B_1 = A, \quad c = b, \quad p = m, \quad q = n, \quad y = x,$$

$$(3.22) \quad \overline{B}_1 = LU, \quad (C_1=L, \ \overline{C}_1=I, \ D_1=U),$$

$$(3.23) \quad B_2 = L^T L, \quad \overline{B}_2 = L_1 D L_1^T, \quad \overline{B}_2^t = (L_1^T)^{-1} D^{-1} (L_1)^{-1} \quad (C_2=L_1, \ \overline{C}_2=D, \ D_2=L_1^T),$$

where $L \in R^{m \times n}$ is a unit trapezoidal matrix, $U \in R^{n \times n}$ is an upper triangular matrix, $L_1 \in R^{n \times n}$ is a unit lower triangular matrix and $D \in R^{n \times n}$ is a diagonal matrix. The matrix that has to be decomposed during the second stage is a positive-definite and symmetric matrix (like the matrix in the second example). It is assumed that symmetry is exploited by using some symmetric form of GE. If the Cholesky method is used to decompose $L^T L$, then $\overline{C}_2 = I$. ∎

Example 3.6. Consider the Golub method, [129]. This algorithm can be represented as a **two-stage** scheme as follows. Set $P_i=I$ and $Q_i=I$ for i=1,2 (i.e. no pivotal interchanges are used during the generalized decomposition). Use the fact that if $rank(A)=n$, then there exist matrices $Q \in R^{m \times n}$, $D \in R^{n \times n}$ and $R \in R^{n \times n}$ (where $Q^T Q = I \in R^{n \times n}$, D is diagonal and R is upper triangular) such that $YAZ = QDR$ (Y and Z being permutation matrices of appropriate orders); see, for example, [299,320-322]. Then the Golub algorithm can be represented by

$$(3.24) \quad B_1 = A^T A, \quad c = A^T b, \quad p = q = n, \quad y = x,$$

$$(3.25) \quad \overline{B}_1 = A^T Y^T Q D R Z^T, \quad (C_1=I, \ \overline{C}_1=A^T Y^T Q, \ D_1=DRZ^T),$$

$$(3.26) \quad B_2 = A^T Y^T Q, \quad \overline{B}_2 = Z R^T D, \quad \overline{B}_2^t = D^{-1} (R^T)^{-1} Z^T \quad (C_2=Q, \ \overline{C}_2=ZR^T, \ D_2=D).$$ ∎

Remark 3.1. All examples given in this section, excepting the first, are applicable for $m \geq n$. Special examples, valid only for $m=n$, can also be considered. This is illustrated by Example 3.1, where GE is represented as a one-stage scheme. ∎

Remark 3.2. Simple calculations show that for the method considered in Example 3.6 the matrix H from (3.3) is given by

$$(3.27) \quad H = ZR^{-1}D^{-2}(R^T)^{-1}Z^T,$$

which means that this matrix does not depend on the matrix Q (the matrix
with orthonormal columns). Therefore there is no need to store Q when the
actual computations with the algorithm from Example 3.6 are carried out.
This is not very important when matrix A is dense, because there are devices
by which information concerning matrix Q can be stored in an economical and
stable way; see, for example, [232]. The situation is different when matrix
A is sparse and the sparsity is to be exploited. This is so because Q as
a rule is much denser than A and, therefore, both the storage of matrix Q
and working with the non-zero elements of Q should be avoided. The fact that
Q is not needed in the actual computations when the algorithm from Example
3.6 is used shows that this algorithm is suitable for sparse matrices and
this has been exploited in the codes discussed in **Chapter 12-Chapter 16**; see
also [296,320-322]. ■

     It has been illustrated in this section that many particular direct
methods for solving $x=A^{t}b$ can be considered as special algorithms within
the general k-stage scheme. It should be emphasized that **the general k-stage
scheme together with the computer-oriented manner of exploiting sparsity
(which has been discussed in the previous chapters) provides a   general
approach that may be used in an attempt to increase efficiency when sparse
matrix techniques are applied with any particular algorithm from the general
k-stage scheme.** The general approach will be sketched in the next section and
several applications of this approach to particular algorithms will be studied
in detail in the next chapters.

3.3. General approach for solving sparse problems by using particular algorithms from the general k-stage scheme

     Assume that matrix A is large and sparse. Then the algorithms from the
general k-stage scheme can be applied together with some sparse matrix
technique. The storage schemes used when sparse matrix techniques are applied
with GE (which is a member of the general k-stage scheme) were described in
the previous chapter. However, it was also pointed out there that GE is used
only to simplify the description of the results. The storage schemes described
in **Chapter 2** can be applied, directly or after some obvious modifications,
to many other algorithms within the general k-stage scheme. Moreover, the
general k-stage scheme together with the computer-oriented manner of
exploiting sparsity can be used to define **a common approach for solving sparse
problems**.

     The computer-oriented manner of exploiting sparsity was described in
Chapter 1 (and illustrated by some numerical examples there and in **Chapter
2**) in the context of GE. It is equally applicable for other methods. The
essential aspects of the computer-oriented manner are: **(i) some   "small"
elements are dropped during the computations** (using an appropriate dropping
criterion) and **(ii) an attempt is made to regain by an iterative method the
accuracy lost because some elements have been dropped.** These operations can
be performed with any particular method within the general k-stage scheme.
Both the dropping criterion and the iterative method can be varied. A relative
drop-tolerance can be applied instead of the simple absolute drop-tolerance
proposed in **Chapter 1.**   Conjugate gradient-type methods ([12-17,45-
46,146,168]) can be applied instead of the simple iterative refinement
discussed in **Chapter 1.**

The ideas sketched above indicate that the following common approach can be used with any particular algorithm within the general k-stage scheme when large and sparse problems $x = A^\dagger b$ are solved.

Common approach for solving large sparse problems

(a) Use some dynamic storage scheme (as, for example, the storage scheme from Chapter 2).

(b) Implement the computer-oriented manner during the generalized decomposition (Step 2 of the general k-stage scheme).

(c) Carry out some iterative process to regain the accuracy lost during the generalized decomposition because some "small" elements are removed.

Remark 3.3 The common approach for solving the problems $x = A^\dagger b$ is only applicable when the problem is **large and sparse**. Such an approach is inefficient when matrix A is dense. This is so because:

(i) there is no meaning in applying a dynamic storage scheme (as that described in **Chapter 2**) for dense matrices,

(ii) removing "small" elements will save neither storage nor computing time when the matrix is dense,

(iii) the use of an iterative process to improve the direct solution will always increase both storage and computing time.

It must be emphasized that (i) - (iii) are true for **any** particular method from the general k-stage scheme. ∎

Remark 3.4. The application of the common approach for solving large and sparse problems $x = A^\dagger b$ is justified if the sparsity of matrix A is not lost during the transformation (Step 1 of the general k-stage scheme). Example 3.2 shows that this may happen: the matrix $A^T A$ may be dense even if matrix A is very sparse. This means that in fact the common approach should be used if matrix B_1 is sparse. For two-stage schemes even this requirement may be insufficient: the sparsity may be lost during the transition from Stage 1 to Stage 2. As an illustration of this statement it should be mentioned that sparsity may be lost when the matrix $B_2 = L^T L$ is calculated in the Peters-Wilkinson method from Example 3.5. ∎

Remark 3.5. If matrix A is sparse and if sparsity is preserved both during Step 1 of the general k-stage scheme and in the transition between any two successive stages when $k \geq 2$, then the common approach, when properly applied, **may** give savings both in storage and computing time. This has already been demonstrated in **Chapter 1** and in **Chapter 2** for the case where m=n and GE with iterative refinement is used. The use of iterative refinement in connection with the general k-stage scheme will be discussed in the following sections of this chapter. ∎

3.4. Application of iterative refinement in connection with the general k-stage scheme

Iterative refinement of the solution y_1 obtained by (3.3) can be carried out as follows:

(3.28) $r_i = c - B_1 y_i$, $d_i = H r_i$, $y_{i+1} = y_i + d_i$, $i = 1, 2, \ldots, \rho-1$.

The vectors r_i are called, as in the case where GE is used, **residual vectors**, while the vectors d_i are called **correction vectors**. Different stopping criteria may be applied in order to terminate the iterative process (3.28) after some $i = \rho-1$ (see, for example, [24,31,172,231,259,271-272]). Then y_ρ (if it satisfies certain acceptability criteria) is used to obtain an approximation to x.

Consider matrix F defined in (3.4). Define $F^0 = I \in R^{q \times q}$. The following results, given here without proof, describe some properties of the iterative refinement process **when it is applied to the general k-stage scheme**. The proof of these results can easily be derived from well-known results; see, for example, [97,341].

Theorem 3.2. Let $\{y_i\}$ be the sequence of vectors calculated by (3.3) and (3.28). Then

(3.29) $$y_i = y + F^{i-j}(y_j - y) + \left(\sum_{\nu=1}^{i-j-1} F^\nu \right) Hs,   \qquad s = c - B_1 y,$$

where $j < i$ is a fixed positive integer. ∎

Theorem 3.3. Let $\{d_i\}$ be a sequence of vectors calculated by (3.20). Then

(3.30) $d_i = F^{i-j} d_j$,

where $j \leq i$ is a fixed integer. ∎

Theorem 3.4. Let λ_i ($i = 1, 2, \ldots, q$) be the eigenvalues of matrix F. Assume that

(3.31) $|\lambda_j| \leq |\lambda_1| < 1$ for ∀ $j \in \{2, 3, \ldots, q\}$.

Then

(3.32) $$y = y_j + \sum_{i=j}^{\infty} d_i - (HB_1)^{-1} Hs,$$

where s is defined as above, see (3.29), and j is a fixed positive integer. ∎

Remark 3.6. Condition (3.31) can be replaced by

(3.33) $\| F \| < 1,$

where $\| \cdot \|$ is any matrix norm induced by the vector norm under considera-
tion.   ∎

Consider (3.32). It implies that

(3.34) $\| d_i \| \to 0$ *as* $i \to \infty$

and, therefore, if $\| d_{i-1} \| > \| d_i \|$, then one could expect the iterative
process to be divergent or only slowly convergent. This condition, combined
with some other conditions, can be used in the stopping criteria (as, for
example, in package **Y12M**, [331,341]). The conclusion is that the above
theorems indicate that the correction vectors d_i can be used in the stopping
criteria for the iterative refinement process applied in connection with the
general k-stage scheme. This fact is well-known for the iterative refinement
process applied in connection with several particular algorithms from the
general scheme ([21-27,231]). The theorems stated above show that this also
holds for **any** particular algorithm from the general scheme.

The correction vectors d_i $(i=1,2,\ldots,\rho)$ can also be used as a measure
of the accuracy achieved (or, in other words, in the acceptability criteria).
This will be discussed in connection with the particular methods for which the
sparse matrix software has been developed, but it is worthwhile to mention
here the fact that this is true for **any** particular method from the general
k-stage scheme.

3.5. Relationship between the convergence of the iterative refinement process and the drop-tolerance used

Assume that the computer-oriented manner of exploiting sparsity is used
and that the decision whether a non-zero element is small is made by criterion
(1.9). Assume also that B_i $(i=1,2,\ldots,k)$ are well-scaled and that the
drop-tolerance T_i (that is to be used at stage i of the general k-stage
scheme) is chosen so that

(3.35) $T_i = w_i b_i,$ $0 \le w_i < 1,$ $i=1,2,\ldots,k,$

where b_i is the size of the non-zero elements in matrix B_i. Assume that at
any stage i the decomposition described by (3.1)-(3.2) is obtained either
by GE or by some orthogonal factorization. Under these assumptions the
following inequality is satisfied by the norm of the perturbation matrix E_i:

(3.36) $\| E_i \| \le f_i(m,n) \; \bar{\epsilon} \; g_i(A),$ $\bar{\epsilon} = max(\epsilon, T_i/b_i),$

where

(i) $f_i(m,n)$ is a function of m and n which depends on the
particular methods used at stage i $(i=1,2,\ldots,k)$,

(ii) ϵ is the machine accuracy (the smallest positive real
number such that $1.0 + \epsilon \neq 1.0$ in the machine arithmetic
on the computer in use),

(iii) $g_i(A)$ $(i=1,2,\ldots,k)$ is a function of the norm of matrix
       A corresponding to the norm used in the left-hand-side of
       (3.36).

For $T_i = 0$ (i.e. if the classical manner of exploiting sparsity is used)
explicit expressions for $f_i(m,n)$ and for $g_i(A)$ for some particular methods
from the general k-stage scheme can be found in [24,110,141,231,270-276].

Let

(3.37) $f(m,n) = \max_{1 \leq i \leq k} \{f_i(m,n)\},$

(3.38) $g(A) = \max_{1 \leq i \leq k} \{g_i(A)\}$

and

(3.39) $\overline{g}(A) = \max_{1 \leq i \leq k} \left(\left\| \left(\prod_{j=1}^{k-1} Q_j D_j^\dagger \right) Q_k \overline{B}_k^\dagger P_k \left(\prod_{j=i}^{k-1} P_j^T C_j \right)^T P_i^T \right\| \ \left\| Q_i^T \left(\prod_{j=1}^{i-1} Q_j D_j^T \right)^T \right\| \right) .$

With this notation the following bound for the matrix F, defined by
(3.4), can be derived:

(3.40) $\|F\| \leq k\, f(m,n)\, \overline{\epsilon}\, g(A)\, \overline{g}(A), \qquad \overline{\epsilon} = \max_{1 \leq i \leq k} \{\overline{\epsilon}_i\} .$

Remark 3.7. For many particular direct methods for solving the algebraic
problem $x = A^\dagger b$ the product $g(A)\, \overline{g}(A)$ can be expressed in the simplest
case (where $k=1$ and $T_1=0$) by the condition number $\kappa(A)$ of matrix A. The
condition number depends on the matrix norm used and is defined by

(3.41) $\kappa(A) = \|A\| \ \|A^\dagger\| .$

In the case of GE, which requires $m=n$, the relationship between $g(A)$
$\overline{g}(A)$ and $\kappa(A)$ is the simplest one:

(3.42) $g(A)\, \overline{g}(A) = \kappa(A) .$

For the method of solving $x = A^\dagger b$ by forming the normal equations (see
Example 3.2) the relationship between $g(A)\, \overline{g}(A)$ and $\kappa(A)$ is given by
the following expression:

(3.43) $g(A)\, \overline{g}(A) = \kappa^2(A) .$

If the method of augmented matrices (see Example 3.3) is used with an
optimal value of parameter α, then the relationship between $g(A)\, \overline{g}(A)$
and $\kappa(A)$ can be expressed in the following way:

(3.44) $g(A)\, \overline{g}(A) = \sqrt{2}\, \kappa(A) .$

It should be mentioned, however, that it is not very easy to find the
optimal value of parameter α, because this value is related to the singular
values of matrix A which are normally not known.

If an orthogonalization method is in use and if A is a square matrix, then the relationship between $g(A)\,\bar{g}(A)$ and $\kappa(A)$ is the same as that for GE; i.e. it is again given by (3.42). While GE can only be used in the case m=n, the case m>n can also be treated with the orthogonalization methods, but the relationship between $g(A)\,\bar{g}(A)$ and $\kappa(A)$ becomes more complicated when the matrix is rectangular.

The relationship between $g(A)\,\bar{g}(A)$ and $\kappa(A)$ is derived (for some of the algorithms discussed in the first four examples in Section 3.2) under the assumption that the spectral matrix norm is in use. This means that the norm of matrix A is defined by

$$(3.45) \qquad \|A\|_2 = \mu \; ,$$

where μ is the square root of the largest eigenvalue of $A^T A$. The condition number obtained by using the spectral matrix norm is called **the spectral condition number.** ■

Remark 3.8. Of course, relationship (3.40) cannot be used for a direct estimation of the norm of matrix F. This is true also in the simplest case with k=1 and T_1=0. In [24,110,231,259,270-273] it is emphasized that the theoretically found values of f(m,n) normally give a very great over-estimation of the norm of F. This is especially true when m and/or n are large (and precisely this is the case when matrix A is sparse). However, at least in the case where $g(A)\bar{g}(A)$ can be expressed as a function of the condition number $\kappa(A)$, the bound (3.40) together with $\|F\| < 1$ (which is needed to ensure a convergent iterative refinement process) reveals an important relationship between the condition number $\kappa(A)$ and the **largest** drop-tolerance T that yields convergence:

> the drop-tolerance can be chosen **larger** when the condition number is smaller.

For the general k-stage scheme the drop-tolerance T is defined by

$$(3.46) \qquad T = \max_{1 \le i \le k} \{T_i\}. \qquad ■$$

Remark 3.9. The choice of T_i ($i \in \{1,2,\ldots,k\}$) depends on the particular method used during the i'th stage of the general k-stage scheme. Assume that a one-stage scheme is used. In this case $T_1 = T$. Consider the method based on forming the normal equations (see Example 3.2) and the method based on the use of the augmented matrices (Example 3.3) with an optimal value of parameter α. In the first case $g(A)\bar{g}(A)=\kappa^2(A)$, see (3.43), while in the second case $g(A)\bar{g}(A)=\sqrt{2}\kappa(A)$, see (3.44). Therefore it is clear (under the assumption that iterative refinement is used to regain the accuracy lost) that the use of the normal equations imposes a very severe restriction on the choice of the drop-tolerance T compared with the use of the augmented matrices with an optimal value of parameter α. This is true even in the case where matrix A is only moderately ill-conditioned. It must be repeated here (see also Remark 3.7 above) that it is not very easy to find the optimal value of α. However, in many experiments good results were obtained with α varied in a quite large interval; even the simplest choice, the choice α=1, gives very often excellent results. Thus, the use of the method based on augmented matrices is often more profitable than the use of the normal equations when it is allowed to drop small elements during the computations. ■

3.6. Implementation of the common approach in the solution of sparse algebraic problems by some particular direct methods

The common approach for solving sparse algebraic problems $x=A^{\dagger}b$, formulated in Section 3.3, has been implemented in several subroutines. A general survey of the software in which this approach is applied is given in this section. Further details will be given in the following chapters, where the particular methods used will be studied separately.

GE is used to obtain the decomposition LU=PAQ in package **Y12M**. The triangular factors L and U are used to obtain an approximation to the solution x of the system of algebraic equations Ax=b (the problem $x=A^{\dagger}b$ is reduced to a system of linear algebraic equations when A is a square matrix, m=n, and when *rank*(A)=n). Both the classical manner of exploiting sparsity and the computer-oriented manner can be specified when package **Y12M** is used. The storage schemes in **Y12M** are very similar to those described in **Chapter 2**. The package and/or different algorithms used in it are discussed [104,105,291,292,295,301,303,306,311,312,323-332,336-341]. In chapters 4, 5, 6, 9, 10 and 11 the following issues will be discussed in detail:

(i) the use of sparse matrix technique during GE,
(ii) the pivotal strategies,
(iii) the use of iterative refinement and of preconditioned conjugate gradient-type methods,
(iv) modifications for parallel computers,
(v) the calculation of an approximation to the condition number of matrix A,

Package **Y12M** can also be used to solve the more general algebraic problem $x=A^{\dagger}b$ (m=n being replaced by m≥n) by forming the augmented matrix from (3.17). This is possible because the problem $x=A^{\dagger}b$ can be reduced to a system of linear algebraic equations $B_1y=c$ after the transformation step (Step 1 in Section 3.1). This observation and formulae (3.17) show that the application of package **Y12M** for the particular method from **Example 3.3** is based on the following three steps:

(a) write a simple subroutine that performs the transformation of the problem solved (Step 1 in the general scheme),
(b) apply package **Y12M** in the solution of the system of linear algebraic equations obtained after Step 1,
(c) form an approximation to the solution x by taking the last n components of the approximation to the solution of the augmented system of linear algebraic equations

The application of package **Y12M** in the solution of $x=A^{\dagger}b$ by forming augmented matrices will be studied in **Chapter 7**. Another application, in the solution of ordinary differential equations, will be discussed in **Chapter 8**.

Orthogonalization methods as those considered in **Example 3.6** are implemented in package **LLSS01**. As in **Y12M**, both the classical and the computer-oriented manners of exploiting the sparsity can be used in **LLSS01**. The dynamic storage scheme described in **Chapter 2** in connection with GE, is applied with some simple modifications also in package **LLSS01**. Package **LLSS01** is discussed in **[104,296,299,320-322]**. Different properties of this package will be studied in chapters 13, 14 and 15.

Iterative refinement, see (3.28), is used in **LLSS01**. A special package, **LLSS02**, in which a conjugate gradient algorithm is applied has also been developed. **LLSS02** will be described and compared with **LLSS01** in **Chapter 16**.

It will be useful to have several subroutines for handling sparse symmetric matrices. The following three classes of problems can be treated by the computer-oriented manner when such subroutines are available: **(1)** systems of linear algebraic equations with symmetric matrices, **(2)** problems $x=A^{\dagger}b$ when the normal equations are formed (Example 3.2) and **(3)** problems $x=A^{\dagger}b$ when the Peters-Wilkinson method is used (Example 3.5) and when the symmetry of $B_2=L^TL$ is exploited during the second stage.

3.7. Testing the subroutines in which the common approach for solving sparse problems $x=A^{\dagger}b$ is implemented

Many rules and algorithms in sparse matrix technique are based on some heuristics. This is also true for many of the rules and algorithms proposed in this chapter (some of which are studied in detail in the following chapters). Therefore it is very often necessary to check carefully how well rules and algorithms work in practice. Test-matrices are to be used in such checks. This shows that the following problem arises in connection with the validation of the rules and algorithms for sparse matrices:

> **How should one choose suitable test-matrices?**

The easiest (in principle at least) way is to collect **matrices which often appear in practice**. This leads to huge files and/or many tapes, and may cause difficulties with data integrity in the transfer between different computers. It should also be mentioned that the number of matrices in such files and tapes is normally not very large. Nevertheless, it is necessary to test the codes on such matrices because the codes must work well on practical problems. During the experiments with the packages **Y12M**, **LLSS01** and **LLSS02** sparse test-matrices from the Harwell set **[76]** and the Harwell-Boeing set **[71,72]** have been used. Many test-matrices connected with: (1) the solution of some partial differential equations which govern the pressure in two-phase flow systems **[59]**, (2) nuclear magnetic resonance spectroscopy **[211-214]**, (3) modeling biological patterns **[152]**, and (4) thermodynamics **[303,325,327]** have also been used in the experiments. Some numerical results obtained in the tests with such matrices have already been demonstrated in the first two chapters.

Several special generators for sparse test-matrices have been developed
in [291,295,330-332]. Some test-matrices obtained by these generators have
already been applied in **Chapter 1** (see Example 1.3). Three generators for
sparse test-matrices will be described in this section and used in the
following chapters. The first two of these generators create square matrices
(the first one symmetric and positive definite matrices). The third generator
can also create rectangular matrices. The generators create matrices dependent
either on some of the parameters listed below (the first two generators) or
on all of these parameters (the third generator).

(1)	parameter m	the number of rows in the desired matrix A can be varied by varying m,
(2)	parameter n	the number of columns in the desired matrix A can be varied by varying n,
(3)	parameter c	the sparsity pattern of the desired matrix A can be varied by varying c,
(4)	parameter r	the number of non-zeros in the desired matrix A can be varied by varying r,
(5)	parameter α	the size of the non-zeros in the desired matrix A can be varied by varying α.

A very large number of test-matrices (in theory an infinite number) can
be obtained by a generator in which all of these five parameters can be
varied, or even if only some of these parameters can be varied. In the first
two generators only the parameters n and c can be varied (moreover, it is
assumed that m=n). The subroutines that are used to create matrices by these
two generators are called **MATRE** and **MATRD**. It is said that matrices of
classes $E(n,c)$ and $D(n,c)$ are created when these generators are in use.
All five parameters are used in the third generator **MATRF2**. The matrices
produced when the third generator is applied are called matrices of class
$F2(m,n,c,r,\alpha)$. All three subroutines, **MATRE, MATRD** and **MATRF2**, are very
simple. The use of these subroutines allows one to carry out a systematic
investigation of the dependence of the performance of the sparse software on
the variation of any of the parameters in the matrix generator under
consideration (several parameters could be varied simultaneously).

Matrices of class $E(n,c)$. The matrices that are created by the first
generator, the matrices of class $E(n,c)$, can be defined as follows (see also
Chapter 1):

(3.47) $a_{ii} = 4$, $i=1,2,\ldots,n$

(3.48) $a_{i,i+1} = a_{i+1,i} = -1$, $i=1,2,\ldots,n-1$,

(3.49) $a_{i,i+c} = a_{i+c,i} = -1$, $i=1,2,\ldots,n-c$,

where $n \geq 3$ and $1 < c < n$ are required. The matrices from this class are
symmetric and positive definite. They are very similar to matrices obtained
by using the five point rule in the discretization of the Laplace operator on

a square domain. An example with n=12 and c=6, i.e. the matrix A=E(12,6), is given on **Fig. 3.1**. ■

	1	2	3	4	5	6	7	8	9	10	11	12
1	4	-1	0	0	0	0	-1	0	0	0	0	0
2	-1	4	-1	0	0	0	0	-1	0	0	0	0
3	0	-1	4	-1	0	0	0	0	-1	0	0	0
4	0	0	-1	4	-1	0	0	0	0	-1	0	0
5	0	0	0	-1	4	-1	0	0	0	0	-1	0
6	0	0	0	0	-1	4	-1	0	0	0	0	-1
c+1 = 7	-1	0	0	0	0	-1	4	-1	0	0	0	0
8	0	-1	0	0	0	0	-1	4	-1	0	0	0
9	0	0	-1	0	0	0	0	-1	4	-1	0	0
10	0	0	0	-1	0	0	0	0	-1	4	-1	0
11	0	0	0	0	-1	0	0	0	0	-1	4	-1
n = 12	0	0	0	0	0	-1	0	0	0	0	-1	4

Figure 3.1
The matrix A = E(12,6).

	1	2	3	4	5	6	7	8	9	10	11	12	13	14	15	16	17	18	19	20
1	■	0	0	0	0	■	■	■	0	0	■	■	■	■	■	■	■	■	■	■
2	0	■	0	0	0	0	■	■	■	0	0	■	■	■	■	■	■	■	■	■
3	0	0	■	0	0	0	0	■	■	■	0	0	■	■	■	■	■	■	■	■
4	0	0	0	■	0	0	0	0	■	■	■	0	0	■	■	■	■	■	■	■
5	0	0	0	0	■	0	0	0	0	■	■	■	0	0	■	■	■	■	■	■
6	0	0	0	0	0	■	0	0	0	0	■	■	■	0	0	■	■	■	■	■
7	0	0	0	0	0	0	■	0	0	0	0	■	■	■	0	0	■	■	■	■
8	0	0	0	0	0	0	0	■	0	0	0	0	■	■	■	0	0	■	■	■
9	0	0	0	0	0	0	0	0	■	0	0	0	0	■	■	■	0	0	■	■
10	0	0	0	0	0	0	0	0	0	■	0	0	0	0	■	■	■	0	0	■
11	0	0	0	0	0	0	0	0	0	0	■	0	0	0	0	■	■	■	0	0
12	0	0	0	0	0	0	0	0	0	0	0	■	0	0	0	0	■	■	■	0
13	0	0	0	0	0	0	0	0	0	0	0	0	■	0	0	0	0	■	■	■
14	■	0	0	0	0	0	0	0	0	0	0	0	0	■	0	0	0	0	■	■
15	■	■	0	0	0	0	0	0	0	0	0	0	0	0	■	0	0	0	0	■
16	■	■	■	0	0	0	0	0	0	0	0	0	0	0	0	■	0	0	0	0
17	0	■	■	■	0	0	0	0	0	0	0	0	0	0	0	0	■	0	0	0
18	0	0	■	■	■	0	0	0	0	0	0	0	0	0	0	0	0	■	0	0
19	0	0	0	■	■	■	0	0	0	0	0	0	0	0	0	0	0	0	■	0
20	0	0	0	0	■	■	■	0	0	0	0	0	0	0	0	0	0	0	0	■

Figure 3.2
The sparsity pattern of matrix **A=D(20,5)**.
The non-zeros are denoted by ■ .

Matrices of class D(n,c). The non-zero elements of a matrix created by the second generator, the matrices of class **D(n,c)**, can been introduced as follows:

(3.50) $a_{ii} = 1, \qquad i=1,2,\ldots,n,$

(3.51) $a_{i,i+c}=i+1, \quad i=1,2,\ldots,n-c, \qquad a_{i,i-n+c}=i+1, \quad i=n-c+1,n-c+2,\ldots,n,$

(3.52) $a_{i,i+c+1}=-i$, $i=1,2,\ldots,n-c-1$, $a_{i,i-n+c+1}=-i$, $i=n-c,n-c+1,\ldots,n$,

(3.53) $a_{i,i+c+2}=16$, $i=1,2,\ldots,n-c-2$, $a_{i,i-n+c+2}=16$, $i=n-c-1,n-c,\ldots,n$,

(3.54) $a_{i,n-11+i+j}=100j$, $i=1,2,\ldots,10$, $j=1,2,\ldots,11-i$,

where $n\geq14$ and $1\leq c\leq n-13$. An example with $n=20$ and $c=5$ is given in **Fig. 3.2.** The matrix in this example is **A=D(20,5).**   ∎

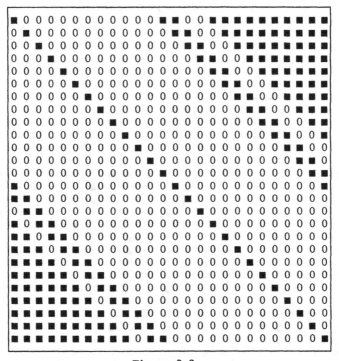

Figure 3.3
The sparsity pattern of matrix $A=F2(26,26,12,3,\alpha)$.
The non-zero elements are denoted by ∎.

    Matrices of class F2(m,n,c,r,α).   The non-zero elements of a matrix created by the third generator are given by the formulae:

(3.55) $a_{i,i-q_n}=1$, $i=1,2,\ldots,m$

(3.56) $a_{i,i-q_n+c+s}=(-1)^s si$, $i=1,2,\ldots,m$, $s=1,2,\ldots,r-1$,

(3.57) $a_{i,n-11+i+j}=j\alpha$, $i=1,2,\ldots,10$, $j=1,2,\ldots,11-i$,

(3.58) $a_{n-11+i+j,i}=j/\alpha$, $i=1,2,\ldots,10$, $j=1,2,\ldots,11-i$,

where

(3.59) $22\leq n\leq m$, $11\leq c\leq n-11$, $2\leq r\leq min(c-9,n-20)$, $1\leq\alpha$,

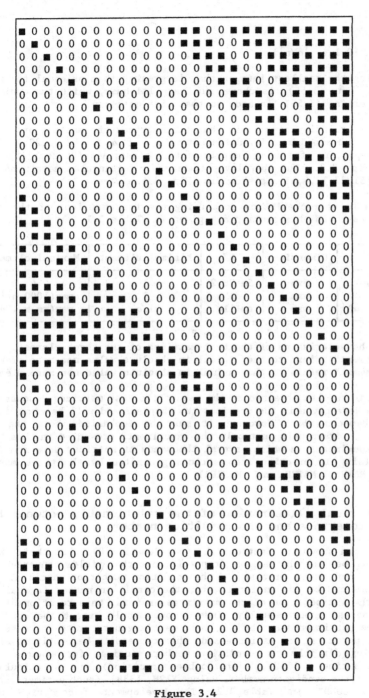

Figure 3.4
The sparsity pattern of matrix A=F2(51,27,12,4,α).
The non-zero elements are denoted by ■.

and

(3.60) $q \in \{0,1,\ldots \lceil m/n \rceil \}$

($\lceil m/n \rceil$ being the smallest integer that is greater than or equal to m/n).
Parameter q in (3.55) is chosen so that $1 \leq i-qn \leq n$, while the choice of
this parameter in (3.56) is made so that $i \leq i-qn+c+s \leq n$.

The above relationships imply that the matrices of smallest order from
this class are obtained by using different values of α in $F2(22,22,11,2,\alpha)$.
The sparsity pattern of a square matrix of this class, $A=F2(26,26,12,3,\alpha)$, is
shown in **Fig. 3.3**. The sparsity pattern of a rectangular matrix of this
class, $A=F2(51,27,12,4,\alpha)$, is shown in **Fig. 3.4**. ∎

Some basic characteristics concerning the test matrices of the three
classes discussed in this section are given in **Table 3.1**.

Class	Dimension	NZ	Minimal element	Maximal element
E(n,c)	nxn	5n-2c-2	1	4
D(n,c)	nxn	4n+55	1	max(1000,n+1)
$F2(m,n,c,r,\alpha)$	mxn	rm+110	$1/\alpha$	max(rm-m,10α)

Table 3.1
Some characteristics of the test-matrices of the three classes.
"Minimal element" refers to the minimal in absolute value non-zero element.
"Maximal element" refers to the maximal in absolute value non-zero element.

Numerical results obtained by the use of the test matrices discussed in
this section were given in the previous two chapters and will be given in the
following chapters also. However, it must be emphasized here that **all
conclusions are drawn using not only the experiments presented in this book,
but using experiments in which several thousands of matrices were run** (both
**matrices arising in practice and matrices generated by the three matrix-
generators MATRE, MATRD and MATRF2**).

3.8. A numerical illustration of the relationship between the drop-tolerance and the condition number

It has been shown in **Section 3.5** (see Remark 3.8) that there exists
a relationship between the largest drop-tolerance that yields convergence of
the iterative refinement process and the condition number of matrix A (the
drop-tolerance can be chosen larger when the condition number is smaller).
Now, after the introduction of the matrix generators, this relationship can
easily be checked experimentally. Indeed, one should expect the condition
number of the matrix produced by the third matrix-generator to increase when
the value of parameter α is increased. This can easily be confirmed either
by calculating directly the condition numbers (using, for example, some
Harwell subroutines, [142], in double precision) or by calculating
estimations of the condition numbers using **Y12M**, [338] (such estimations are
given under "COND" in **Table 3.2**). The development of condition number
estimators for a sparse matrix code (not necessarily **Y12M**) will be studied
in detail in **Chapter 9**.

An experiment, performed to verify the conclusion made in **Remark 3.8**, is described below. Matrices of class $F2(m,n,c,r,\alpha)$ are used. The first four parameters are kept fixed: $m=n=22$, $r=2$, $c=11$. The starting value of the fifth parameter is $\alpha=1$. After every successful run, α is multiplied by 2. The maximal values of α, by which an acceptable solution can be obtained, have been found for four different drop-tolerances. The results are given in **Table 3.2**. It is clearly seen that the results confirm the conclusion drawn in Section 3.5: **if the condition number is larger, the largest drop-tolerance that yields convergence is smaller.**

It should be mentioned here that a slightly modified version of package **Y12M** has been used in this experiment. In this version all inner products are calculated and accumulated in double precision, while in the original **Y12M** some inner products are calculated in double precision and then rounded and accumulated in single precision; [331,341]. By using this procedure, a compatibility with the method applied in package **LLSS01** is achieved. The performance of the iterative refinement when the GE process is used (implemented in **Y12M**) and when the Givens algorithm is used (implemented in **LLSS01**) will be discussed in detail in **Chapter 5** and **Chapter 15**, respectively.

Drop-	Package Y12M			Package LLSS01		
tolerance	k	COND	$\|x-y\|$	k	COND	$\|x-y\|$
0	34	4.4E+14	8.6E-14	15	1.7E+12	7.1E-11
1.0E-4	16	6.8E+12	2.8E-12	9	4.0E+08	7.5E-10
1.0E-3	13	1.0E+11	6.5E-12	7	2.4E+07	4.7E-10
1.0E-2	10	1.7E+09	3.4E-12	4	2.4E+05	5.2E-10

Table 3.2
An experiment with matrices produced by the matrix-generator MATRF2 with $m=n=22$, $r=2$, $c=11$ and with different values of α. When the drop-tolerance T is larger, then the maximal value α^* of parameter α for which the iterative refinement process is convergent becomes smaller for both codes ($\alpha^*=2^k$ and k is given in the table). COND is an estimate of the condition number found by package Y12M, [338]. The exact solution of Ax=b is denoted by x and the calculated one by y.

The calculations have been carried out on an **IBM 3033** computer usung single precision, where the machine accuracy is approximately equal to $\epsilon=10^{-6}$. It is seen (Table 3.2) that the accuracy actually obtained is nearly twice as high when the iterative refinement process converges, though all matrices are stored in single precision and all arithmetic operations in the decomposition step are carried out in single precision. The high accuracy that can be achieved when certain inner products are calculated and kept in double precision will be discussed in **Chapter 5** and in **Chapter 15**.

3.9. Concluding remarks and references concerning the general scheme for solving linear algebraic problems

Many direct methods for solving linear algebraic problems have some similar properties. Therefore it would seem to be useful to unite the direct methods in a common scheme. This has been done here.

The general k-stage scheme for solving linear algebraic problems, which has been defined in **Section 3.1**, was first introduced in [292]. The same definition is also used in [299,301].

Many well-known direct methods can be represented as special cases within the general scheme for solving linear algebraic problems. Six examples are given in **Section 3.2**. However, it is not clear whether **all** direct methods can be represented as special cases within the general k-stage scheme. For example, the question whether the algorithm developed by **Björck and Duff** [28] can be represented as a member of the general scheme is open.

The common approach for solving **sparse** linear algebraic problems, which has been defined in **Section 3.3**, is an illustration of the usefulness of the general k-stage scheme. Some applications of this common approach for particular algorithms of the general k-stage scheme will be discussed in the following chapters. The common approach for solving sparse linear algebraic problems can be considered as a more formal description of the computer-oriented manner for exploiting sparsity.

The iterative refinement process is perhaps the simplest algorithm that can be used in part **(c)** of the common approach. Some features of the iterative refinement are described in **Section 3.4**. The use of iterative refinement with some special direct methods will be discussed in **Chapter 5 and Chapter 15**. The application of other iterative processes, conjugate gradients and conjugate gradient-type algorithms, will also be studied (**Chapter 11 and Chapter 16**).

The rate of convergence of the iterative refinement process depends both on the drop-tolerance used and on the condition number of matrix A. The relationship between the largest drop-tolerance that yields convergence of the iterative refinement and the condition number is discussed in **Section 3.5**. This is a common discussion for all direct methods from the general k-stage scheme. The conclusion is that **the drop-tolerance should be chosen smaller when the condition number is larger**. This means that it may be useful to estimate the condition number of matrix A. This is especially true when a long simulation process containing many problems with the same matrices (or even with similar matrices) has to be carried out. If this is the case, then one might be interested in estimating the condition number in the beginning of the computations and, after that, in performing the whole simulation process with a proper value of the drop-tolerance. Some algorithms for estimating the condition number of a **sparse** matrix will be discussed in **Chapter 9**.

Several implementations of sparse matrix techniques for particular direct methods within the general k-stage scheme have been briefly discussed in **Section 3.6**. These implementations will be studied in detail in the following chapters. It should be stressed here that the dynamic storage scheme described in **Chapter 2** is used (with some obvious modifications when necessary) in all implementations discussed in this chapter. Both the classical manner of exploiting sparsity and the computer-oriented manner can be used in these implementations.

It is very important to check the efficiency of the subroutines in a proper way. The test matrices used in the experiments are described in Section 3.7. Both matrices that appear in practice ([59,71,72,76,152,211, 303,325-327]) and artificially created test matrices have been used. Three generators for sparse test-matrices are described. These generators were first discussed in detail in [330], but some of them have been used earlier (for example in [324]). It must be emphasized here that it is necessary to establish some balance between the use of two groups of sparse test matrices:

Group 1	Artificially created matrices.
Group 2	Matrices that appear in scientific and engineering problems.

The first group of matrices allows a systematic investigation of the behaviour of the software as a function (1) of the dimensions of the matrix, (2) of the density of its non-zero elements, (3) of the locations of the non-zero elements, and (4) of the magnitude of the non-zero elements. The second group of matrices provides an extra (and very often valuable) check of the results obtained during the investigations performed with test matrices from the first group. It should be emphasized here that some of the experiments carried out by using artificially created sparse test matrices cannot be carried out (or, at least, it is very difficult to carry them out) when matrices of the second group are to be used. As an illustration of this the experiment described in Section 3.8 can be mentioned. Such an experiment can be performed only with a long sequence of test matrices that have the following properties: (1) the same order, (2) the same density of the non-zero elements and (3) the same locations of the non-zero elements. Moreover, the condition number should gradually become larger, i.e. if A_i and A_j are two matrices in the sequence and if $j > i$, then $cond(A_i)$ should be larger than $cond(A_j)$. There is no sufficiently large set of sparse test matrices from the second group that satisfies these conditions.

A very simple and clear illustration of the relationship between the largest drop-tolerance that yields convergence of the iterative refinement and the condition number is given in Section 3.8 (see also [292,341]). Some other examples will be given in the following chapters. It should be mentioned that some experiments with small, but very ill-conditioned, test-matrices (proposed by Zielke [288,289]) have also been performed. These experiments also confirm the conclusion made in this chapter: for very ill-conditioned matrices the drop-tolerance should be small. Finally, it should be emphasized that the relationship between the largest drop-tolerance that yields convergence and the condition number is valid not only when iterative refinement is used but also when other iterative methods are applied.

4. PIVOTAL STRATEGIES FOR GAUSSIAN ELIMINATION

The discussion of some special algorithms within the general k-stage scheme will begin with the simplest example; Gaussian elimination (**GE**). If GE is used in the solution of linear algebraic equations with general sparse matrices, then the pivotal strategy plays a very important role. The pivotal strategy is a powerful tool that can efficiently be used during the efforts to preserve as well as possible the sparsity of the original matrix and, at the same time, to keep the rounding errors as small as possible. Normally these two requirements (preserving both sparsity and stability) cannot be satisfied similtaneously. Therefore a compromise is needed in order to obtain a balance between sparsity and accuracy. Some simple examples, where the lack of such a balance leads to catastrophic results, will be demonstrated in this chapter. Several pivotal strategies will be described and tested on many systems with sparse matrices. Some recommendations concerning the choice of a pivotal strategy for a sparse matrix code will be given.

4.1. Pivoting for dense matrices

Assume that classical GE is used to represent a **general** square matrix $A \in R^{n \times n}$ as a product of a lower triangular matrix $L \in R^{n \times n}$ and an upper triangular matrix $U \in R^{n \times n}$. Denote by $a_{ij}^{(1)} = a_{ij}$ the elements of A. The elimination process (see **Chapter 1**), which leads to the factorization (or decomposition) $A = LU$, is carried out by

$$(4.1) \qquad a_{ij}^{(s+1)} = a_{ij}^{(s)} - a_{is}^{(s)} (a_{ss}^{(s)})^{-1} a_{sj}^{(s)},$$

where $s=1,2,\ldots,n-1$, $i=s+1,s+2,\ldots,n$, $j=s+1,s+2,\ldots,n$. It is assumed that $a_{ss}^{(s)} \neq 0$ for all $s \in \{1,2,\ldots,n-1\}$.

It is well-known, [231,259,270-277], that in many cases pivotal interchanges must be applied during the GE in an attempt both to ensure that the computations will not break down for a non-singular matrix A and to obtain an accurate decomposition LU. For any $s \in \{1,2,\ldots,n-1\}$ consider the three sets:

$$(4.2) \qquad A_s = \{a_{ij}^{(s)} \ / \ i=s,s+1,\ldots,n, \ j=s,s+1,\ldots,n\},$$

$$(4.3) \qquad R_{is} = \{a_{ij}^{(s)} \ / \ j=s,s+1,\ldots,n\}, \qquad i=s,s+1,\ldots,n,$$

$$(4.4) \qquad C_{js} = \{a_{ij}^{(s)} \ / \ i=s,s+1,\ldots,n\}, \qquad j=s,s+1,\ldots,n.$$

It is said that:

    (i) A_s is the **active part of matrix A at stage s**,

    (ii) R_{is} is the **active part of row i at stage s**,

    (iii) C_{js} is the **active part of column j at stage s**.

Pivotal interchanges at stage s are normally performed by one of the following two rules.

Rule 1 - Partial pivoting. Let $a_{is}^{(s)}$ ($a_{sj}^{(s)}$) be one of the largest in absolute value elements in the active part R_{is} of row i (in the active part C_{js} of column j) at stage s. Bring this element into position (s,s) by interchanging rows i and s (columns j and s). ■

Rule 2 - Complete pivoting. Let $a_{ij}^{(s)}$ be one of the largest in absolute value elements in the active part A_s of matrix A at stage s. Bring this element into position (s,s) by interchanging rows i and s as well as columns j and s. ■

The interchanges required by the above two rules can be described by permutation matrices, say P and Q. If this is done, then the decomposition process performed by using complete pivoting can be represented in a matrix form by

(4.5) $LU = PAQ$, $P \in R^{n \times n}$, $Q \in R^{n \times n}$.

Let I be the identity matrix in $R^{n \times n}$. Then the GE process with partial pivoting may be considered as a special case of (4.5) with Q=I when only row interchanges are used or with P=I when only column interchanges are carried out. GE without any pivoting may also be considered as a special case of (4.5) with P=Q=I. The GE process is stable without any pivoting for some special matrices (see [270]). No pivoting for stability is needed for the matrices of class **E(n,c)**. However, it will be shown that the attempt to use pivotal interchanges in order to preserve sparsity **may** cause stability problems when matrices of class **E(n,c)** are factorized (**Table 4.4**).

It can be proved (see, for example, [231]) that if the following conditions are satisfied:

   (i) matrix A is non-singular (A^{-1} exists),

   (ii) the computations can be performed without
        rounding errors,

   (iii) neither overflow nor underflow occurs,

then the GE algorithm (4.1) with either of the two rules for pivoting will not break down and some non-singular factors L and U will be calculated. This means, in particular, that **all** pivotal elements (i.e. the elements brought in position (s,s), s=1,2,...,n-1 as well as the element which is in position (n,n) at the end of GE) will be **non-zero elements**.

However, rounding errors are in general unavoidable when the calculations are performed on a computer. Therefore

(4.6) $LU = PAQ + E$

is satisfied instead of (4.5). $E \in R^{n \times n}$ is a perturbation matrix, whose elements e_{ij} (i=1,2,...,n, j=1,2,...,n) satisfy ([231], p. 151):

(4.7) $|e_{ij}| \leq n\pi b_n \epsilon$,

where ϵ is the machine precision (the smallest positive number such that $1.0 + \epsilon \neq 1.0$ in the floating-point arithmetic on the computer under consideration), π is a constant of order O(1) that is independent of matrix A, and for   k=1,2,...,n:

(4.8) $b_k = \max\limits_{1 \leq s \leq k, s \leq i \leq n, s \leq j \leq n} (|a_{ij}^{(s)}|)$.

It is well-known (see [231] again) that

(4.9) $b_n \leq [n(2^1 3^{1/2} 4^{1/3} ... n^{1/(n-1)})]^{1/2} b_1$

for complete pivoting and

(4.10) $b_n \leq 2^{n-1} b_1$

for partial pivoting. Moreover, examples where the latter bound is achieved, can be constructed ([271-272]). However, experience indicates that this happens very seldom. Therefore it is commonly accepted that GE with partial pivoting is a stable algorithm that normally provides sufficiently accurate results. This fact, together with the fact that complete pivoting is rather expensive, explains why partial pivoting is used in nearly all standard subroutines for solving systems of linear algebraic equations with dense matrices. Of course, it is worthwhile to check the **growth factor** b_k/b_1 (k=1,2,...,n) and the magnitude of the pivots during the GE process. Large values of the growth factor as well as small values of $a_{ss}^{(s)}$ (s=1,2,...,n) should be considered as a signal of inaccurate results.

In the following sections of this chapter it will be shown that all pivotal strategies for sparse matrices are generalizations of the pivotal strategies for dense matrices.

4.2. Common principles in the construction of pivotal strategies for sparse matrices

The GE process is carried out by (4.1) also in the case when A is sparse. It has been mentioned already (in **Chapter 1**) that if $a_{ij}^{(s)} = 0$ but both $a_{is}^{(s)} \neq 0$ and $a_{sj}^{(s)} \neq 0$, then a new non-zero element, **a fill-in**, $a_{ij}^{(s+1)} \neq 0$ is created. It is clear that one should attempt to keep the number of **fill-ins** small. It has been shown in **Chapter 2** that it is difficult to handle fill-ins when a dynamic storage scheme is in use. If many fill-ins are created, then both the storage requirements and the computer time requirements may be excessive. The requirement to keep the number of fill-ins small is equivalent to a requirement to preserve the sparsity of matrix A as well as possible during the calculation of the factors L and U. The pivotal strategy is a useful tool in the efforts to avoid many fill-ins. This means that not only is the pivotal strategy for sparse matrices a device by which an attempt to calculate an **accurate** LU-decomposition is made, but it can also be used to calculate an LU-decomposition which is as sparse as possible. The requirements of sparsity and accuracy work in opposite directions and, therefore, it is not a surprise that a pivotal strategy for sparse matrices is usually a compromise between sparsity and accuracy.

The pivotal strategies for sparse matrices can, roughly speaking, be divided into three groups.

| Group 1 - Pivotal strategies with a priori interchanges. |
| Group 2 - Pivotal strategies based on the use of the Markowitz cost-function. |
| Group 3 - Pivotal strategies based on a local minimization of the number of fill-ins |

A pivotal strategy belonging to any of these three groups can formally be obtained from some pivotal strategy for dense matrices by the use of two main principles:

| (i)   relax the accuracy requirements, |
| (ii) introduce some additional criteria by which an attempt to preserve sparsity better is carried out. |

It should be emphasized here that it is necessary to keep carefully the balance between (i) and (ii) because **(a)** the results will be inaccurate if the accuracy requirement is relaxed too much, and **(b)** the preservation of sparsity will be obtained at too high a cost price when complicated additional criteria are introduced.

Denote by $r(i,s)$ and $c(j,s)$ the numbers of elements in the active part R_{is} of row i at stage s and in the active part C_{js} of column j at stage s. Let

$$(4.11) \quad \alpha_{is} = \max_{a_{ij}^{(s)} \in R_{is}} (|a_{ij}^{(s)}|)$$

and

$$(4.12) \quad \beta_{js} = \max_{a_{ij}^{(s)} \in C_{js}} (|a_{ij}^{(s)}|)$$

The accuracy requirements are usually relaxed by introducing a **stability factor** $u \geq 1$ and by choosing the pivotal elements at any stage s $(s=1,2,\ldots,n-1)$ among the non-zero elements of a certain **stability set** $B_s \subseteq A_s$. The stability set B_s depends on the pivotal strategy chosen and on the stability factor u. The elements of this set satisfy at least one of the following two relations (the choice being dependent on the pivotal strategy selected):

$$(4.13) \quad u|a_{ij}^{(s)}| \geq \alpha_{is}, \qquad u|a_{ij}^{(s)}| \geq \beta_{js}.$$

Sparsity is preserved in a different way in the three groups of pivotal strategies. However, the following common principle can be formulated. For each pivotal strategy some subsets $C_s^* \subseteq B_s$ $(s=1,2,\ldots,n-1)$ are determined by

$$(4.14) \quad C_s^* = \{a_{ij}^{(s)} \ / \ a_{ij}^{(s)} \in B_s \ \wedge \ a_{ij}^{(s)} \text{ satisfies some sparsity criterion}\}$$

and one of the elements of set C_s^* is selected as a pivot at stage s of the GE. Set C_s^* will be called the **set of candidates**.

The general description of the sparse pivotal strategies given in this section is sufficient to explain why the accuracy requirements are normally relaxed. In this way the number of elements in the stability set is in general increased and, thus, the sparsity criterion is applied on a larger set.

4.3. Pivotal strategies with a priori interchanges

A pivotal strategy from this group can be described by the following definition.

Definition 4.1. Carry out the calculations in two parts as follows

> **Part 1** - Order the rows (columns) of matrix A in an increasing number of non-zero elements, i.e. $r(1,1) \leq r(2,1) \leq \ldots \leq r(n,1)$ after a priori interchanges of the rows, while $c(1,1) \leq c(2,1) \leq \ldots \leq c(n,1)$ after a priori interchanges of the columns.

> **Part 2** - At stage s (s=1,2,...,n-1) of the GE choose as a pivot an element $a_{sj}^{(s)}$ $(a_{is}^{(s)})$ in row (column) s which has a minimal number of non-zero elements in C_{js} (R_{is}) and which satisfies the first (second) condition (4.13). Bring this element into position (s,s) by column (row) interchanges. ■

If **row** interchanges are carried out in **Part 1**, then

$$(4.15) \quad B_s = \{ a_{sj}^{(s)} \ / \ a_{sj}^{(s)} \in R_{ss} \ \wedge \ u|a_{sj}^{(s)}| \geq \alpha_{ss} \}$$

$$(4.16) \quad C_s^* = \{a_{sj}^{(s)} \ / \ a_{sj}^{(s)} \in B_s \ \wedge \ (a_{sj}^{(s)} \in C_s^*, \ a_{sm}^{(s)} \notin C_s^*, \ a_{sm}^{(s)} \in B_s) \Rightarrow c(m,s) \geq c(j,s)\}. \ ■$$

If **column** interchanges are carried out in **Part 1**, then

$$(4.17) \quad B_s = \{ a_{is}^{(s)} \ / \ a_{is}^{(s)} \in C_{ss} \ \wedge \ u|a_{is}^{(s)}| \geq \beta_{ss} \}$$

$$(4.18) \quad C_s^* = \{a_{is}^{(s)} \ / \ a_{is}^{(s)} \in B_s \ \wedge \ (a_{is}^{(s)} \in C_s^*, \ a_{ms}^{(s)} \notin C_s^*, \ a_{ms}^{(s)} \in B_s) \Rightarrow r(m,s) \geq r(i,s)\}. \ ■$$

From (4.15)-(4.18) it follows that any pivotal strategy among the strategies introduced by **Definition 4.1** is a natural generalization of Rule 1 (partial pivoting). Indeed, observe that:

> **(i)** if matrix A is dense, then Part 1 of Definition 4.1 is not needed (one may assume that the identity matrix I is used in the a priori interchanges when A is dense),

> **(ii)** if matrix A is dense, then Rule 1 is obtained from Definition 4.1 by setting u=1.

The pivotal strategies from Definition 4.1 have several advantages over the pivotal strategies from the other two groups (see **Section 4.2**):

---

(1) Only column (row) interchanges are carried out during the GE. The a priori interchanges can easily and efficiently be implemented by using standard algorithms; see, for example [136,137].

---

(2) The development of a code based on a pivotal strategy from Definition 4.1 is more straightforward than the development of a code based on a pivotal strategy from Group 2 or Group 3.

---

(3) The amount of additional information that has to be kept and updated during the factorization is smaller than that for the pivotal strategies from the other two groups.

---

(4) The code will normally be shorter than that for the strategies from the other two groups.

There is only one drawback when such strategies are in use: **the code may produce many fill-ins in the decomposition of some matrices.** Unfortunately, this is a very serious drawback, because both the storage and the computing time are increased when many fill-ins are created during the GE; see, for example, the results presented in **Table 4.2**. Nevertheless, subroutines based on a strategy from the first group should be included in a package for sparse matrices. The reason for this can be explained as follows. Very often many problems with quite similar matrices have to be solved. If this is so, then it may be worthwhile to test the efficiency of the available pivotal strategies for the class of matrices under consideration. If it happens that a pivotal strategy with a priori interchanges works well, then this strategy may efficiently be used in the later runs.

It should be mentioned that pivotal strategies based on a priori interchanges and different from those introduced by **Definition 4.1** can be applied; see, for example, [73,199]. Such pivotal strategies will not necessarily have all the advantages **(1)-(4)**. However, more important is the fact that there is always a danger that many fill-ins will be produced during the decomposition of matrix A by the use of a pivotal strategy of the first group. This is so because any strategy from this group is based on the expectation that some sparsity criterion by which the number of fill-ins produced at stage s will be small when this criterion is applied just before stage s (s=1,2,...,n-1), will also be small when the sparsity criterion is applied before the beginning of the GE (or, in other words, before stage 1 and taking into account the original sparsity pattern only). If the class of matrices is such that this expectation holds at each stage of the GE, then a pivotal strategy from this group is useful. If the class of matrices is such that this expectation does not hold, then it is better to try a pivotal strategy from the other two groups. This explains why a pivotal strategy from the first group may work quite satisfactorily in some situations. However, if a long sequence of systems with similar matrices is to be treated, then one should be careful when deciding whether a pivotal strategy from the first group should be used.

The conclusions drawn above are valid for the case where all non-zero elements are kept during the whole elimination process; i.e. when the classical manner of exploiting sparsity is in use. If the computer-oriented manner of exploiting sparsity is applied (if some elements $a_{ij}^{(s+1)} \neq 0$ are removed from the dynamic storage scheme when these become small in some sense), then the situation **may** change. This possibility will be discussed in the next chapter. Here it should only be noted that removing "small" non-zero elements could change dramatically the nature of the computational process, and a pivotal strategy which performs rather badly without neglecting "small" elements can sometimes become the best choice when removing "small" elements is allowed (when the computer-oriented manner of exploiting sparsity is applied). As an illustration of this fact, compare the numerical results given in **Table 4.2** with those given in **Table 4.5**.

4.4. Pivotal strategies based on the use of the Markowitz cost-function

The following definitions can be used in order to introduced pivotal strategies of the second group.

Definition 4.2. Let $a_{ij}^{(s)} \neq 0$ be an arbitrary non-zero element belonging to set A_s defined in **Section 4.1**. The product

(4.19) $M_{ijs} = [r(i,s)-1][c(j,s)-1]$

of the other non-zero elements in the active part of row i at stage s and the other non-zero elements of column j at stage s is called the **Markowitz cost of element** $a_{ij}^{(s)}$. The integer

(4.20) $M_s = \min_{a_{ij}^{(s)} \in B_s} (M_{ijs})$

is called the **optimal Markowitz cost at stage s**. ∎

Remark 4.1. The optimal Markowitz cost at stage s depends on the stability requirements imposed (by the stability factor u; see **Section 4.2**). Therefore it must be emphasized here that the optimal Markowitz cost at stage s may be (and very often is) greater than the **minimal Markowitz cost at stage s**, which is defined by $M_s^* = min(M_{ijs})$, the minimum being taken over all $a_{ij}^{(s)} \in A_s$ and all $a_{ij}^{(s)} \neq 0$. ∎

Definition 4.3. Let $s \in \{1,2,\ldots,n-1\}$. Consider a set of $p(s)$ rows with $1 \leq p(s) \leq n-s+1$. Define set I_s as follows:

(4.21) $I_s = \{i_1, i_2, \ldots, i_{p(s)}\}$ ($i_m \in \{s, s+1, \ldots, n\}$, $m=1,2,\ldots,p(s)$),

where

(4.22) $r(i_1,s) \leq r(i_2,s) \leq \ldots \leq r(i_{p(s)},s)$,

(4.23)   ($i \notin I_s \wedge s \leq i \leq n$) $\Rightarrow$ $r(i_{p(s)},s) \leq r(i,s)$.

Let the stability set B_s and the set of candidates C_s^* (defined in **Section 4.2**) be given by

(4.24) $B_s = \{ a_{ij}^{(s)} / a_{ij}^{(s)} \in A_s \ \wedge \ u|a_{ij}^{(s)}| \geq \alpha_{is} \ \wedge \ i \in I_s \}$

and

(4.25) $C_s^* = \{ a_{ij}^{(s)} / a_{ij}^{(s)} \in B_s \ \wedge \ M_{ijs} = M_s \}$.

The class of **generalized Markowitz strategies, GMS's,** contains all pivotal strategies in which an **arbitrary** element of set C_s^* is chosen as a pivot at stage s (s=1,2,...,n-1).

The class of **improved generalized Markowitz strategies, IGMS's,** contains all pivotal strategies in which **one of the largest (in absolute value) elements** of the set of candidates, C_s^*, is chosen as a pivot at stage s (s=1,2,...,n-1). ∎

Remark 4.2. Column numbers ($j_1, j_2, \ldots, j_{p(s)}$) can be considered in (4.21) instead of row numbers. If this is done, then the relations:

(4.26) $c(j_1,s) \leq c(j_2,s) \leq \ldots \leq c(j_{p(s)},s)$,

(4.27) $(j \notin I_s \ \wedge \ s \leq j \leq n) \ \Rightarrow \ c(j_{p(s)},s) \leq c(j,s)$

and

(4.28) $B_s = \{ a_{ij}^{(s)} / a_{ij}^{(s)} \in A_s \ \wedge \ u|a_{ij}^{(s)}| \geq \beta_{js} \ \wedge \ j \in I_s \}$

are to be used instead of (4.22), (4.23) and (4.24) respectively. ∎

Remark 4.3. The class of IGMS's is a subclass of the class of GMS's. ∎

Remark 4.4. Consider a particular pivotal strategy P^*. The expressions: "P^* is a **GMS**" and "P^* is a **IGMS**" will often be used as abbreviations of "P^* belongs to the class of generalized Markowitz strategies" and "P^* belongs to the class of improved generalized Markowitz strategies" respectively. ∎

The pivotal strategies introduced by **Definition 4.3** and **Remark 4.2** depend on two parameters: **the stability factor u** and **the number p(s) of rows (columns) among the non-zero elements of which a pivotal element is to be selected.**

The original Markowitz strategy, **[169]**, is a GMS with $u = \infty$ (this means that no stability restriction is imposed and, thus, $B_s = A_s$ for all $s \in \{1,2,...,n-1\}$) and with p(s) = n-s+1 (this means that the pivotal search is carried out on the whole A_s for all $s \in \{1,2,...,n-1\}$). Both the value of u and the value of p(s) chosen in the original Markowitz strategy may cause difficulties during the GE. This will be discussed later in this section. It is sufficient to mention here that because of the bad choice of the parameters the original Markowitz strategy is not used in the sparse codes at present.

The values of u and p(s) used in the pivotal strategies of three well-known codes for solving systems of linear algebraic equations with sparse matrices are given in **Table 4.1** below.

The implementation of a strategy from the second group in a package for sparse matrices is much more complicated than the implementation of a strategy from the first group. The major difficulty arises from the necessity to arrange the rows or the columns (or both the rows and the columns in some implementations as [63]) in increasing numbers of non-zero elements in their active parts at the beginning of each stage s of the GE; see (4.21) and (4.26). This process requires both extra computing time and extra storage. However, the number of fill-ins is often reduced considerably and, therefore, both the total computing time needed to solve Ax=b and the total storage are reduced in comparison to those for the pivotal strategies from the first group. This is demonstrated numerically in Table 4.2.

C o d e	Type of the strategy	u	p(s)
MA18 (Curtis and Reid,[48]) MA28 (Duff, [63]) Y12M (Zlatev et al., [231])	GMS GMS IGMS	u=4 u=10 u∈[4,16]	n-s+1 n-s+1 min(n-s+1,q)

Table 4.1
Basic characteristics of the pivotal strategies in three codes. The values of u given in the table are the values that are recommended in the codes, i.e. the user may specify other values. The values of p(s) in the first two codes are fixed. In Y12M q=3 is recommended (there are three different pivotal strategies in Y12M; the pivotal strategy for general sparse matrices is considered here).

Density	A priori ordering of the columns			Markowitz cost-function		
parameter	COUNT	Time	Accuracy	COUNT	Time	Accuracy
r = 2	1015	0.09	1.02E-2	750	0.10	1.15E-3
r = 3	15793	2.30	2.33E-3	1941	0.21	3.07E-3
r = 4	19075	3.28	2.56E-4	4066	0.49	2.14E-3
r = 5	23271	4.94	5.72E-4	6066	0.88	4.73E-3
r = 6	26838	8.11	3.23E-3	7409	1.18	3.49E-3
r = 7	27942	10.42	1.88E-3	10320	2.03	3.01E-3
r = 8	28069	11.41	1.98E-3	13505	3.23	3.20E-3
r = 9	29479	12.37	2.69E-3	14619	3.64	2.89E-3
r = 10	29823	12.43	2.40E-4	14569	3.50	2.34E-3
r = 11	30275	13.60	1.01E-4	17816	5.07	5.64E-3
r = 12	30730	13.29	8.08E-5	21853	7.49	3.86E-3

Table 4.2
Comparison of results obtained by running two versions of package Y12M in the solution of systems of algebraic equations with matrices F2(300,300,100,r,100.0) with NZ=300r+110. COUNT is the largest number of elements found in array ALU during the GE. This experiment has been carried out on an IBM 3081 in single precision. The times are given in seconds.

The algorithm for ordering rows (columns) in increasing number of non-zero elements at stage s (s=1,2,...n-1) will not be discussed here (but it will be discussed in detail in **Chapter 6**, where the implementation of GE for sparse matrices in package **Y12M** will be described). Only the pivotal search (carried out under the assumption that some ordering has already been made in some way) will be studied in this section for two typical cases.

Case A - p(s) is small. The pivotal strategy for general matrices in package **Y12M** ([331]) is an example of a pivotal strategy with a small p(s). All non-zero elements in the rows (columns) of set I_s are searched. The number of searched elements is normally small, because p(s) is small and matrix A is sparse. Moreover, a pivotal strategy of the class of **IGMS's** can easily be specified when p(s) is small (this could lead to an essential improvement of the accuracy of the computed solution in some cases). ■

Case B - p(s) is large. It is clear that it is not efficient to search for a pivot among all non-zero elements in the rows (columns) of I_s in this situation. A very elegant device, by which searching all non-zero elements in the rows (columns) of I_s is often avoided, has been developed for the case p(s)=n-s+1 by **Curtis and Reid [48]** and improved by **Duff [63]**. Both rows and columns are searched by this device (in order of increasing number of non-zero elements in their active parts). The treatment of rows is much easier than that of columns, because the non-zero elements in the code in which this device is applied, **MA28**, [63,77], are ordered by rows. Therefore, if there are both rows and columns with the same number of non-zero elements in their active parts, then the rows are searched before the columns. It is expected that a pivot will be found while rows are searched. If a row (column) with k non-zero elements in its active part is searched at stage s of the GE and if the Markowitz cost M_{ijs} of the non-zero element $a_{ij}^{(s)}$ currently searched is less than $(k-1)^2$ (k(k-1)), then the pivotal search is terminated. ■

The efficiency of the second device (where large p(s) is used) depends crucially on the mean number of rows (columns) searched during the GE. If this number is small, then the device performs well. If this number is large, then the device fails to find pivots quickly and it may be better to apply a pivotal strategy with a small p(s). The following examples show that for some matrices the use of a small p(s) is much more profitable.

Example 4.1. Let $B \in R^{n \times n}$. Assume that r(i,1)=c(j,1)=k for all i and for all j. Let $C \in R^{n \times n}$ have the same sparsity pattern as B. Consider

$$(4.29) \quad A = \begin{vmatrix} \alpha I & B \\ C & 0 \end{vmatrix}, \quad \alpha \in R, \quad A \in R^{(2n) \times (2n)}, \quad I \in R^{n \times n},$$

where I is the identity matrix.

The number of non-zero elements in any of the first n rows of matrix A is k+1, while the number of non-zero elements in the last n rows of A is k. The number of non-zero elements in any of the first n columns of A is k+1, while the corresponding number in the last n columns of A is k.

Assume that **all** non-zero elements of matrix A satisfy the stability requirement at stage 1 of GE (or, in other words, belong to B_1). Set $p(s)=n-s+1$ and use the device described in **Case B**. The minimal number of non-zero elements in the active parts of a row (column) at stage 1 is k. Moreover, there are n rows with k non-zero elements and n columns with k non-zero elements. The rows will be searched first. For any element $a_{ij}^{(1)} \neq 0$ with $i \in \{n+1,n+2,\ldots,2n\}$ the Markowitz cost is

$$(4.30) \qquad M_{ijs} = (k-1)k$$

and the stopping criterion stated above is not satisfied, because

$$(4.31) \qquad M_{ijs} > (k-1)^2 .$$

On the other hand, any $a_{ij}^{(1)} \neq 0$ with $i \in \{n+1,n+2,\ldots,2n\}$ belongs to the set of candidates for pivots, C_1^*, and may be chosen as a pivot, because it is assumed that all non-zero elements satisfy the stability requirements. This proves that the **pivotal strategy based on the second searching device fails to find quickly a pivot at stage 1 for this example, not because the pivot is not among the searched non-zero elements but because the pivotal strategy is not able to establish the simple fact that any of the searched elements can be chosen as a pivot.** ∎

The above example **proves** that a pivotal strategy with a large $p(s)$ may fail to find quickly a pivot when $c(j,s) > r(i,s)$ holds for many non-zero elements. Such relations occur often when linear least squares problems are solved by the use of augmented matrices (see **Example 3.3**). However, the long pivotal search is not the whole story when large $p(s)$ is in use. The stability requirements may also cause difficulties for the pivotal strategies with a large $p(s)$. In these strategies the quick termination of the pivotal search is related to the requirement that at least one element with a minimal Markowitz cost belongs to the set of candidates C_s^*. If the stability requirement is such that the optimal Markowitz cost M_s is greater than the minimal Markowitz cost M_s^*, then a pivotal strategy with a large $p(s)$ may fail to find a pivot quickly. A typical situation where this occurs is again the solution of linear least squares by augmentation. An illustration is given by **Example 4.2**.

Example 4.2. Consider the first stage of the decomposition of matrix A from (4.29). Let now $B \in R^{m \times n}$ with say $m \geq 2n$. Set $C = B^T$. Assume that there are q rows with r non-zero elements among the rows of matrix B. Assume also that (i) q is large and (ii) r is the smallest number of non-zero elements in a row of B. Finally, assume that the numbers of non-zero elements in all columns of B that are involved in the pivotal search of the q rows with r non-zeros is greater than r+1. The q diagonal elements in αI (corresponding to the q rows in B with r non-zeros) have a minimal Markowitz cost at stage 1; $M_1^* = r^2$ (take into account the symmetry of A). It is well-known, however, that the choice of diagonal elements of αI as pivots should be avoided during the first m stages (because this leads effectively to forming the matrix $B^T B$ in the right-hand side lower corner of A when $\alpha=1$). Therefore it is natural to exclude the diagonal elements of αI from the stability set B_1 (and also for the next stages $s=2,3,\ldots,m$). This can be achieved either by an appropriate choice of α and u or by incorporating a special device for rejecting the diagonal elements of αI as pivots. If an element on the diagonal of αI cannot be selected as a pivot and if $p(s)=m+n-s+1$, then at least q rows and q

columns will be searched before the determination of the first pivot. The
situation will be similar at many of the next stages. A numerical demonstration of this situation is presented in **Table 4.3**. It is seen, from **Table
4.3**, that the search of many rows and columns leads to a great increase of
the computing time. The accuracy achieved is rather poor; the reasons for this
will be described in the second part of this section.

Compared characteristics	Small value of $p(s)$ Package Y12M; Zlatev et al.,[331]	Large value of $p(s)$ Package MA28; Duff [63]
Accuracy	3.5E-5	1.7E-2
Computing time	130.15	1466.04
Storage (COUNT)	255889	271356

Table 4.3
Numerical results obtained in the solution of a system
$Ax=b$, where A is an augmented matrix of order 2000
with $B=F2(1500,500,150,6,100.0)$. COUNT is the largest
number of non-zero elements found during the LU-decomposition of A. The number NZ of non-zero elements in
A is 19720. The computer used is an IBM 3081 and the
time is given in seconds. ■

By **Example 4.1** and **Example 4.2** it is proved that there are at least
two cases in which the pivotal strategy based on the use of a large $p(s)$ may
fail to find a pivot quickly:

(i) $c(j,s) > r(i,s)$ for many $a_{ij}^{(s)} \neq 0$ in the rows
searched at stage s.

(ii)   the minimal Markowitz cost is not optimal (the
non-zero elements with minimal Markowitz cost do
not belong to the stability set B_s).

Consider now the choice between a pivotal strategy from the class of
GMS's and a pivotal strategy from the class of IGMS's. Since the elements
of the sets of candidates for pivots C_s^* are also elements of the stability
sets B_s, the change made to obtain an IGMS (from the corresponding GMS)
does not seem to be very important for the accuracy of the results. However,
it can be verified both theoretically and experimentally that there exist
classes of matrices for which the use of a pivotal strategy from the class of
IGMS's is clearly superior over the use of a pivotal strategy of the class
of GMS's.

The theorem proved below, **Theorem 4.1**, shows that there exist classes
of matrices for which the use of any IGMS ensures stable results, while
there is no guarantee that this will happen when a GMS is applied.

Theorem 4.1. Let matrix $A \in R^{n \times n}$ be diagonally dominant (the absolute
value of any diagonal element is greater than the sum of the absolute values
of the other elements in its row) and symmetric in structure ($a_{ij} \neq 0 \Rightarrow a_{ji}$
$\neq 0$ for all $i \in \{1,2,\ldots,n\}$ and for all $j \in \{1,2,\ldots,n\}$). Then the GE
process is stable when any pivotal strategy from the class of IGMS's is
used.

Proof. Let only diagonal elements be chosen as pivotal elements during the first $s-1$ stages of the GE. By this choice the symmetry in structure is preserved. Moreover, submatrix A_s (the active part of matrix A at stage s) is diagonally dominant too (see [270]). Therefore $c(j,s)=r(j,s)$ ($j \in \{s,s+1,\ldots,n\}$) and $M_{ijs} = [r(i,s)-1][r(j,s)-1]$. Let $r(i_1,s)=r(i_2,s)=\ldots=r(i_k,s)$ where $k \leq p(s)$. Then the diagonal elements in rows $i_1,i_2,\ldots,i_k$ are elements of set C_s^* and the largest (in absolute value) element of C_s^* will be one of these elements. It follows that if an **IGMS** is used then the pivotal element will be one of the diagonal elements. Thus, if an **IGMS** is used, then only diagonal elements can be chosen as pivots. Therefore the symmetry in structure is preserved, and it can be shown (as in [270], pp. 288-289) that $b_n \leq 2b_1$, where b_k for $k=1,2,\ldots,n$ is defined by (4.8). Hence the GE process is stable. ∎

Remark 4.5. If the pivotal strategy used belongs to the class of **GMS's**, then the above result does not hold. This can be proved as follows. Let $r(k,s)=r(m,s)$, where $k=i_1$ and $m=i_2$, i.e. row k and row m contain the smallest number of non-zero elements in their active parts at stage s (among the rows of A_s). Then the off-diagonal element $a_{km}^{(s)}$ can be chosen as a pivot when it satisfies the stability condition (it is easily seen that $a_{km}^{(s)} \in C_s^*$). In this way both the symmetry in structure and the diagonal dominance will be destroyed, and the stability of the GE process is not ensured when this happens. ∎

Remark 4.6. If matrix A satisfies the conditions of **Theorem 4.1**, then no pivoting for numerical stability is needed, but pivoting for preservation of sparsity is needed. The usefulness of the pivotal interchanges can be demonstrated by the following example. Consider the matrix whose $3n-2$ non-zero elements are a_{ii} ($i=1,2,\ldots,n$), a_{1j} and a_{j1} ($j=2,3,\ldots,n$). Assume that this matrix is diagonally dominant. Then the conditions of **Theorem 4.1** are satisfied. The computations will be stable without any pivoting. However, if no pivoting is used, then the non-zero elements after the first stage will become n^2 (this means that n^2-3n+2 fill-ins are created). If any **IGMS** is applied, then no fill-in will appear during the factorization. The same result (no fill-in during the factorization) will also be obtained when any **GMS** is applied in the treatment of this particular example (because only diagonal elements are candidates for pivots also when a **GMS** is used). ∎

Remark 4.7. It is easy to prove that when matrix A satisfies the conditions of **Theorem 4.1** and when an **IGMS** is used, the selection of pivots does not depend on the stability factor u (or, in other words, any value of u will produce the same sequence of pivots). But many matrices which are encountered in practice do not belong strictly to the above class and nevertheless retain the behaviour of b_n (b_n does not become very large). However, pivoting for numerical stability must be carried out in the latter case. The stability factor u should be chosen carefully: some balance between the following two extreme cases must be found.

(a) A very small value of u **may** save some computations in the search for pivotal elements (because the number of non-zero elements which are to be examine becomes smaller), but this will in general lead to a bad preservation of sparsity (i.e. many fill-ins will often be created).

(b) The use of large values of u **may** cause instability. ∎

   Remark 4.8. The requirement for diagonal dominance can be slightly
relaxed. In **Theorem 4.1** it is required that **every** diagonal element is
greater (in absolute value) than the sum of the absolute values of the other
elements in its row. One can require that the absolute value of every diagonal
element is greater than or equal to the sum of the absolute values of the
other elements in its row, but at least one diagonal element is greater (in
absolute value) than the sum of the absolute values of the other elements in
its row (at least one strict inequality can be found).

   Some results concerning the stability of the **IGMS's** for symmetric and
positive definite matrices can be established (see **[341]**, **p. 51**). This issue
will not be pursued here, because there are special methods (**[82,115]**) for
this type of matrices which are rather efficient. ■

   The second theorem proved in this section, **Theorem 4.2**, indicates that
some of the unacceptable pivotal strategies which a **GMS** may entail are
removed when the corresponding **IGMS** is used.

   Theorem 4.2. Assume that:

   (i) system Ax=b is solved,

   (ii) y is the solution found

and

   (iii) $\|x-y\| < \delta$ is required.

   Let $G(\delta)$ be the set of all possible pivotal sequences from the class
of the **GMS's** that are not acceptable (i.e. any pivotal strategy from set
$G(\delta)$ may be chosen during the factorization of matrix A, and if this
happens then $\|x-y\| \geq \delta$ will be produced). Let $G^*(\delta)$ be the corresponding
set when pivotal strategies from the class of **IGMS's** with the same
parameters u and p(s) (s=1,2,...n-1) are used. Then $G^*(\delta) \subseteq G(\delta)$.

   Proof. Let $g(\delta)$ be any unacceptable pivotal sequence found by a **GMS**;
i.e. $g(\delta) \in G(\delta)$. If at any stage s of the GE there is an element which
is greater (in absolute value) than the absolute value of the s'th component
of $g(\delta)$ and which belongs to C_s^*, then this element will be chosen as
pivotal by the corresponding **IGMS** and therefore $g(\delta)$ does not belongs to
$G^*(\delta)$. It is clear that there is no element og $G^*(\delta)$ which is not an element
of $G(\delta)$. This implies the assertion of the theorem. ■

   Remark 4.9. The proof of **Theorem 4.2** shows that if the matrix is
large, then many unacceptable pivotal sequences which are possible if a **GMS**
is used may be removed from the set of the possible sequences when the
corresponding **IGMS** is applied (indeed, any unacceptable pivotal strategy in
a **GMS** that is caused by the choice of a non-zero element which is not among
the largest elements in the set of candidates at the stage under consideration
is not present in the set of unacceptable pivotal strategies of the
corresponding **IGMS**). Many numerical examples which illustrate that this
happens in practice are given in **[291,295,301,332,341]**; see also **Table 4.4**.
All experiments show that if n is large, then the use of an **IGMS** will
normally give more accurate solutions (and sometimes **much more** accurate
solutions), because many pivotal sequences which cause large errors in the
solution are removed; see again **Table 4.4**. ■

The above theorems, Theorem 4.2 and Theorem 4.2, as well as many experiments with different classes of matrices indicate clearly that the use of an **IGMS** should be preferred for matrices that satisfy the following conditions (even the use of these conditions with s=1 is often worthwhile; for s=1 the conditions could easily be checked):

(a) $\left|a_{i,k(i,s)}^{(s)}\right| = max \left(\left|a_{ij}^{(s)}\right|\right),$
$s \le j \le n$

(b) $\left|a_{i,k(i,s)}^{(s)}\right|/\left|a_{ij}^{(s)}\right| \approx u^* < u$ (only non-zero elements $a_{ij}^{(s)}$
with $s \le i, j \le n$ and with $j \neq k(i,s)$ being used here),

(c) $a_{i,k(i,s)}^{(s)} \in C_s^*,$

(d) $p_{is} > 1$ (where p_{is} is the number of non-zero elements in row i, i=s,s+1,...,n, which belongs to the set of candidats for pivots C_s^*).

The test-matrices of class E(n,c) (see **Section 3.7**) satisfy conditions **(a)-(d)** for s=1 with $u^* = 4$. Therefore it is not surprising that any matrix $A \in E(n,c)$ is difficult for the **GMS's** when n is sufficiently large and when the stability factor u is greater than 4. This is so because the choice of an off-diagonal element as a pivot may cause a growth by a factor $1+u^*$ at the stage where this happens. On the other hand, only diagonal elements will be chosen as pivots when any **IGMS** is applied, and this causes no growth of the non-zero elements of A_s (s=1,2,...,n-1). Therefore poor (or even very bad) results should be expected when a **GMS** is chosen, while very accurate results are to be expected when a **IGMS** is used. This is demonstrated in **Table 4.4**. It should be noted that the fact that the computed solution is quite wrong can (for these matrices, at least) easily be detected by checking the growth factor b_n/b_1. It should also be noted that for this kind of matrices the choice of off-diagonal elements as pivots leads to a very poor preservation of the sparsity (i.e. many fill-ins are created).

c	The pivotal strategy is a GMS				The pivotal strategy is an IGMS			
	COUNT	Time	GROWTH	Accuracy	COUNT	Time	GROWTH	Accuracy
20	162788	164	1.9E5	2.3E0	105074	16	1.0	8.7E-3
40	256747	791	2.4E5	6.9E1	123720	24	1.0	3.0E-3
60	301889	840	1.5E5	1.8E0	148439	41	1.0	1.4E-3
80	329007	1348	3.9E6	7.0E1	131682	31	1.0	8.4E-4
100	308611	1384	2.5E5	7.8E0	138913	35	1.0	5.0E-4
120	271453	909	5.5E5	6.2E0	119119	24	1.0	3.6E-4
140	238675	705	1.9E5	1.8E0	111747	20	1.0	1.7E-4

Table 4.4
Comparison of a pivotal strategy from the class of the CMS's and the corresponding pivotal strategy from the class of IGMS's. Systems of linear algebraic equations whose matrices are of class E(3600,c) with c=20(20)140 are solved. The number of non-zero elements in matrix A is NZ=5n-2c-2=17998-2c. GROWTH = b_n/b_1 is the growth factor. The experiment is carried out on an IBM 3081 and the computing times are given in seconds.

The analysis of the pivotal strategies as well as the numerical experiments (not only these given here; see [291,301,306,323,327,331,332,341]) show that a small p(s) should be used in the pivotal strategy. This choice has the following advantages when compared with the use of a large value of p(s):

(A) It is not necessary to store and update, at each stage of the GE process, information about the order of columns in increasing number of non-zero elements.
(B) The additional storage needed in the computational process is reduced.
(C) No pivotal search of columns is needed.
(D) The number of computational operations connected with the pivotal search is reduced considerably.
(E) A pivotal strategy from the class of IGMS's can easily be implemented. The use of such a strategy may result in much much accurate solutions than those obtained by the use of a pivotal strategy from the class of GMS's (see Table 4.4).

Since only a few rows are searched, the pivotal strategy with a small p(s) **may** produce more fill-ins than the corresponding pivotal strategy with a large p(s). The results of many experiments indicate, however, that this will happen very seldom. Indeed, in many situations the pivotal strategy with a small p(s) produces even less fill-ins than that with a large p(s); see **Table 4.4**). Nevertheless, an option where a large p(s) could be specified may be useful. In the new version of **MA28** both small and large values of p(s) may be chosen (see [66,69,142,202]).

4.5. Pivotal strategies based on a local minimization of fill-ins

The following definitions can be used to introduce pivotal strategies of the third group.

Definition 4.4. Let $a_{ij}^{(s)}$ be an arbitrary non-zero element in set A_s and let F_{ijs} be the number of **fill-ins** created at stage s of the GE when $a_{ij}^{(s)}$ is chosen as a pivot. Then the integer

$$(4.32) \quad F_s = min(F_{ijs}) \quad (\ s \in \{1,2,\ldots,n\text{-}1\}, \quad s \leq i \leq n, \quad s \leq j \leq n, \quad a_{ij}^{(s)} \in B_s \ )$$

is called **the optimal fill-in cost at stage s**. If the stability requirement, $a_{ij}^{(s)} \in B_s$, is removed from (4.32), then **the minimal fill-in cost at stage s**, F_{ijs}^*, will be found. It is clear that $F_{ijs} \geq F_{ijs}^*$. ∎

Definition 4.5. Let I_s and B_s be the same as in Definition 4.3 or in Remark 4.2. Let

$$(4.33) \quad C_s^* = \{ \ a_{ij}^{(s)} \ / \ a_{ij}^{(s)} \in B_s \quad \wedge \quad F_{ijs} = F_s \ \}.$$

The class of **generalized minimum fill-in strategies (GMFS's)** contains **all** pivotal strategies in which an **arbitrary** element of the set of candidates C_s^* is chosen as pivot at stage s (s=1,2,...,n-1).

The class of **improved generalized minimum fill-in strategies (GMFS's)** contains **all** pivotal strategies in which an **one of the largest in absolute value** elements of the set of candidates C_s^* is chosen as pivot at stage s (s=1,2,...,n-1). ∎

It is rather difficult to implement a pivotal strategy of this class in a code for sparse matrices. **Duff and Reid** [73] have carried out some experiments using a pivotal strategy from the third group. Their conclusion is that "although this algorithm is the most computationally expensive of those under consideration its cost is not totally prohibitive". Therefore if a pivotal strategy from this group is included in a package for sparse matrices, then it should not be the only available pivotal strategy. Such a strategy may be useful for some special matrices. Consider, for example, the matrix whose sparsity pattern is given on **Fig. 4.1**. The application of any strategy based on the Markowitz cost-function will give **three times more fill-ins** than the application of a strategy based on a local minimization of fill-ins (the numbers of fill-ins being 27 and 9 respectively).

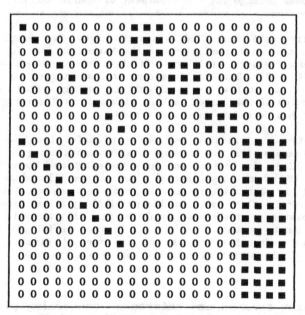

Figure 4.1
The sparsity pattern of a 22x22 matrix such that 9 fill-ins are produced when a pivotal strategy based on a local minimization of fill-ins is used, while the corresponding figure for a pivotal strategy based on the Markowitz cost-function is 27. The non-zero elements are denoted by ■.

One can attempt to apply some ideas as those in **Section 4.4** and develop pivotal strategies with a small p(s) (i.e. one restricts the search for a pivot in a set of a few "best" rows or columns). Such a strategy may perform well for matrices that give a considerably smaller number of fill-ins than the strategies from the other two groups.

4.6. Choice of a pivotal strategy in a package for sparse matrices

The survey made in this chapter indicates that no pivotal strategy will be the best one in all situations. This is so because in all of these pivotal strategies one attempts to preserve sparsity by some criterion that preserves (or at least attempts to preserve) sparsity **locally** at each stage of GE. The attempt to preserve sparsity globally leads to a very big combinatorial problem and, therefore, seems to be prohibitive in general. It is well-known that any pivotal strategy based on some attempt to preserve sparsity locally may fail to preserve it globally. The conclusion is that:

> several different pivotal strategies have to be implemented in a
> general purpose package for the treatment of sparse matrices.

This requirement is satisfied by some packages. Two examples are given below. Both pivotal strategies based on the Markowitz cost-function with a small p(s) and with large p(s) are available in the new version of package **MA28** ([66,69,142,202]). Three different pivotal strategies (two of them for special matrices; see below) are available in package **Y12M** ([331,341]). Moreover, a version of the latter package with a pivotal strategy based on a priori ordering has been developed and tested.

Special pivotal strategies may be very useful for some classes of matrices. For example, the choice of pivotal elements among the elements of the main diagonal is the best choice for some matrices. There is an option provided in **Y12M** where only diagonal elements are allowed as pivots. For some matrices it is worthwhile to supress entirely the pivotal interchanges. There is an option in **Y12M** by which this possibility is exploited. It should be stressed, however, that the user should be sure that the matrix that is to be decomposed can be successfully treated by the special pivotal strategy he/she intends to specify. The use of any special pivotal strategy is efficient only if the matrix is suitable for such a strategy.

4.7. Concluding remarks and references concerning pivotal strategies for sparse Gaussian elimination

An attempt to show how any pivotal strategy for sparse matrices can be obtained from some pivotal strategy for dense matrices has been carried out in this chapter. Therefore the discussion started with a short description of the pivotal strategies for dense matrices in **Section 4.1**. The pivotal strategies for sparse matrices are divided into three groups in **Section 4.2**. Representatives of the pivotal strategies of each group are described in **Section 4.3** - **Section 4.5**. The description is complete for the pivotal strategies based on the Markowitz cost-function, while some basic ideas concerning the pivotal strategies from the other two groups are sketched. Pivotal strategies based on the Markowitz cost are very efficient and commonly used in practice. Many details about pivotal strategies from the other two groups as well as for pivotal strategies in general can be found in [5,33,48,69,73,78,82-86,115,117,155,198-202,218-219,237-245]. The choice of a pivotal strategy for a general-purpose packages for sparse matrices is briefly discussed in **Section 4.6**.

If the matrix in the system of linear algebraic equations solved is **symmetric and positive definite**, then much simpler pivotal strategies may be used. In this case:

(i)   no pivoting for numerical stability is needed
(ii) all pivotal interchanges can be performed before the numerical computations.

The exploitation of the fact that the matrix is symmetric and positive definite has not been considered in this chapter. The interested reader is advised to consult **[82,115]**. It is only necessary to point out here that the pivotal elements for symmetric and positive definite matrices must be chosen on the main diagonal (otherwise both the symmetry and the positive definiteness will be destroyed). This observation shows that the simplicity of the pivotal strategies for such matrices is due not only to the fact that no pivoting for stability is needed, but also because a few elements only (the diagonal elements) are to be searched. The simplicity is sometimes hidden behind concepts of graph theory (commonly used in this situation). However, if the graph theory concepts are translated into matrix theory concepts (which is always possible and should always be done), the simplicity becomes apparent.

If the matrix of the system is symmetric but not positive definite, then this fact can be exploited in the design of pivotal strategies. The exploitation of the fact that the matrix is only symmetric (but not positive definite) has not been discussed here. Sparse codes in which this property is exploited are described in **[78,175,201]**. More details can be found in these references. It should only be noted here that special care should be taken to preserve symmetry during the factorization process (by choosing the so-called **two-by-two** pivots, see, for example, [175]).

Many other special properties of the matrices treated can be exploited to develop special pivotal strategies. This is, for example, true for many classes of matrices that arise after the sparse discretization of partial differential equations by finite elements; as an illustration of this see [255-256].

It must be emphsized here that it has been implicitly assumed in nearly the whole of **Chapter 4** that the classical manner of exploiting sparsity is used. Of course, all pivotal strategies discussed in this chapter can also be applied when the computer-oriented manner of exploiting sparsity is applied. However, **the conclusions concerning the efficiency of the different strategies made here are not necessarily true when the computer-oriented manner is specified**. All conclusions made in this chapter are valid when all non-zero elements are kept and updated during the whole factorization. If neglecting "small" elements is allowed (i.e. if the computer-oriented manner of exploiting sparsity is in use), then the comparison between the different pivotal strategies **may** lead to other conclusions. This will be illustrated in the next chapter, where the use of iterative refinement in connection with the GE process for sparse matrices will be described.

The influence of the pivotal strategy on the accuracy of the results is an important issue when general sparse matrices are treated. In the latter case it seems to be necessary to relax the stability requirements in order to preserve better the sparsity of the matrix under consideration. Results for some classes of matrices are given in **Section 4.5**. However, results for general matrices are also needed. A general result is given in **[197]**. It is proved in **[197]** that the constant $\pi=2.01$ in the classical estimate of the error in the LU factorization of dense matrices should be replaced by $\pi=3.01$ when sparse matrices are decomposed by using a stability factor u. Unfortunately, the constant π is not very important in the error bound. Much more important is the multiplier b_n, see (4.7) and (4.8). When dense matrices are decomposed by using partial pivoting the bound $b_n \leq 2^{n-1}b_1$ can be obtained; see (4.10). The corresponding bound for general sparse matrices that are decomposed by using some Markowitz pivotal strategy with stability factor $u \geq 1$ is $b_n \leq (1+u)^{n-1}b_1$. Therefore the error is unacceptable (in both cases) even if $b_n \approx 2^k b_1$ with $k \ll n$ (assuming that the sparse pivotal strategy is used with u=1). It is believed that partial pivoting gives accurate results for dense matrices, but this is so only because all experiments support an assumption that the growth factor b_n is small. For sparse matrices the situation is not very clear. Some examples where b_n is very large are given in **Table 4.4**, see also **Chapter 9**. Moreover, large growth factors do imply inaccurate solutions (for the particular examples in **Table 4.4**, at least). This means that more efforts towards obtaining stability results for general sparse matrices are needed. Monitoring the growth factor is the simplest tool that can be used to check the stability of the computations. One can do this either directly during the factorization or use the bound derived in **[92]**. The first device is more robust (the actual growth factor is calculated when it is used), but also more expensive. The second device works rather well for some matrices (first and foremost for well-scaled matrices), but in many cases the estimate of the growth factor calculated by the second device differs too much from the actual growth factor. On the other side, the second device is much cheaper than the first one. The second device is implemented in **MA28** (**[63]**).

Many classical results for dense matrices, **[35]**, could probably be extended for general sparse matrices and this should be done in the near future.

CHAPTER 5

USE OF ITERATIVE REFINEMENT IN THE GE PROCESS

It is well-known that if Gaussian elimination (GE) with iterative refinement (**IR**) is used in the solution of systems of linear algebraic equations Ax=b whose coefficient matrices are **dense**, then the accuracy of the results will usually be greater than the accuracy obtained by the use of Gaussian elimination without iterative refinement (**DS**). However, both more storage (about 100% because a copy of matrix A is needed) and more computing time (some extra computing time is needed to perform the iterative process) must **always** be used with the **IR** process.

Assume that matrix A is **sparse** and that some sparse matrix technique is used. Then the accuracy of the solution computed with the **IR** will normally still be greater than the accuracy of the solution computed by the **DS**. However, even more important is the fact that in the latter case the use of the **IR** **may** also lead to a reduction of both the computing time and the storage needed (as illustrated by **Table 1.3**). The effect of applying **IR** when sparse systems of linear algebraic equations are solved will be studied in detail in this chapter.

IR will normally be used together with applying a positive drop-tolerance during GE. This means that **IR** is used in connection with the computer-oriented manner for exploiting sparsity introduced in **Chapter 1**. Since GE is a particular method within the general k-stage scheme (see **Example 3.3**), it is clear that **IR** can also be considered as a particular case in the common approach for solving sparse problems $x = A^{-1}b$ that has been described in **Section 3.3**.

5.1. Convergence of the IR process

Let us assume (as in the previous chapter, **Chapter 4**) that systems of linear algebraic equations

(5.1) $Ax = b$ ($A \in R^{n \times n}$, $x \in R^{n \times 1}$, $b \in R^{n \times 1}$, $rank(A) = n$)

are solved by GE. The application of GE during the factorization of matrix A (together with some pivotal strategy; see again **Chapter 4**) leads to

(5.2) $LU = PAQ + E$ ($L \in R^{n \times n}$, $U \in R^{n \times n}$, $P \in R^{n \times n}$, $Q \in R^{n \times n}$, $E \in R^{n \times n}$),

where, as explained in the previous chapters, L and U are triangular matrices, P and Q are permutation matrices, while E is a perturbation matrix. In this case E is caused not only by rounding errors made during the calculations on the computer used, but also because some "small" non-zero elements are removed during the GE. The criterion defined by (1.9) will be used in this chapter in the decision whether an element is small or not. Another criterion will be defined and used in the **Chapter 11**.

Assume that the factors L and U are calculated. Then an approximation x_1 to the exact solution of (5.1) can be found by

(5.3) $x_1 = QU^{-1}L^{-1}Pb$.

If no "small" elements have been removed during the computation of (5.2) (or, in other words, if the classical way of exploiting sparsity is used), then x_1 is the **DS** (the direct solution) and the solution process is terminated by (5.3). If it is allowed to remove "small" elements (i.e. if the computer-oriented manner of exploiting sparsity is used), then the triangular factors L and U will normally be inaccurate, in which case the approximation x_1, the first solution, will also be inaccurate. Therefore it is worthwhile to attempt to regain the accuracy lost during the computation of (5.2) and (5.3) by performing the well-known iterative refinement (**IR**) process ([100,172,179,231,271-272]):

(5.4) $r_i = b - Ax_i$,   i=1,2,...,q-1   (compute the residual),

(5.5) $d_i = QU^{-1}L^{-1}Pr_i$,   i=1,2,...,q-1   (compute the correction),

(5.6) $x_{i+1} = x_i + d_i$,   i=1,2,...,q-1   (update the solution).

Some stopping criteria are to be applied in order to terminate the **IR** when one of the following conditions is satisfied:

(i)	**further calculations are not justified because the changes that are to be made by (5.6) can not be represented on the computer used (the correction is so small that the updated solution is equal to the previous approximation),**
(ii)	**the iterative process is either not convergent or is very slowly convergent,**
(iii)	**the accuracy required is achieved.**

If **(i)** is satisfied, then **(iii)** is normally also satisfied. However, the greatest accuracy that can be achieved (when the iterative process converges) depends on the computer used. If the user requires greater accuracy than the greatest accuracy that can be achieved on the computer used, then the iterative process may be terminated by **(i)** without achieving the accuracy that is requested.

The stopping criteria by which the three conditions are checked in package **Y12M** are

(5.7) $\|x_q - x_{q-1}\| \leq \epsilon \|x_q\|$,

(5.8) $\|d_q\| > \|d_{q-1}\|$ $\wedge$ $q > 2$,

(5.9) $q = MAXIT$,

where the machine precision is denoted by ϵ, $\|\cdot\|$ is the vector norm used and MAXIT is the prescribed maximal number of iterations allowed in the iterative process.

Conditions (i) and (iii) are checked by (5.7). An attempt to achieve the greatest accuracy on the computer used is made in **Y12M**. This requirement can be relaxed by introducing an additional stopping criterion in which ϵ replaced by some $\epsilon < PREC < 1$. This is the choice made in the new version of **MA28**. An attempt to terminate the iterative process when the convergence rate is slow is carried out by (5.9). The following theorems can be used to justify (heuristically at least) the introduction of (5.8) by which an attempt to check whether the iterative process is convergent or not is made.

Theorem 5.1.   Let x be the exact solution of (5.1). Assume that all computations with (5.3)-(5.6) can be performed without rounding errors. If

(5.10) $F = U^{-1}L^{-1}E$,

then the following two equalities are satisfied:

(5.11) $x_{i+1}-x = QF^iQ^T(x_1-x) = -QF^{i+1}Q^Tx$ (i = 1,2,...),

(5.12) $d_{i+1} = QF^iQ^Td_1$ (i = 1,2,...).

       Proof.   The assertions of this theorem are simple corollaries of the assertions of **Theorem 3.2** and **Theorem 3.3** (which once again illustrate the usefulness of the general k-stage scheme). Indeed, replace in (3.29) and (3.30) i by i+1 and j by 1. Also take into account that vector s from (3.29) is equal to zero for square matrices of full rank. Finally, use the representation of GE as a member of the general k-stage scheme (see **Example** 3.1). In this way (5.12) and the first part of (5.11) will be obtained from (3.30) and (3.29) respectively. The second part of (5.11) can easily be obtained by using (5.2), (5.3) and (5.10). ■

       Theorem 5.2.   Let λ_k (k=1,2,...,n) be the eigenvalues of matrix F   and let

(5.13) $|\lambda_1| \geq |\lambda_k|$ (k=2,3,...,n).

       Under the same assumptions as in the previous theorem the following equality holds:

(5.14) $x = x_j + \sum_{i=j}^{\infty} d_i$ (j=1,2,...)

when

(5.15) $|\lambda_1| < 1$.

       Moreover, if (5.15) holds, then the following two equalities also hold:

(5.16) $x = \lim_{i \to \infty}(x_i)$,

(5.17) $\lim_{i \to \infty}(d_i) = 0$.

       Proof. The proof of Theorem 5.2 follows from the proof of Theorem 3.4 (see also the remarks in the proof of the previous theorem). ■

       Corollary 5.1.   All assertions of **Theorem 5.2**, i.e. (5.14), (5.16) and (5.17), hold also if (5.15) is replaced by

(5.18) $\|F\| < 1$.

   Proof.   See the corresponding result concerning the general k-stage scheme (**Remark 3.6**).   ■

   Definition 5.1. If λ_1 satisfies (5.13), then the number $|\lambda_1|$ is called **the spectral radius of matrix F.** It is often denoted by $\rho(F)$:

(5.19) $\rho(F) = |\lambda_1|$.   ■

   Definition 5.2. The non-negative number

(5.20)   RELEST $= \|d_{q-1}\|/\|x_q\|$

will be called the estimated relative error.   ■

   Using of the spectral radius of F and the results proved above the following behaviour of the stopping criteria (5.7)-(5.9) should be expected.

   (A) If $\rho(F) \ll 1$, then **IR** will converge quickly and will typically be stopped by the second criterion, (5.8), or possibly by the first one, (5.7), if rounding errors dominate.   RELEST will give a good estimate of the relative error in the computed solution, x_q, assuming that $b \neq 0$.   ■

   (B) If $\rho(F) < 1$ but close to one, then the convergence rate will as a rule be very slow and **IR** will typically be stopped by (5.9).   RELEST will provide a fair error estimate (assuming again that $b \neq 0$).   ■

   (C) If $\rho(F) \geq 1$, then the **IR** will probably not converge and (5.17) will not hold. This will be detected by (5.8), normally with q=3. The value of RELEST gives information about the accuracy of the approximation.   ■

   It must be emphasized here that the theorems on which the above interpretation of the behaviour of the stopping criteria is based hold under the assumption that (5.3)-(5.6) are performed without rounding errors (an error analysis in which the rounding errors are taken into account is given in [172]; see also [100]). Of course, **rounding errors do appear** when the computational process is performed on a computer. The computation of the residual vectors, r_i, is very often carried out in extended precision and there is a good experimental evidence that this is sufficient for the **IR** process to converge within machine accuracy (see also **Section 5.5**). It should also be mentioned that the rounding errors made during the back substitution are usually much smaller than those made during the factorization (although the rigorous bounds for the rounding errors during the back substitution are larger than those for the factorization; [231, pp. 155-159]).

5.2.   The drop-tolerance

   In package **Y12M** the **IR** process is used in connection with the computer-oriented manner of exploiting sparsity, [331,341]. This implies that some "small" elements are removed during the elimination process (5.2), and **IR** is a tool by which an attempt to regain the accuracy lost during the computation of L and U is made. In this section first the effect of using **IR** for dense and sparse matrices will be discused, and then the choice of the drop-tolerance in package **Y12M** will be described.

Assume that A is **dense** and consider IR and DS. More storage is needed when IR is applied (because a copy of A is used in the calculation of the residual vectors). Also more computing time is needed for IR (because the iterative process is to be carried out after obtaining the first solution). DS will normally ensure a sufficiently accurate solution. It is true that the IR process provides cheap error estimations. However, this is not a serious advantage of IR when matrix A is dense, because error estimates can be obtained by other means; see, for example, [42-44,55,99-100], where some cheap devices for calculating an estimate of the condition number of a dense matrix are given. The estimate of the condition number can be used to obtain information concerning the sensivity of the solution of the system of linear algebraic equations to rounding errors during the factorization. The conclusion is that DS should preferred when A is dense, but possibly one should try to evaluate the errors made (directly or indirectly).

The situation changes when A is **sparse**. Now one of the most important tasks in the efforts to achieve a high degree of efficiency is the attempt **to keep the number of fill-ins small.** Often many of the fill-ins are small (compared with the original elements). Therefore it is natural to throw away fill-ins that are smaller than a certain level defined by a special parameter called **the drop-tolerance.** Of course, not only fill-ins but also other elements, which becomes small during the factorization, can be dropped. An attempt to justify the usefulness of removing small elements when sparse matrices are treated was probably first made by **Evans** [94], but other authors have also considered this possibility ([41,198,244,278]). The first code for general sparse matrices in which a drop-tolerance is used together with IR is SIRSM ([318,319]). This possibility is also exploited in Y12M and in LLSS01 ([320-322,331,341]). The idea is fairly general and can be applied to a large class of direct methods (**the common approach of solving sparse linear algebraic problems**; see **Section 3.3** or [299]). An answer to the question why this approach may be profitable when sparse matrices are treated will be answered in the next two sections.

An absolute drop tolerance, introduced in (1.9), will be used in this chapter. This means that an element is removed when it becomes smaller in absolute value than the drop-tolerance T chosen by the user. It would be desirable to apply a relative drop tolerance, so that an element is removed when

$$(5.21) \qquad |a_{ij}^{(s)}| \leq T \ min[\ \underset{s \leq k \leq n}{max} \ (|a_{ik}^{(s)}|), \ \underset{s \leq k \leq n}{max} \ (|a_{kj}^{(s)}|)].$$

Unfortunately, it is not easy to implement such a criterion in an efficient way, because the maximal elements in the rows and columns would have to be kept and updated during the GE, which is a rather expensive procedure. Experiments were carried out with both criteria. In general, the simple use of an absolute T, the criterion defined by (1.9), performed better than (5.21). However, it should be pointed out that this is true when IR is used and something about the magnitude of the elements of A is known. If more advanced methods (as, for example, preconditioned conjugate gradient-type methods) are applied, then in general one can apply larger values of T than the values of T for which the IR converges. The use of larger values of T requires a more careful choice of the dropping criteria. This problem will be discussed in the **Chapter 11.** In this chapter only the use of a criterion based on an absolute drop-tolerance will be assumed.

It is important to answer the question: **how can an optimal value of the drop-tolerance be found?** If a long sequence of systems with similar matrices is to be treated, then it may be profitable to try to determine an optimal or nearly optimal value of the drop-tolerance. Long sequences of systems of algebraic equations appear often in practice (when time-dependent problems are solved). Therefore the important problem of finding an optimal drop-tolerance in this situation deserves a special discussion and such a special discussion will be presented in **Chapter 8.** The problem of finding an optimal value of the drop-tolerance in the case where one system of linear algebraic equations is to be solved is open. If the system is well-scaled and not very ill-conditioned, then the use of drop-tolerances in a rather wide range gives good results. This has been demonstrated in [295,301,303,323,332,341]. Many numerical examples will also be given in this chapter. The conclusion is that, although it is not clear how to find an optimal drop-tolerance in the case where only one system is solved, often it is rather easy to find a good value of the drop-tolerance that can efficiently be used during the factorization.

5.3. Storage comparisons

The following abbreviations will be used in this section: **DMT** is some dense matrix technique code, **SMT-DS** is a sparse matrix technique code with direct solution and **SMT-IR** is a sparse matrix technique code with iteratively refined solution. The choice of the **DMT** code is not critical (one can use, for example, the subroutines **DECOMP/SOLVE** from [99]). Y12M is used here as an **SMT** code, but the ideas are fairly general and could be applied to other codes also.

An attempt to answer the following two questions will be carried out in this section:

(A) When is the **SMT** code (either **SMT-DS** or **SMT-IR**) more efficient than the **DMT** code with regard to the storage requirements?
(B) When is the **SMT-IR** code more efficient than the **SMT-DS** code with regard to the storage requirements?

Assume that the integers occupy the same space in the computer memory as the reals.

The storage needed by a typical **DMT** code is given by

(5.22) $S_1 = n^2 + 3n.$

Since the **SMT-DS** is in fact the case where the classical manner of exploiting sparsity is in use, the three arrays in the input storage scheme can be overwritten by the three large arrays of the dynamic storage scheme (see **Remark 2.1**). Moreover, it is reasonable to assume that $NZ \leq NN1 = 0.6NN$. Therefore the storage needed for the **SMT-DS** is

(5.23) $S_2 = 2.6NN + 13n.$

The computer-oriented manner of exploiting sparsity is applied in the SMT-IR. Thus, the three arrays of the input storage scheme cannot be overwritten by the arrays of the dynamic storage scheme (matrix A will be needed in the iterative process). Two extra arrays are needed: to keep the residual vectors r_i and the correction vectors d_i. The storage needed is

(5.24) $S_3 = 2.6NN + 3NZ + 15n$.

Assume that

(5.25) $NN = \upsilon NZ$.

From (5.22), (5.23) and (5.25) it can be deduced that

$$(5.26) \quad S_2 < S_1 \Leftrightarrow NZ/n^2 < g(\upsilon,n) \overset{\text{def}}{=} (1-10/n)/(2.6\upsilon) .$$

It is easily seen that

$$(5.27) \quad g(\upsilon,n) < g(2,n) < \lim_{n\to\infty}[g(2,n)] = 1/5.2 \approx 0.1923077$$

and the following criterion concerning the storage requirements of the DMT and the SMT-DS can be formulated.

Criterion 5.1 If more of 20% of the elements of matrix A are non-zero elements, then the DMT will use less storage then the SMT-DS. ∎

It is difficult to formulate a converse of Criterion 5.1; i.e. a criterion which will answer the question: **when is the SMT-DS more effecient (with regard to the storage used) than the DMT?** The difficulties arise because the number of locations needed for fill-ins is normally not known at the beginning of the GE process. If this information (parameter COUNT; see Section 2.7) is available, then the choice NN = COUNT + 2n is as a rule a good one when the storage requirements only are taken into account (but this choice may be time-consuming, because many garbage collections may be needed when the length of the main arrays is so small; see Table 2.2). If COUNT is not known, then information about the performance of the SMT-DS with regard to the storage requirements can be obtained by studying the function $g(\upsilon,n)$. Selected values of this function are given in Table 5.1.

n	$\upsilon=2$	$\upsilon=3$	$\upsilon=4$	$\upsilon=5$	$\upsilon=6$	$\upsilon=7$
50	0.154	0.103	0.0769	0.0615	0.0513	0.0440
100	0.173	0.115	0.0865	0.0692	0.0577	0.0495
1000	0.190	0.127	0.0952	0.0762	0.0635	0.0544
∞	0.192	0.128	0.0962	0.0769	0.0641	0.0549

Table 5.1
Selected values of function $g(\upsilon,n)$.

The function $g(\upsilon,n)$ varies slowly with n when n is sufficiently large (and this is the case when sparse matrices are handled numerically). Therefore the efficiency of the SMT-DS with regard to the storage requirements depends in practice only on parameter υ (or, in other words,

on the length of the main arrays in the dynamic storage scheme). **Table 5.1**
proves in fact that even if many locations are reserved for fill-ins (if
parameter v is large), then the **SMT-DS** will still be more efficient than
the **DMT** when less than 5% of the elements of the matrix are non-zeros
(which is normally satisfied when n is large).

The comparison of the storage requirements of the **DMT** and the **SMT-IR**
can be carried out as follows. From (5.22), (5.24) and (5.25) it follows
that the relationship

$$(5.28) \quad S_3 < S_1 \; \Leftrightarrow \; NZ/n^2 < g^*(v,n) \;\overset{def}{=}\; (1-12/n)/(3+2.6v) \; .$$

has to be satisfied when the **SMT-IR** is more efficient than the **DMT** with
regard to storage requirements. The function $g^*(v,n)$ is such that

$$(5.29) \quad g^*(v,n) < g^*(2,n) < \lim_{n\to\infty}[g^*(2,n)] = 1/8.2 \approx 0.12119512 \; .$$

Criterion 5.2. If more than 12% of the elements of matrix A are non-
zeros, then the **DMT** will use less storage than the **SMT-IR**. ∎

Again it is difficult to express the converse of Criterion 5.2; i.e. a
criterion which will answer the question: **when is the SMT-IR more efficient
(with regard to the storage used) than the DMT?** The reasons are the same as
those given above (where the **SMT-DS** and the **DMT** are compared). Some
information could be obtained by studying the values of function $g^*(v,n)$.
Selected values of this function are given in **Table 5.2.**

n	$v=2$	$v=3$	$v=4$	$v=5$
50	0.0976	0.0741	0.0597	0.0500
100	0.110	0.0833	0.0672	0.0563
1000	0.121	0.0917	0.0739	0.0619
∞	0.122	0.0926	0.0746	0.0625

Table 5.2
Selected values of function $g^*(v,n)$.

It is seen from **Table 5.1** and **Table 5.2** that the two functions
$g(v,n)$ and $g^*(v,n)$ have similar properties. Also the function $g^*(v,n)$
varies slowly with n when n is sufficiently large. This means that also
the efficiency of **SMT-IR** (with regard to the storage requirements) depends
in practice only on v, but now the length of the arrays in the dynamic
storage scheme can as a rule be chosen considerably smaller than in the case
where **SMT-DS** is in use. An illustration of this fact is given by **Table 2.1**:
$v=9$ is needed when **SMT-DS** is used, while $v=2$ is quite sufficient when
SMT-IR is applied. **Table 5.2** proves that if $v=3$ is chosen (which often is
more than enough; see **Table 2.1** and **Table 2.3**) and if n is sufficiently
large, then **SMT-IR** is more efficient than **DMT** when the number of non-zeros
is less than 9% (and this condition is usually satisfied by sparse matrices).

Remark 5.1. The comparison between **SMT-IR** and **DMT** is not quite fair
for **SMT-IR**, because it has been assumed that the smallest value of v is
2. This is only true when the value of NZ1, the non-zeros that are larger

in absolute value than the drop-tolerance at the beginning of the GE, is equal to NZ (i.e. no non-zero element has been removed during the initialization of the dynamic storage scheme; see **Remark 2.2**). However, experiments indicate that for some matrices that arise in practice NZ1 is considerably smaller than NZ (see, for example, the results given in **Table 2.3**). Therefore v can be smaller than 2. This shows that **SMT-IR** may be more efficient than **DMT** even if more than 12% of the elements of matrix A are non-zeros, provided that many "small" elements are removed during the initialization of the dynamic storage scheme. However, this property depends on the particular problem, and Criterion 5.2 is rather good as a general rule (because the number of non-zeros in a large sparse matrix is normally less than 12%). ■

Let us try to compare (with regard to the storage requirements) **SMT-DS** and **SMT-IR**. The lengths of the main arrays in the dynamic storage scheme are different for these two options. Let NN_2 be the length of the arrays in the row ordered list when **SMT-DS** is used, and let NN_3 be the corresponding length when **SMT-IR** is chosen. Assume that

(5.30) $NN_2 = v_2 NZ$, $NN_3 = v_3 NZ$.

The following relation can be derived from (5.23), (5.24) and (5.30):

(5.31) $S_3 < S_2 \Leftrightarrow 3 + 2n/NZ < 2.6(v_2 - v_3)$.

Assume that $NZ \geq 5n$ (this inequality is very often satisfied). Then the following criterion can be introduced.

Criterion 5.3. **SMT-IR** is more efficient than **SMT-DS** (with regard to storage requirements) when $v_2 - v_3 > 1.31$. ■

The results in this section are derived under several assumptions. Several remarks concerning these assumptions are necessary.

Remark 5.2. It is assumed that integers and reals occupy the same storage in the computer memory. Some obvious modifications must be performed when this is not the case (when, for example, double precision is used). ■

Remark 5.3. Criterion 5.1 is defined by using the assumption made in **Y12M**: NN ≥ 2NZ. In general such an assumption is not needed (see **Remark 2.1** and **[341]**). The condition in Criterion 5.1 can be relaxed if the requirement is NN ≥ NZ. However, all experiments show that in practice NN must very often be considerably greater than 2NZ when the classical manner of exploiting sparsity is in use (see the results presented in **Table 1.2**, **Table 1.3**, **Table 2.1** and **Table 2.3**). Therefore the requirement NN ≥ 2NZ is normally not a restriction. ■

Remark 5.4. Consider Criterion 5.3. If a more economical scheme, consisting of a copy of arrays ALU and CNLU as well as of the first and the third column of array HA, is used instead of the input storage scheme, then the condition $v_2 - v_3 < 1.31$ can be replaced by $v_2 - v_3 < 1$ (see **[341]**, p. 66). However, the difference is not very great. The input storage scheme is very convenient for some large practical problems, as for example, problems arising from the space discretization of partial differential equations or problems arising in the discretization of systems of ordinary differential equations. ■

The numerical results presented in the previous chapters confirm many of the conclusions made in this section. Another experiment was carried out by the use of some matrices of class $E(n,44)$, $n=650(50)1000$. The results of this experiment are given in **Table 5.3**.

n	NZ	SMT-DS (F01BRE + F04AXE)		SMT-IR (package Y12M)	
		COUNT	COUNT/NZ	COUNT	COUNT/NZ
650	3160	22246	7.04	7697	2.44
700	3410	24286	7.12	8453	2.48
750	3660	26932	7.36	9174	2.51
800	3910	30424	7.78	9882	2.53
850	4160	35290	8.48	11643	2.80
900	4410	37230	8.44	12360	2.80
950	4660	40488	8.69	12551	2.69
1000	4910	45850	9.34	14082	2.87

Table 5.3
The storage used with SMT-DS and SMT-IR for matrices E(n,44).

The results shown in **Table 5.3** indicate, once again, that for some matrices the difference v_2-v_3 is much greater than 1.31 (in fact, for all eight matrices $v_2-v_3 > 4$). It should be mentioned that the **NAG** subroutines (see [178]) are used in the **SMT-DS** option. This is not the best choice for these matrices, because a Markowitz pivotal strategy with a large value of $p(s)$ is implemented in these subroutines. The use of a small value of $p(s)$ will reduce the values of COUNT and COUNT/NZ, but the new values will still be much larger than the values of COUNT and COUNT/NZ for the **SMT-IR** option.

5.4. Computing time

Several factors play an important role when computing times spent by different linear equation solvers are compared. For **dense** matrices the situation is fairly straightforward: the computing time is directly proportional to the number of simple arithmetic operations and this number is $2n^2/3 + 2n^2 + O(n)$ for GE.

The situation is much more complicated when sparse matrix techniques are in use. Then the computing time spent depends on the following quantities:

(i)   the order n of the matrix,
(ii) the number NZ of the non-zero elements,
(iii) the number of fill-ins created,
(iv) the pivotal strategy used.

The order n and the non-zeros NZ are kept fixed when the two options, **SMT-DS** and **SMT-IR**, are compared. However, even when the numbers of fill-ins are roughly the same for two pivotal strategies, the computing

times can be very different. Therefore in this section not only are SMT-DS
and SMT-IR compared, but also the use of different pivotal strategies is
discussed. First, two pivotal strategies based on the Markowicz cost-function
will be used in the comparisons, and thereafter some other pivotal strategies
will be tested.

The two Markowicz strategies are implemented in the NAG Library [178]
and in Y12M. The first of them is a GMS with a large p(s), the second one
is an IGMS with a small p(s); see the previous chapter. The abbreviations
DS and IR will be used in this section (instead of SMT-DS and SMT-IR),
because only codes in which sparse matrix technique is implemented will be
compared.

   A). Varying the order of the matrix. The combined effect of IR (with
drop-tolerance T=0.001) and a better pivotal strategy is shown in Table
5.4. It is seen that the speed-up (DS comp. time / IR comp. time) becomes
larger when the order of the matrix becomes larger, which is a very nice
property. ∎

   B). Comparison with a code for special matrices. Matrices of class
E(n,c) are used in this experiment, but similar results have been found for
many other matrices. The matrices of class E(n,c) are symmetric, positive
definite band matrices with bandwidth   2c+1 (see Section 3.7). It is
interesting that the IR option of Y12M combined with a proper value of the
drop-tolerance is competitive (when   n   is sufficiently large) with some
subroutines that are written for symmetric, positive definite band matrices.
The results given in Table 5.5 are obtained for matrices of class E(n,c)
with $n=c^2$ (similar matrices appear in the discretization of certain partial
differential equations on a square space domain). The NAG subroutine F04ACE
([178,277]), which exploits the symmetry, positive definiteness and bandedness
of the matrix, has been selected for this experiment. For small values of n
F04ACE is more efficient. However, the IR option of Y12M becomes better
when   n   is greater than   1521; for n=2304 IR is considerably better
(although Y12M is written for general matrices and does not exploit
symmetry, positive definiteness and bandedness). ∎

n	NZ	DS (F01BRE + F04AXE)		IR (package Y12M)	
		COUNT	Comp. time	COUNT	Comp. time
650	3160	22246	23.47	7697	4.55(5.1)
700	3410	24286	28.46	8453	5.27(5.4)
750	3660	26932	38.29	9174	6.12(6.2)
800	3910	30424	53.67	9882	6.88(7.8)
850	4160	35290	58.54	11643	6.97(8.4)
900	4410	37230	57.35	12360	7.47(7.7)
950	4660	40488	115.58	12551	8.07(14.3)
1000	4910	45850	152.31	14082	8.50(17.9)

Table 5.4
The storage and computing times for matrices A=E(n,44).
IR is used with T=0.001. The speed-up for IR is given in brackets.
The computing times (on a UNIVAC 1100/82 computer) are given in seconds.

n	c	Band subroutine - F04ACE	Package Y12M - IR
900	30	5.17	8.15
961	31	5.90	8.84
1024	32	6.59	9.41
1089	33	7.40	10.02
1156	34	8.29	10.57
1225	35	9.25	11.23
1296	36	10.28	11.86
1369	37	11.41	12.65
1444	38	12.62	13.33
1521	39	13.93	14.23
1600	40	15.35	14.76
2304	41	30.66	21.45

Table 5.5
Comparison of the IR option of package Y12M with a subroutine
which exploits the symmetry, the positive definiteness and the bandedness
of the matrices of class $E(n,c)$. The computing times are given in
seconds on a UNIVAC 1100/82 computer.

n	DS - F01BRE+F04AXE	DS - Y12M	IR - Y12M
650	47.11	25.59	13.16
700	59.63	33.81	13.00
750	57.57	34.99	13.77
800	65.69	36.22	14.55
850	69.04	26.63	15.16
900	77.36	41.41	16.00
950	90.91	40.71	16.28
1000	82.92	42.13	17.85

Table 5.6
Computing times obtained in runs with matrices of class $D(n,c)$.
The figures are the sums of the computing times obtained in the solution
of six systems with different values of c (c=4(40)204) for each n.
The time is measured in seconds on a UNIVAC 110/82 computer.
The DS option of Y12M is used with $T=10^{-12}$, while $T=10^{-2}$ is used
with the IR option of Y12M.

C). Varying both the order and the sparsity pattern. Many experiments
have been carried out in order to compare the performance of different options
and different pivotal strategies (based on the Markowicz cost-function) when
the parameters n and c are varied (i.e. both the order of matrix A and
its sparsity pattern are varied in this experiment). Results obtained in runs
with test-matrices of class $D(n,c)$ with n=650(50)1000 and c=4(40)204 are
given in **Table 5.6**. The same experiment has been performed with matrices of
class $E(n,c)$; the results are given in **Table 5.7**. The figures in these two
tables are sums of the computing times (in seconds) for the six systems with
different values of c solved for each n. It is seen that the IR option
of **Y12M** is the most efficient option for these two classes of matrices. It
should be mentioned that this is also true for many other classes of matrices
(including many matrices from the Harwell-Boeing set of sparse test-matrices;
[71,72]). ■

n	DS - F01BRE+F04AXE	DS - Y12M	IR - Y12M
650	52.45	25.53	16.43
700	69.31	29.09	19.10
750	86.68	31.82	21.55
800	109.38	36.52	23.06
850	131.07	42.68	25.33
900	143.69	46.31	27.45
950	227.23	52.04	29.51
1000	253.87	58.62	31.79

Table 5.7
Computing times obtained in runs with matrices of class E(n,c).
The figures are the sums of the computing times obtained in the solution
of six systems with different values of c (c=4(40)204) for each n.
The time is measured in seconds on a UNIVAC 110/82 computer.
The DS option of Y12M is used with $T=10^{-12}$, while $T=10^{-2}$ is used
with the IR option of Y12M.

c	DS - F01BRE+F04AXE	DS - Y12M	IR - Y12M
4	3.45	2.42	2.37
44	11.85	6.50	2.09
84	14.15	6.97	2.43
124	14.19	7.34	2.44
164	12.92	6.49	2.75
204	9.13	6.50	2.47

Table 5.8
Computing times obtained in runs with matrices of class D(800,c).
The time is measured in seconds on a UNIVAC 110/82 computer.
The DS option of Y12M is used with $T=10^{-12}$, while $T=10^{-2}$ is used
with the IR option of Y12M.

c	DS - F01BRE+F04AXE	DS - Y12M	IR - Y12M
4	3.24	2.11	2.45
44	52.52	14.76	6.88
84	24.99	9.19	4.43
124	12.69	4.52	3.67
164	7.79	4.11	3.15
204	6.15	2.83	2.48

Table 5.9
Computing times obtained in runs with matrices of class E(800,c).
The time is measured in seconds on a UNIVAC 110/82 computer.
The DS option of Y12M is used with $T=10^{-12}$, while $T=10^{-2}$ is used
with the IR option of Y12M.

D). Varying the sparsity pattern only (the order is fixed). In Table
5.8 and Table 5.9 the dependence of the computing time on parameter c
(i.e. on the sparsity pattern of the matrix; see Section 3.7) is illustrated
for n=800. Small and large values of c seem to be easiest for all three

codes, but the difference is not great for the IR option of Y12M.
Intermediate values of c are especially tough for the NAG subroutines
F01BRE and F04AXE, although these values put also a certain strain on the
DS option of package Y12M. This is the reason for the great differences in
the performance of the three codes in Table 5.6 and Table 5.7. But in all
cases the DS option of package Y12M is better than the NAG subroutines
and in turn the IR option of package Y12M is better still, excepting the
case with A = E(800,4). ∎

   E). Varying the density of the non-zero elements. In Table 5.10 the
NAG subroutines F01BRE and F04AXE are compared with the IR option of
package Y12M by using matrices of class $F2(m,n,c,r,\alpha)$; see Section 3.7.
The values of parameters m, n, c and α are kept fixed: m=n=500, c=20,
α=100.0. Parameter r is varied: r=5(5)40. By varying r the density of
the non-zero elements in the matrix is varied; NZ=rm+110; see Table 3.1.
Thus, the problem becomes harder when the value of parameter r is larger.
It is seen from Table 5.10 that the IR option of Y12M is 3-5 times
faster than the NAG subroutines also when the density of the non-zero
elements is varied. ∎

r	NZ	NZ/(N*N)	DS - F01BRE+F04AXE	IR - package Y12M
5	2610	0.01	9.91	2.22
10	5110	0.02	32.96	6.16
15	7610	0.03	56.84	11.60
20	10110	0.04	59.32	14.84
25	12610	0.05	131.39	25.59
30	15110	0.06	97.89	34.32
35	17610	0.07	144.16	50.76
40	20110	0.08	288.03	62.81

Table 5.10
Comparison of computing times obtained in runs with matrices of
class F2(500,500,20,r,100.0). The computing times are given in
seconds on a UNIVAC 1100/82 computer. IR is used with $T=10^{-2}$.

   F). Markowitz strategy versus a priori ordering: varying the density of
the non-zero elements. The results in experiments (A) - (E) have been
obtained by using pivotal strategies based on the Markowitz cost-function.
Some experiments, where a pivotal strategy based on an a priori ordering was
used, have also been carried out. In the first of these experiments the
density of the non-zero elements of the matrix has been varied. Some results,
which correspond to the results obtained with the DS option and presented
in Table 4.1, are given in Table 5.11. The results obtained with the
strategy based on an a priori ordering are compared with the IR option of
Y12M (where a Markowitz pivotal strategy is used). It should immediately be
emphasized that these results are quite unexpected (at least for r>3). It was
expected that for the pivotal strategy with the a priori ordering the values
of COUNT would be larger than the corresponding values for the Markowitz
strategy, but that the difference would be smaller than that for the case
where DS is used (see Table 4.1). It was expected that this may lead to
slightly smaller computing times for the a priori ordering (because it is
considerably cheaper than the Markowitz pivotal search). However, this is
true, for the matrices tested at least, only for r≤3. For r>3 the values of
COUNT obtained when the a priori ordering is used are considerably smaller

than those obtained when the Markowitz strategy is used and, therefore, also
the computing time for the former pivotal strategy are considerably smaller.
Compare the results in **Table 4.1** with the results in **Table 5.11**. The
Markowitz strategy performs better when the **DS** option (with T=0.0) is used,
while the a priori ordering performs better for the **IR** option (with T>0.0).
Of course, this will not be true in general. However, the results given in
Table 5.11 indicate that the use of a large drop-tolerance may influence in
a different way the performance of the pivotal strategy and that the influence
of a positive drop-tolerance on the pivotal strategies deserves further investigation. ∎

Density	A priori ordering the columns			Markowitz cost-function		
parameter	COUNT	Time	Accuracy	COUNT	Time	Accuracy
r= 2	784	0.09(6)	2.91E-12	710	0.12(8)	1.81E-12
r= 3	1331	0.15(7)	6.43E-13	1203	0.19(8)	3.38E-13
r= 4	1778	0.18(7)	5.61E-13	2027	0.26(7)	6.71E-13
r= 5	1982	0.23(9)	9.56E-13	2928	0.40(9)	8.92E-13
r= 6	2331	0.24(7)	4.97E-13	3531	0.51(9)	4.65E-13
r= 7	2707	0.30(8)	1.01E-12	4249	0.57(7)	1.02E-12
r= 8	3098	0.31(6)	7.43E-13	4923	0.75(10)	7.85E-13
r= 9	3497	0.42(10)	9.02E-13	5639	0.84(8)	8.78E-13
r=10	3827	0.41(7)	7.39E-13	6057	0.95(8)	7.71E-13
r=11	4191	0.45(7)	8.21E-13	6859	1.14(10)	7.96E-13
r=12	4549	0.47(6)	5.26E-13	7247	1.18(9)	7.53E-13

Table 5.11
Comparison of two IR options based on two different pivotal strategies.
The density of the non-zero elements is increased when r is increased.
The matrices are F2(300,300,100,r,100.0). The numbers of iterations are
given in brackets. The times are measured in seconds on an IBM 3081.
These results should be compared with the results given in Table 4.1.

G). **Performance of the IR option for augmented matrices**. Some results,
obtained with two **DS** implementations for augmented matrices, were presented
in **Chapter 4 (Table 4.3)**. The corresponding results that are obtained with the
IR option are given in **Table 5.12**. The storage used is reduced by a factor
approximately equal to five, while the time is reduced by a factor greater
than 16 (for the old **MA28** the corresponding factor is 181). The use of
IR in connection with augmented matrices desrves a special discussion. The
treatment of augmented matrices will be studied in **Chapter 7**. ∎

Compared charecteristics	IR - package Y12M; small p(s)
Accuracy	1.03E-6
Computing time	8.11
Iterations	9
Storage used (COUNT)	55412

Table 5.12
Numerical results obtained in the solution of system By=c,
where B is the augmented matrix of A=F2(1500,500,150,6,100.0).
The results in this table should be compared with those in Table 4.3.

H). **Performance of IR in connection with another code (MA28).** In the previous experiments the implementation of an **IR** option in package **Y12M** has been compared with **DS** implementation. Of course, **IR** can be implemented in other sparse codes also, and has been implemented in **MA28** [66,69]. Some results obtained with the **IR** option of **MA28** are given in **Table 5.13**; the corresponding results (found with two **DS** options in **MA28**) are given in **Table 4.4**. The same conclusions as above can be drawn by comparing the results in **Table 5.15** with those in **Table 4.4**: both the storage and the computing time are reduced when the **IR** option is in use. ∎

c	COUNT	Time	Iterations	Accuracy
20	74462	30.23	114	7.92E-5
40	76572	18.57	53	7.53E-5
60	74390	12.87	27	2.67E-5
80	71521	10.23	17	5.34E-5
100	69789	8.94	12	3.05E-5
120	67260	7.87	9	6.87E-5
140	65743	7.44	8	2.00E-5

Table 5.13
Numerical results obtained with the IR option of package MA28.
The new version of MA28 is used with T=0.001 and with a small p(s).
The matrices are E(3600,c). The code stops the computations when
estimated accuracy becomes smaller than a prescribed parameter PREC
(PREC=10^{-4} is selected for this experiment). The results in this table
should be compared with the results in Table 4.4.

The results presented in this section indicate that very often it is efficient to apply **IR** when large sparse matrices are treated numerically. These results as well as many other runs show that the use of an **IR** option in a sparse code is worthwhile and as a rule leads to a reduction of both the storage needed and the computing time spent:

(i)   for values of n varying in a wide range,
(ii) for different sparsity patterns (different values of c),
(iii) for different densities of the non-zeros (different r's),
(iv) in connection with different codes.

It has also been demonstrated that the **IR** process can be used in several different ways. The different ways of using the **IR** process are the topic of the discussion in the next section.

5.5. Different ways of using the IR process

Three different ways of using the **IR** process in the solution of linear algebraic equations have been proposed in the literature.

A). The classical or the English way. The residual vectors r_i in (5.4) are accumulated in extended precision and then rounded to single precision. All other computations are performed in single precision. This strategy has been analyzed in several works of **Wilkinson** [271-272]; see also [172,179,231,259]. Assume that the length of the mantissa of the reals is t_1 digits when single precision is in use. If **IR** converges and if the residual vectors are calculated as above, then the accepted approximation x_q will be very near the exact solution x rounded to t_1 digits. ■

B). The revolutionary or the Polish way. In a paper written by **Jankowski and Wozniakowski** [154] it is shown that under certain conditions the extended precision is not needed even in the computation of the residual vectors. Therefore all computations can be carried out in one precision only. This is an advantage of the method (using extended precision may cause difficulties on some computers). On the other hand, one normally will not achieve full machine accuracy of the approximate solution as in the classical way ([154,221-223]). It should be noted, however, that in many scientific and engineering problems full machine accuracy is not needed; the accuracy actually required is much lower than the machine accuracy, but one needs a reliable error evaluation of the the the accuracy obtained. ■

C). The cautious or the Scandinavian way. The vectors r_i, d_i and x_i are stored in extended precision and all inner products in (5.3)-(5.6) are accumulated in extended precision. Everything else is performed in single precision. Assume that the lengths of the mantissa of the reals are t_1 digits when single precision is used and t_2 digits when extended precision is used. Assume also that $2t_1 \leq t_2$ (this will be satisfied when double precision is used, for example, on IBM and UNIVAC computers). If these assumptions are satisfied and if the **IR** process converges, then the resulting approximation x_q will be very near to the exact solution x rounded to $2t_1$ digits. This result was shown in [21,22,24] for an algorithm developed in [30], and all experiments indicate that it holds for GE as well; see, for example, the results given in **Table 5.13**.

The price that has to be paid with the Scandinavian way of using the **IR** process is:

(i) some extra storage (the arrays in which r_i, d_i and x_i are stored have to be declared in double precision),

(ii) some extra time per iteration,

(iii) a few (usually 3-4) extra iterations for the high accuracy.

The gain is not only a higher accuracy, but also a more reliable error estimation and possibly a more robust algorithm, which may converge in some cases where the classical way of using the **IR** process does not converge.

It should be mentioned here that in a version of package **Y12M**, used in the experiments, all three arrays involved in the **IR** process are stored in double precision and all inner products involving these three arrays (r_i, d_i and x_i) are calculated in double precision. This is probably the simplest choice: all matrices are stored in single precision, while the vectors are stored in double precision. However, the same accuracy can be obtained if the correction d_i is stored in single precision. ■

Two remarks are necessary with the three ways of using the IR process.

Remark 5.5. All conclusions concerning the accuracy of the approximation calculated by the IR process are drawn under the assumption that the input data (the non-zero elements of matrix A and the components of vector b) are represented exactly in the computer memory. ∎

Remark 5.6. It may be worthwhile to attempt to stop the IR process when a certain degree of accuracy is achieved. There is a device in the new option of package **MA28** by which such an attempt is carried out. The use of this device is illustrated in **Table 5.13**. Such a device is very useful when, for example, systems of ordinary differential equations are solved numerically. The development of a device for stopping the iteration when the required accuracy is achieved will be discussed in **Chapter 8** and in **Chapter 11**. ∎

The classical way of using the IR process is applied in package **Y12M**. The Polish way of using the IR process is applied in the new version of **MA28** [66,69] and in a modification of package **Y12M** attached to a package for solving linear ordinary differential equations [335-337]. There exists an experimental version of **Y12M** in which the Scandinavian way of using the IR is applied. This version has been run with many systems of linear algebraic equations; some results will be presented in the following chapters.

5.6. When is the IR process efficient?

The examples given in the previous sections of this chapter demonstrate the fact that the IR process combined with a good pivotal strategy and a large drop-tolerance can be very efficient. However, the use of the IR process is not always better than the use of the corresponding **DS** options. Some examples have already been given (see **Table 1.1**, **Table 1.2** and the first examples in **Table 5.9** and **Table 5.13**). Several other examples are given in **Table 5.14**. These examples are test-matrices from the Harwell set [76]. These particular matrices are not typical because **(i) no fill-in is produced during the GE** and **(ii) some of them are merely permutations of triangular matrices.** However, for the purposes of this section these matrices are really very good because they indicate what is to be expected in the worst (for the IR process) case.

Consider the extra storage for the IR option relative to the storage used with **DS** options (see **Section 5.3**):

(5.32) $(S_3-S_2)/S_2 = (3NZ+2n)/(2.6\upsilon NZ+13n)$.

Assume that $\upsilon=1$ can be used (this will practically never happen; again the worst case for the IR option is considered). Since $NZ \geq n$,   (5.32) indicates that the storage used with the IR option is, in the worst case, approximately twice as large as the storage needed for the DS options. However, it must be emphasized that here it is also assumed that the contents of the original matrix (the three main arrays in the input storage scheme) can be overwritten by the contents on the triangular factors (the contents of the three main arrays in the dynamic storage scheme) when the DS options are in use. Overwriting cannot be done when some large scientific and engineering problems are solved, because the original matrix is needed also after the decomposition. This is the case when systems of ordinary differential equations are solved by the use of implicit Runge-Kutta methods (see **Chapter**

8). When overwriting cannot be done then the storage used with the **IR** option will, in the worst case, be approximately the same as that needed for the **DS** options.

For the problems in **Table 5.14** the **IR** option uses from 20% to 29% extra computing time. If the **IR** process converges very slowly, then the amount of extra time needed for the **IR** option will be greater. In the first example in **Table 5.13** the computing time for the **IR** option is increase by a factor approximately equal to two (compare the results in **Table 5.13** with those in **Table 4.4**).

Name of the				Computing time		Extra time
matrix	n	NZ	Density	DS - Y12M	IR - Y12M	in percent
SHL 0	663	1687	0.004	0.89	1.11	25%
SHL 200	663	1726	0.004	0.97	1.16	20%
SHL 400	663	1712	0.004	0.93	1.12	20%
STR 0	363	2454	0.019	0.68	0.86	26%
BP 0	822	3276	0.005	1.24	1.60	29%

Table 5.14
Numerical results obtained in runs with five Harwell test-matrices.
No fill-ins are produced during the factorization of these matrices.

	DS - Y12M	IR - Y12M		
n	T = 1.0E-12	T = 1.0E-4	T = 1.0E-3	T = 1.0E-2
250	8.83	5.87	5.37	5.35
300	16.52	8.31	7.33	6.52
350	19.00	9.07	7.48	7.37
400	21.11	9.96	8.85	7.76
450	30.13	11.56	9.88	9.09
500	24.11	11.93	10.44	9.48
550	38.47	13.61	11.66	10.24
600	36.52	14.52	13.00	11.16
Total	194.69	84.83	74.02	66.97

Table 5.15
Comparison of computing times for matrices of class D(n,c) obtained with
the DS option and with the IR option applied with three drop-tolerances.
For each value of n the numbers given in the table are sums of the
computing times obtained in the solution of six systems with different
values of parameter c (c=4(40)204).

The numerical results indicate that as a rule both the computing time and the storage needed for the **IR** option will in the worst case be increased by a factor of two in comparison with the same characteristics for the **DS** option. However, the **IR** option may be preferred even in this case, because it will normally give higher accuracy and a rather reliable error estimate. Nevertherless, it must be emphasized here, and this has also been done in the beginning of this study, that if the matrix is such that no fill-ins or a few fill-ins only are produced during the GE, then the DS option is more efficient than the IR option both with regard to the storage used and the computing time spent.

Assume now that the matrix is such that many fill-ins are created in the course of the factorization. Then the IR option may be efficient (and often very efficient if the drop-tolerance is chosen in an optimal way). Unfortunately, it is not very easy to determine an optimal value of the drop-tolerance for an arbitrary matrix. On the other hand, it is often not very difficult to find a good value for the drop-tolerance. Assume that the matrix that is to be factorized is not very ill-conditioned and that its non-zero elements are of the same order of magnitude. Then $T \in [10^{-5}a, 10^{-2}a]$, where a is the order of magnitude of the non-zero elements, is normally a good choice. This has been demonstrated by many numerical examples in [295,332]. Some numerical examples are given in **Table 5.15** and **Table 5.16**. More numerical illustrations will be given in **Chapter 7** - **Chapter 9**. It should be noted that this rule can be applied even if the matrix is not well-scaled (after some kind of scaling; scaling will be discussed in **Chapter 11**).

	DS - Y12M	IR - Y12M		
n	T = 1.0E-12	T = 1.0E-4	T = 1.0E-3	T = 1.0E-2
250	3.45	3.66	3.54	3.32
300	4.49	4.72	4.47	4.42
350	6.38	6.31	5.84	5.68
400	8.51	8.11	7.32	6.96
450	10.51	10.25	9.11	8.49
500	13.49	12.70	11.00	10.03
550	15.67	15.23	12.83	11.50
600	20.01	20.00	14.73	12.94
Total	82.51	80.98	68.84	63.34

Table 5.16

The same as Table 5.15, but the matrices are of class E(n,c); c=4(40)204.

Sometimes a nearly optimal value of the drop-tolerance can be determined by using a special device. This will be demonstrated in **Chapter 8**, where large and sparse systems of linear ordinary differential equations will be treated by exploiting sparsity. However, the device is applicable not only for systems of ordinary differential equations, but for any problem that contains a long sequence of systems of linear algebraic equations.

5.7. Large drop-tolerance versus large stability factor

The large drop-tolerance T has been introduced in order to limit the number of fill-ins. In the previous chapter it has been explained that the use of a **large stability factor** u leads to increasing the candidates for pivotal elements in set C_s^* ($s \in \{1,2,\ldots,n-1\}$) and, possibly, to a better preservation of sparsity (i.e. to a reduction of the number of fill-ins). Therefore it is interesting to investigate the combined effect of T and u on the efficiency of the computational process. This effect is discussed below by using some experimental results shown in **Table 5.17**. Many other experiments indicate that the results presented in **Table 5.17** are typical. Other tests are discussed in [295], pp. 391-392.

T	Stability factor u = 4			Stability factor u = 512		
	COUNT	Iterations	Time	COUNT	Iterations	Time
0.0	3376	7	5.68	3044	7	4.99
0.001	1790	10	2.44	2218	11	3.47
0.1	1475	12	2.25	1947	11	3.01
1.0	1120	13	1.79	1333	15	2.34
10.0	860	11	1.27	946	21	2.32

Table 5.17
The drop-tolerance T versus the stability factor u.
The matrix used is F2(125,125,15,6,4.0) with NZ=860.
A CDC CYBER 173 is used and the times are given in seconds.

When $T = 0.0$ (and the same is also true when T is very small, say, $T=10^{-20}$ as in [295]), the use of a large value of the stability factor u does lead to a reduction of both the storage and the computing time. However, when T is large, then the results are better when the stability factor is smaller. This very surprising result is probably due to the fact that rather small pivotal elements can be selected when the stability factor is large, and small pivotal elements tend to produce larger fill-ins. Thus, although fewer fill-ins are produced, most of them are retained despite the use of a large value of the drop-tolerance T. The conclusion is that **the stability factor u should not be chosen too large when the drop-tolerance T is large.**

5.8. Concluding remarks and references concerning the IR process

The IR described in this chapter is a particular case of the common approach for solving sparse algebraic problems $x = A^t b$ for the case where:

(i) matrix A is a square non-singular matrix, so that the problem $x = A^t b$ is reduced to a system of linear algebraic equations $Ax = b$,

(ii) the simple GE method is used to factorize matrix A,

(iii) IR is applied in step (c) of the common approach (see **Section 3.3**).

It is possible to obtain an **IR** option that is more efficient than the simple **DS** option **only** together with using a positive drop-tolerance T during the calculation of the triangular factors by GE. Therefore the method described in this chapter can also be considered as **an implementation of the computer-oriented manner of exploiting sparsity (see Chapter 1) in connection with the GE process.**

The idea of using a positive drop-tolerance has been suggested in
[41,94,198,244,278]. The combination of this idea with IR has probably been
implemented first in code SIRSM, [318-319]. Now it is used in several
codes; as, for example, MA28 ([66,69,142]) and Y12M ([323,329,331,341]).

For "cheap" matrices (matrices that are factorized without fill-ins or
with a few fill-ins only) the DS option gives better results with regard
both to the storage and the computing time. However, the gain is as a rule not
very big. The discussion in this chapter shows that the greatest reductions
are by a factor of two both in storage and in computing time when the DS
option is used instead of the IR option for "cheap" matrices. Moreover, it
is clear that a reduction by a factor close to two can be achieved very
seldom. Normally the reductions both in storage and computing time are by
factors that are considerably smaller than two and, thus, the IR option can
sometimes be preferred even for "cheap" matrices, because both a more accurate
solution and a reliable error estimation can be achieved by this choice.

For "expensive" matrices the use of the IR option is normally very
efficient. The matrix is called "expensive" here if many fill-ins are
created during its factorization by the GE. Experiments show that many
matrices that appear in scientific and engineering problems are often very
"expensive" (see Table 1.3, Table 2.1 and Table 2.3; many other examples
are given in [295,301,303,323,327,332,338,341]). Moreover, in some cases the
use of the IR option with large drop-tolerance is the only option by which
the problem under consideration can be made tractable numerically on the
computers that are available at present (see Table 2.3 and [336]). This
situation will be discussed in Chapter 8.

> The main conclusion is that if the IR option is specified
> for a cheap matrix, then the penalty is not large, while
> there are matrices that are so expensive that the choice of
> the DS option is catastrofic and reasonable results (in
> terms of storage and computing time) could be achieved only
> with the IR option (and a large value of the drop-tole-
> rance). Thus, an option based on IR (or another iterative
> method) with a large drop-tolerance is absolutely necessary
> in a general purpose package. On the other hand, a DS
> option may be useful for some matrices and, therefore,
> such an option should also be included in the package.

A remark concerning storage is needed here. It has been mentioned already
that it may be necessary to keep a copy of the original matrix when many
problems arising in science and engineering are solved. One such problem,
leading to the solution of large systems of linear ordinary differential
equations, will be discussed in detail in Chapter 8. Nevertheless, in the
comparison of the storage needed for the two options it was assumed that the
original matrix is overwritten by the triangular factors when the DS option
is used. If the IR option is used, then both the original matrix and the
triangular factors are to be kept. However, even in this situation, one can
reduce the storage needed when IR is applied because the original matrix and
its LU factorization are never used simultaneously. Such an algorithm, which
exploits some special properties of the UNIVAC computers, is discussed by
Wasniewski et al., [264]. The algorithm may also be applied on some other
computers.

CHAPTER 6

IMPLEMENTATION OF THE ALGORITHMS

After the foregoing discussion of the pivotal strategies that can be used during the GE process and of the iterative refinement (**IR**) process, everything is prepared to describe the main principles that can be applied in a sparse matrix code for solving systems of linear algebraic equations. The particular code used to illustrate the implementations is **Y12M**. Hence the two main storage schemes described in **Chapter 2**, the input storage scheme and the dynamic storage scheme, are applied. However, similar ideas can be implemented if other codes and/or other storage schemes are used.

An attempt to answer the following questions will be carried out in this chapter:

(a) How can the Gaussian transformations be applied in a sparse code?

(b) How can the pivotal search be organized in a sparse code?

(c) How can the IR process be incorporated?

Parallel codes and conjugate gradient-type algorithms will be discussed later, in **Chapter 10** and **Chapter 11**, but some implementation issues concerning these two topics will briefly be discussed in the end of this chapter.

Some parts of the computational process that are valid for any particular method within the general k-stage scheme for solving linear algebraic problems, $x=A^\dagger b$ (see **Chapter 3**), such as **(i)** the initialization of the dynamic storage scheme, **(ii)** the insertion of fill-ins in the dynamic storage scheme, and **(iii)** the performance of garbage collections in the dynamic storage scheme, have already been discussed in detail in **Chapter 2**. Therefore, these will not be considered in this chapter.

Assume that: **(1)** only one system Ax=b is to be solved, **(2)** the direct solution (**DS**) option is applied, and **(3)** certain operations on the right-hand side vector b are carried out during the GE process. Then the lower triangular matrix L is not needed and storage and/or computing time can be saved if L is not stored. An option, in which matrix L is not stored, will also be discussed in this chapter.

6.1. Implementation of Gaussian elimination

Assume that the Gaussian transformations at stage s, $s \in \{1,2,\ldots,n-1\}$, are to be performed. Assume also that the pivotal element is already found and that the interchanges needed to bring it to position (s,s) are made. Then the Gaussian transformations at stage s are carried out by the use of the following formula:

$$(6.1) \quad a_{ij}^{(s+1)} = a_{ij}^{(s)} - a_{is}^{(s)}[a_{ss}^{(s)}]^{-1}a_{sj}^{(s)}, \quad a_{ss}^{(s)} \neq 0,$$

$$i=s+1,s+2,\ldots,n, \quad j=s+1,s+2,\ldots,n .$$

Since the matrix is sparse, the calculations with (6.1) are carried out only if both $a_{is}^{(s)} \neq 0$ and $a_{sj}^{(s)} \neq 0$. The first of these two inequalities shows that only **the target rows**, the rows that contain non-zero elements in the pivotal column, are modified during the calculations at stage s of the GE. The row numbers of the target rows are stored in array **RNLU** of the column ordered list of the dynamic storage scheme from position HA(s,5) to position HA(s,6); see **Section 2.4**.

Assume that $i \in (HA(s,5), HA(s,5)+1, \ldots, HA(s,6))$. First it is necessary to find the location of the element $a_{is}^{(s)}$ by searching in array **CNLU** from position HA(i,2) to position HA(i,3). It must be noted here that **(1)** the search is terminated when the element is found, and **(2)** this step could be carried out (and is carried out in the new version of **Y12M**) together with the column interchanges. The element $a_{is}^{(s)}$ is interchanged with the element stored in position HA(i,2) in array **ALU**, and similar interchanges are performed with the column numbers of these two elements in array **CNLU**. Finally the contents of HA(i,2) is increased by one, and the number

(6.2) $t = a_{is}^{(s)}[a_{ss}^{(s)}]^{-1}$

is calculated and stored in ALU[HA(i,2)-1].

After this introductory step the main formula (6.1), by which the GE at stage s is carried out, is simplified,

(6.3) $a_{ij}^{(s+1)} = a_{ij}^{(s)} - t a_{sj}^{(s)}$,

and it is clear that the operation that is to be performed is: **multiply a row (the pivotal row s) by a scalar, -t, and add the result to another row (the target row i)**. This is a very well-known **basic linear algebra operation**. For dense matrices there are many standard subroutines that perform this operation very efficiently (in many computing centres these subroutines are available in machine language, which makes them even more efficient). For sparse matrices, however, the situation is much more complicated. The device described below is based on ideas used in **[77,324,341]**.

Since the pivotal row s plays a special role (it participates in **all** basic linear algebra operations at stage s), it is natural to attempt to exploit this property. This can be done in the following way. Assume that all locations of array **PIVOT** from position s to position n contain 0.0 at the beginning of stage s. Then before the beginning of the calculations at stage s the non-zero elements of the pivotal row are stored in array **PIVOT** in positions corresponding to their column numbers; i.e. if $a_{sj}^{(s)} \neq 0$ (for $j \in (s, s+1, \ldots, n)$), then $a_{sj}^{(s)}$ is stored in PIVOT(j). This means that the active part of the pivotal row is **scattered**; i.e. it is stored as a dense vector in array **PIVOT** between positions s and n. Note that the pivots found during the first s-1 stages are stored in the first s-1 locations of array **PIVOT**. The s'th pivot is stored in the right position when the above operations are performed.

Two sweeps are performed for each target row (each row i with i > s that has a non-zero element in the pivotal column):

Sweep 1. Go from position HA(i,2) to position HA(i,3) of array **CNLU** (the column numbers of the non-zero elements of the target row i are located in this part of CNLU). For each j=CNLU(K), K∈[HA(i,2),HA(i,3)], check whether PIVOT(j)$\neq$0.0. If this is so, then modify element $a_{ij}^{(s)}$ using (6.3) and set PIVOT(j)=0.0. It should be mentioned that $a_{ij}^{(s)}$ is stored in ALU(K), while $a_{sj}^{(s)}$ is stored in PIVOT(j) when (6.3) is used to obtain $a_{ij}^{(s+1)}$ from $a_{ij}^{(s)}$. It should also be mentioned that the modified non-zero element is always kept when the classical manner of exploiting sparsity is used, but this element is removed and the relevant locations in the dynamic storage scheme are freed when the computer-oriented manner of exploiting sparsity is applied and when $a_{ij}^{(s+1)}$ is "small" according to the dropping criterion chosen. ■

Sweep 2. Go from position HA(s,2) to position HA(s,3) in CNLU and for each j=CNLU(K), K∈[HA(i,2),HA(i,3)], check whether PIVOT(j)=0.0. If this is so, then restore the contents of this location; in other words, set PIVOT(j)=ALU(K). If PIVOT(j)$\neq$0.0, then a new non-zero element, fill-in, $a_{ij}^{(s+1)} = - ta_{sj}^{(s)}$, is to be created and inserted in the dynamic storage scheme (according to the rules described in **Section 2.5**). All fill-ins are stored when the classical manner of exploiting the sparsity is used, while only fill-ins that are not "small" according to the dropping criterion chosen are stored when the computer-oriented manner of sparsity is applied. ■

At the end of stage s (when the modifications have been performed for all target rows) all locations from position s+1 to position n in array **PIVOT** are set to zero (using the information about the sparsity pattern of the pivot row that is stored in the row ordered list of the dynamic storage scheme). After this operation array **PIVOT** is ready for the next stage, stage s+1, of the GE process.

6.2. Implementation of a pivotal strategy

The basic principles used in the implementation of a pivotal strategy based on the Markowitz cost-function with a small value of p(s) are described in this section. Many of these principles can also be applied in connection with other pivotal strategies described in **Chapter 4**.

It is not efficient to search all n-s+1 rows in the active part A_s of matrix A at stage s in order to find the p(s) "best" rows and to arrange them in increasing number of non-zero elements in their active parts; see (4.22). Therefore the rows are ordered by increasing number of non-zero elements before the beginning of the GE, and this information is updated at the end of every stage s. Three integer arrays of length at least equal to n are needed for the efficient storage and use of the information concerning the pivotal strategy. Columns seven, eight and eleven of array HA are used for this purpose.

The row numbers, ordered by an increasing number of non-zero elements in their active parts, are held in the seventh column of **HA**. The relevant information is stored, at stage s, from position HA(s,7) to position HA(n,7), and the information in these locations has to be updated during stage s because the following operations must be performed: **(i)** the number of the pivotal row should be removed from the seventh column of **HA**, **(ii)** the non-zero elements of the pivotal column should be moved to the begining of the active parts of the target rows, **(iii)** fill-ins are to be added in the active parts of the target rows, and **(iv)** if the computer-oriented manner of exploiting sparsity is used, then all "small" elements are to be removed from the active parts of the target rows. These operations may change the number of elements in the target rows and, therefore, the order of the rows in seventh column of array   HA.

The position of row i, in the seventh column of array HA, is stored in HA(i,8). The position, again in the seventh column of HA, of the first row with j non-zeros in its active part is stored in HA(j,11). If there is no row with j non-zeros in its active part, then HA(j,11)=0.

By the use of these three arrays (3n locations) it is possible to keep track of the number of non-zero elements in the active parts of the rows in A_s (s=1,2,...,n-1) and to keep the rows ordered in a proper way. Moreover, since only a few rows are altered during stage s of the GE, this is done in an efficient way. The number of operations needed in this process is, roughly speaking, about $O(npq)$, where p is the average number of target rows per stage and q is the average difference between the number of non-zeros per target row before and after the computation at stage s. The corresponding number is $O(n^2)$ when all rows in A_s are searched to find the p(s) "best" rows needed in (4.22). Assume: (i) that the order of matrix A is n=1000, (ii) that the average number of target rows per stage is p=10, and (iii) that q=5 (i.e. the number of the non-zeros per target row is, on the average, either 5 more or 5 less after stage s; in comparison with the non-zeros before stage s). Then $O(npq)=c_1*5*10^4$, while $O(n^2)=c_2*10^6$ in the case where all rows of A_s are searched to find the "best" rows. Although $c_2 < c_1$, it is clear that the complexity of the second algorithm is much greater. On sequential machines this will lead to some reduction of the computing time used. On vector and/or parallel machines the situation is not very clear; for more details see **Chapter 10.**

The first s locations of columns 7 and 8 of **HA** are not needed after stage s of the GE. This fact is exploited to store information about the row and column interchanges in these two columns. More precisely, the row and column number of the pivotal element selected by the pivotal strategy at stage s are stored in HA(s,7) and HA(s,8) respectively.

The p(s) "best" rows are stored from position s to position s+p(s)-1 in the seventh column of **HA**. All elements in these rows are searched to find the elements of set C_s^* defined by (4.25). Therefore a small value of p(s) is recommended; $p(s) = min(n-s+1,RPIV)$ where RPIV<4 is appropriate. It is easy to select one of the largest, in absolute value, elements of C_s^* as a pivot; i.e. an **IGMS** (see **Chapter 4**) can be applied when this device is used.

The use of the seventh, eigth and eleventh columns of array **HA** in the pivotal strategy is demonstrated in **Fig. 6.1** and **Fig. 6.2**. The small test-matrix from **Section 2.8** is used in these illustrations.

The contents of the arrays involved in the pivotal strategy before stage 1 of GE are given in **Fig. 6.1**. In the seventh column of **HA** first the row numbers of the four rows with two non-zero elements are stored and then the row number of the row with three non-zero elements. HA(5,8)=1 shows that the first row number in the seventh column of array **HA** is 5. HA(3,11)=5 shows that the row number of the first row with 3 non-zero elements is stored in position 5 of the seventh column of array HA, while HA(4,11)=0 shows that there are no rows with four non-zero elements.

The contents of the same arrays after the first stage of the GE are given in **Fig. 6.2.** It is assumed that a_{11} is the first pivot. This choice is possible for the ordering given in **Fig. 6.1** if $p(s)=4$. The row and column number of the pivot are stored in HA(1,7) and HA(1,8) respectively. The other information in **Fig. 6.2** is presented following the same rules as in **Fig. 6.1**. Note, however, that row 5 and column 4 have changed places with row 1 and column 1.

It should also be noted that the same ideas have been applied in the implementation of a pivotal strategy from the first group (see **Section 4.3**).

Position	1	2	3	4	5
HA(■,7)	5	4	2	1	3
HA(■,8)	4	3	5	2	1
HA(■,11)	0	1	5	0	0

Figure 6.1
**Contents of the columns of array HA that are used in the pivotal strategy
before stage 1 of the GE for the small matrix from (2.2).**

Position	1	2	3	4	5
HA(■,7)	1	4	2	5	3
HA(■,8)	1	3	5	2	4
HA(■,11)	0	2	5	0	0

Figure 6.2
**Contents of the columns of array HA that are used in the pivotal strategy
after stage 1 of the GE for the small matrix from (2.2).**

6.3. On the storage of the lower triangular matrix L

It is natural to store the elements of the lower triangular matrix L (without its diagonal elements) when systems of linear algebraic equations with dense coefficient matrices are solved, because the necessary locations are available in the lower triangular part of the two-dimensional array ALU(n,n) used in this situation. When several systems of linear algebraic equations with the same coefficient matrix are to be solved, maybe one after another, computing time is saved by using the same LU factorization. Even when only one system is to be solved no extra time or space is needed for the storage of matrix L.

With sparse matrices the situation is different. The matrices are often large and, thus, it could be desirable to reserve as little storage in the computer memory as possible. If this is so, then one can save space and/or reduce the number of garbage collections (see **Section 2.6**) by removing the non-zero elements of matrix L. Whenever an element below the diagonal is eliminated, the space occupied by it is freed and can be used, for example, to store a fill-in. Even if a copy of the row is still needed, it is only necessary to copy the non-zero elements above the diagonal (the non-zero elements of matrix U). When a garbage collection is to be performed, the structure can be compressed more tightly than in the case where the non-zeros

of L are kept. Also, the computing time can be reduced slightly because
fewer elements have to be handled. It must be emphasized here that some
operations involving the right-hand side b of the system Ax=b have to be
carried out during the factorization when the non-zero elements of matrix L
are removed during the GE. This can be done in a fairly efficient way. There
is an option in package **Y12M** in which the elements of L are removed.

If several systems with the same matrix are to be solved, then it is
better to keep L and to use the same LU-factorization in the solution of all
of the systems. Several systems with the same matrix must be solved during the
iterative improvement (when both **IR** and conjugate gradient-type methods are
in use). Therefore L must be kept in this case.

The reduction of the storage, measured by parameter **COUNT** (the maximal
number of non-zeros kept in array ALU during the GE) is demonstrated for
some matrices of order 1000 in **Table 6.1**. Similar results were obtained
in many other tests. Some other results will be presented in the next chapter.

c	Matrices of class D(n,c)			Matrices of class E(n,c)		
	L kept	L removed	Percentage	L kept	L removed	Percentage
4	8719	5564	64%	8126	6128	75%
44	16131	9823	61%	27658	14289	52%
84	16263	9724	60%	21411	11123	52%
124	16734	9902	59%	17456	9934	57%
164	16277	9803	60%	14621	8602	59%
204	15319	9625	63%	12111	7575	63%

Table 6.1
**The storage needed (measured by parameter COUNT) in the cases where L is
stored and L is removed for some matrices of order 1000 (n=1000).**

6.4. Solving systems with triangular matrices

Assume that the factorization process is successfully completed and two
triangular matrices L and U such that

(6.4) $LU = PAQ + E$

are available. Then the vector

(6.5) $x_1 = QU^{-1}L^{-1}Pb$

has to be calculated. The use of the dynamic storage scheme in the calculation
of x_1 will be described in this section.

The first step in the calculation of x_1 is to permute b in order to
obtain Pb. Let vector b is stored in array B (of length at least equal
to n). Information about the row interchanges during the GE are stored in
the seventh column of array HA (see **Section 6.2**). More precisely, the row
number of the pivotal element selected at stage s of GE is stored in
HA(s,7). A piece of code that can be used in the calculation of Pb is given

in **Fig. 6.3**. It is assumed that the components of vector Pb overwrite the components of vector b (i.e. vector Pb is stored in array B).

```
DO 10 I=1,N-1
   T=B(HA(I,7))
   B(HA(I,7))=B(I)
   B(I)=T
10 CONTINUE
```

Figure 6.3
Calculation of vector Pb.

The second step in the calculation of x_1 is to compute vector $L^{-1}Pb$, where vector Pb is already calculated and stored in array B by the code given in **Fig. 6.3**. The non-zeros in row i of L (without its diagonal element) are stored, in array **ALU**, from position HA(i,1) to position HA(i,2)-1, i=1,2,...n. HA(i,1) > HA(i,2)-1 indicates that the diagonal element is the only non-zero element in row i of L. The column numbers of the non-zero elements are stored in array **CNLU** (at the same positions); see more details in **Section 4.2**. The code for calculation vector $L^{-1}Pb$ is given in **Fig. 6.4** (assuming that this vector is stored in array B).

```
DO 20 I=1,N
   DO 15 J=HA(I,1),HA(I,2)-1
      B(I)=B(I)-ALU(J)*B(CNLU(J))
15    CONTINUE
20 CONTINUE
```

Figure 6.4
Calculation of vector $L^{-1}Pb$.

The third step in the calculation of x_1 is to compute vector $U^{-1}L^{-1}Pb$, where $L^{-1}Pb$ is already calculated by the code in **Fig. 6.4** (and stored in array B). The non-zero elements of row i of U (without its diagonal element) are stored, in array **ALU**, from position HA(i,2) to position HA(i,3). Their column numbers are stored at the same positions in array CNLU. The relationship HA(i,2) > HA(i,3) indicates that the diagonal element is the only non-zero element in row i of U; the diagonal element is stored in PIVOT(i). More details are given in **Section 2.4**. The code that calculates $U^{-1}L^{-1}Pb$ is given in **Fig. 6.5**. Vector $U^{-1}L^{-1}Pb$ is stored in array B.

```
DO 40 I=N,1,-1
   DO 30 J=HA(I,2),HA(I,3)
      B(I)=B(I)-ALU(J)*B(CNLU(J))
30    CONTINUE
   B(I)=B(I)/PIVOT(I)
40 CONTINUE
```

Figure 6.5
Calculation of vector $U^{-1}L^{-1}Pb$.

The fourth, and last, step in the calculation of x_1 is to permute the vector obtained after the third step, the vector $U^{-1}L^{-1}Pb$ which is stored in array B. Information about the column interchanges (about matrix Q by which $U^{-1}L^{-1}Pb$ is to be permuted) is stored in the eighth column of array HA. More precisely, the column number of the pivotal element selected at stage s of GE is stored in HA(s,8). The code that can be used to calculate vector x_1 is given in **Fig. 6.6** (assuming that x_1 is stored in array B).

```
      DO 50 I=N-1,1,-1
         T=B(HA(I,8))
         B(HA(I,8))=B(I)
         B(I)=T
   50 CONTINUE
```

Figure 6.6
Calculation of $x_1 = U^{-1}L^{-1}Pb$.

6.5. Implementation of iterative improvement

If iterative refinement, **IR**, of the first solution x_1, calculated by (6.5), is required (and this is usually the case when the computer-oriented manner of exploiting the sparsity is used), then the following computations

$$(6.6) \quad r_i = b-Ax_i, \quad d_i = QU^{-1}L^{-1}Pr_i, \quad x_{i+1} = x_i + d_i,$$

are to be carried out for $i=1,2,\ldots,p-1$ until some stopping criterion is satisfied. The **IR** process is described in **Chapter 5**. In this section the implementation of the **IR** process in a sparse code will be discussed.

Consider (6.6). The performance of the last operation is trivial, while the implementation of the second operation has been described in the previous section. Therefore it is necessary only to explain how the first operation, the calculation of the residual vectors r_i, can be performed when some sparse matrix technique is used. A very similar operation, the operation $z=y+Ax$, has been described in **Section 2.2** (and the code by which this operation can be performed, under the assumption that the input storage scheme is used, is given in **Fig. 2.1**).

Some norms of vectors are to be used in the stopping criteria. Also some vectors are to be copied (see the code in **Fig. 2.1**). Some **BLAS** (Basic Linear Algebra Subprograms, [165]) may be applied in the code to perform these operations.

If any preconditioned conjugate gradient-type algorithm is used, then again, as in (6.6), two operations, matrix-vector multiplication and solving systems with triangular matrices, are the most time consuming parts of the computations during the iterative improvement of x_1. The latter operation was described in the previous section, while the code given in **Fig. 2.1** can easily be modified to perform the first operation.

6.6. Implementation of parallel algorithms

Some parallel codes for **general sparse matrices** will be described in **Chapter 10** and **Chapter 11**. These have been developed first and foremost for the **ALLIANT** and tested on the **ALLIANT**. However, this does not mean that the codes are only applicable on **ALLIANT** computers. Assume that a part of the code can be run concurrently, but that the parallelism has not been detected by the compiler. The parallelism of any such part of the code should be considered as non-trivial and written as a separate subroutine. This subroutine is then called concurrently in a loop. This means that the main task is **(1)** to identify the non-trivial parts of the code that could be run in parallel, **(2)** to prepare separate subroutines for these parts, and **(3)** to call these subroutines concurrently. This strategy is not the most efficient one, but it makes the implementation of the codes on other parallel machines easier. If concurrent calls of subroutines are allowed, then the implementation is nearly trivial. Several illustrations of this approach, which will be discussed in more detail in **Chapter 10** and **Chapter 11**, will be given in this section.

Consider first the pivotal search. Assume, for the sake of simplicity, that the row with minimal number of non-zeros (or one of these rows if there are several) is to be used in the pivotal search. The main part of this search, and the most expensive one, is to find at every stage s the "best" row (the row with minimal number of non-zeros in its active part). The routine by which the "best" row could be found concurrently is given in **Fig. 6.7.**

```
      SUBROUTINE SEARCH(N,ISTART,IEND,MINROW,RROW,HA2,HA3)
C
      INTEGER N, ISTART, IEND, LROW, MINROW, RROW, HA2, HA3, K
      DIMENSION HA2(*), HA3(*)
C
      MINROW=N+1
cvd$1 nosync
      DO 10 K=ISTRART,IEND
         LROW=HA3(K)-HA2(K)+1
         IF(LROW.LT.MINROW) THEN
            RROW=K
            MINROW=LROW
         END IF
   10 CONTINUE
C
      RETURN
      END
```

Figure 6.7
The subroutine used in the determination of the "best" row
at each stage of the GE.

The subroutine **SEARCH** should be compiled (on the **ALLIANT**) by the option **"recursive"**. ISTART, IEND, MINROW and RROW are integer arrays of length at least equal to NPROC (NPROC being the number of processors) in the program that calls **SEARCH**. ISTART and IEND contain the first and last rows from the part of rows that are to be searched by the processor under consideration. HA2(K) and HA3(K) contain the first and last non-zero element in the active part of row K. N is the order n of the matrix. On exit MINROW and RROW contain the number of non-zero elements in the active part of the locally "best" row and the row number of the locally "best" row found among the rows searched by the processor under consideration. The globally "best" row can easily be found after the loop where SEARCH is called by studying the values of arrays MINROW and RROW in the program that calls SEARCH. These two arrays are short dense arrays and apropriate BLAS's (see [165]) could be called to find the globally "best" row.

The loop in subroutine SEARCH is performed in a vector mode; a compiler instruction, "cvd$l nosync" has to be used in order to achieve vectorization. Note that all rows from row s to row n are examined to find the "best" row. On parallel computers this strategy, which is more expensive on sequential machines, often performs better than the strategy from Section 6.2, where a list of the "best" rows is updated at the end of every stage of GE. For more details see Chapter 10.

As a second example, one of the subroutines used to perform pivotal interchanges will be considered. It should be mentioned here that in some of the parallel codes the actual (physical) interchanges are not performed, and the subroutine given below is not used when this is so. Assume that stage I of the GE (I=1,2,...,n-1) is to be performed. Assume also that RCOLL is the column number of the pivotal element found at stage I. Finally, assume that ALU(1) and CNLU(1) contain the first non-zero element and its column number in a target row (a row that has a non-zero element in the pivotal column RCOLL). The following subroutine searches through the elements of a target row until it finds the element with column number RCOLL. Then it puts this element (which is needed in the following computations: it will be divided by the pivotal element and used to modify the other non-zeros in the active part of the target row) in the front of the active part of the target row. The subroutine that performs this work and that could be called concurrently over the target rows is given in Fig. 6.8.

```
          SUBROUTINE COLMN1(I,RCOLL,A,CNLU)
    C
          INTEGER I, RCOLL, CNLU, R
          DOUBLE PRECISION A, T
          DIMENSION A(*), CNLU(*)
    C
    cvd$r noconcur
          R=0
      10  R=R+1
          IF(CNLU(R).NE.RCOLL) GO TO 10
          T=A(1)
          A(1)=A(R)
          A(R)=T
          CNLU(R)=CNLU(1)
          CNLU(1)=RCOLL
    C
          RETURN
          END
```

Figure 6.8
The subroutine that finds the multipliers and moves them
in the front of the active parts of the target rows.

Since the subroutine has to be called concurrently (in a loop whose length is equal to the number of target rows), concurrency in the performance of its loops is not needed and it is suppressed by the compiler instruction "cvd$r noconcur". This is, strictly speaking, not needed and is done only in order to avoid any confusion. If this instruction is not inserted in the code, then the compiler gives a message that loop 10 is performed concurrently. This is certainly not true, because the whole subroutine **COLMN1** is called concurrently.

It should be emphasized that these two examples are the shortest subroutines that could be called concurrently. There are many other sub-routines. It is not necessary to give more examples. It is much more important to point out, once again, that the main principle used in the development of parallel code for general sparse matrices is: **all non-trivial parts of the code that can be performed in parallel should be written as separate subroutines that are called concurrently.**

6.7. Concluding remarks and references concerning the implementation of the algorithms

The **sequential** algorithms described in this chapter are implemented (with modifications in some cases) in the codes **ST** ([324]), **SIRSM** ([319], **SSLEST** ([311,312]) and **Y12M** ([329,331,341]) developed at the Technical University of Denmark and at the University of Copenhagen during the period 1975-1978. Some of the ideas used in these algorithms were described and/or applied in other works also (as, for example, [5,63,69,83,136,137,194,244]).

The **parallel** algorithms, to be studied in detail in **Chapter 10**, are implemented in three codes, **Y12M1**, **Y12M2** and **Y12M3** (see also [104,105]), developed at the Center for Supercomputing Research and Development at the University of Illinois at Urbana-Champaign in 1988-1990. The work on these codes and the underlying algorithms is still continuing.

The algorithm described in **Section 6.1** was proposed in a slightly modified form in [63,77].

The algorithm from **Section 6.2** was probably first used in this form in **SSLEST** ([312]). However, the idea of ordering the rows and/or the columns in an increasing number of non-zero elements in their active parts is not new. The same idea is applied in **MA28** ([63,77]), but linked lists are used there for the arrays corresponding to the seventh, eighth and eleventh columns of array HA.

The idea of removing the non-zero elements of the lower triangular matrix L, discussed in **Section 6.3**, when these are not needed in the further calculations, was first implemented in code **ST** ([324]) and then in the other codes developed at the Technical University of Denmark and at the University of Copenhagen. To the author's knowledge, **there is no other code for general sparse matrices that has such an option.**

The main difficulty in solving sparse systems of linear algebraic equations by the GE is the factorization of the coefficient matrix. When this is done, the remaining problems can easily be solved. This has been demonstrated in **Section 6.4** and in **Section 6.5** in connection with solving systems with triangular matrices and with the implementation of different iterative improvement processes (iterative refinement and conjugate gradient-type methods).

Two illustrations of the implementation of some parallel algorithms (to be discussed in the **Chapter 10** and **Chapter 11**) are given in **Section 6.6**. The main idea used in the development of parallel codes is to identify the parts of the code that could be done in parallel and then to write a separate subroutine for each of these parts. Two such subroutines, **SEARCH** and **COLMN1**, are given.

The algorithms described in this chapter, and especially the sequential algorithms, are not the only possible algorithms that could be applied during the solution of sparse systems of linear algebraic equations. Other algorithms (including here algorithms by which some special properties of the coefficient matrices, as, for example, symmetry and positive defineteness, are exploited) are discussed in   [33,48,53,62-86,126,155-156,166-168,170,175-178,185-189, 192,194,198-202,237-248,251-252,260].

CHAPTER 7

SOLVING LEAST SQUARES PROBLEMS BY AUGMENTATION

The solution of linear least squares problems by orthogonalization will be studied in **Section 12-Section 16**. However, one of the methods for solving such problems, which may be rather efficient (for some classes of problems, at least), is closely connected to the Gaussian elimination (GE) process. This is the method where the augmented matrix is used (see **Example 3.3**). It is quite natural to describe this method in the part of the book where GE is studied. It will be shown in this chapter that codes for solving systems of linear algebraic equations can very easily be used to solve linear least squares problems by augmenting the coefficient matrix. Y12M will be used, but this is done only to facilitate the exposition (the ideas could also be used in other codes).

The numerical experiments indicate that the augmented systems of linear algebraic equations, which have to be solved when the method of augmentation is used, are very often expensive, both in terms of computing time and storage. Therefore the use of the computer-oriented manner of exploiting sparsity is often more profitable than the classical manner. However, **care should be taken when the augmented matrix is formed** in order to avoid instability of the decomposition process.

7.1. Forming the augmented matrix

Consider the linear least squares problem:

(7.1) $x = A^{\dagger}b$, $A \in R^{mxn}$, $b \in R^{mx1}$, $x \in R^{nx1}$, $m \geq n$, $rank(A) = n$.

One of the particular methods within **the general k-stage scheme for solving linear algebraic problems** is based on an augmentation of matrix A and the reduction of the problem defined by (7.1) to a system of linear algebraic equations (see **Example 3.3**). The system of linear algebraic equations (henceforth called **the augmented system**) can be written as:

(7.2) $By = c$ ($B \in R^{pxp}$, $y \in R^{px1}$, $c \in R^{px1}$, $p = m+n$),

where

(7.3) $B = \begin{vmatrix} \alpha I & A \\ A^T & 0 \end{vmatrix}$, $c = \begin{vmatrix} b \\ 0 \end{vmatrix}$, $y = \begin{vmatrix} r/\alpha \\ x \end{vmatrix}$, $\alpha \neq 0$.

Assume that the input storage scheme from **Section 2.1** is used. This means that the non-zero elements of matrix A from (7.1) are stored, in an arbitrary order, in the first NZ locations of array AORIG and their row and column numbers are stored in the same positions in arrays RNORIG and CNORIG (NZ being the number of non-zero elements in matrix A). Assume that the length of these three arrays is at least 2NZ+m. Assume also that the components of vector b from (7.1) are stored in the first m locations of array B which contains at least p=m+n locations. Matrix B and vector c can be obtained by using the code given in **Fig. 7.1**.

In the first loop of the code in **Fig. 7.1** the transposed matrix A^T is prepared. At the same time the column numbers of the elements of A are increased by m (M being used in the code instead of m). The cost of this part of the work is 4NZ operations (but it must be noted that there are only assignment statements and additions of integers in this loop; i.e. no floating point operations).

```
        DO 10 I=1,NZ
           AORIG(NZ+I)=AORIG(I)
           CNORIG(NZ+I)=RNORIG(I)
           CNORIG(I)=CNORIG(I)+M
           RNORIG(NZ+I)=CNORIG(I)
     10 CONTINUE
C
        NZ=2*NZ
        DO 20 I=1,M
           AORIG(NZ+I)=ALPHA
           CNORIG(NZ+I)=I
           RNORIG(NZ+I)=I
     20 CONTINUE
C
        NZ=NZ+M
        NP=M+N
        NSTART=M+1
        DO 30 I=NSTART,NP
           B(I)=0.0
     30 CONTINUE
```

Figure 7.1
Preparation of matrix B and vector c from the augmented system (7.3).

In the second loop the block αI (α being replaced in the code by ALPHA) is set up. The cost of this part is 3M operations (only assignment statements are used).

The cost of the third loop (in which the last N locations of array B are set equal to zero) is N operations. NP is used instead of p. Vector c is stored in array B. NSTART is the starting point (in array B) of the zero part of vector c.

The total cost of the work needed to prepare the augmented system is 4NZ+3M+N operations (but no floating point operations). This cost is negligible in comparison with the work needed in the other parts of the solution process.

The cost is very small because the input storage scheme from **Section 2.1** is used. The great advantage of this scheme is that the elements can be stored in an arbitrary order. Thus, new elements can very simply be added to the end of the list, and this is exploited in the code given in **Fig. 7.1**. The use of any other storage scheme will cause some extra work. Assume, for example, that **(i)** the elements are ordered by rows in array AORIG and their column numbers are stored at the same positions in array CNORIG and **(ii)** pointers for row starts and ends are stored in the first and third columns of array HA. If such an economical storage scheme is used, then copies of the rows are needed to store the diagonal elements of matrix αI (because there are no free locations between the rows of matrix A). However, even more troublesome is the fact that the last n rows of matrix B from (7.2) are columns in matrix A. It is not very easy to prepare this part of matrix B because the economical storage scheme is ordered row-wise. This example illustrates some of the difficulties one may have with other input storage schemes and justifies (once again) the choice of the input scheme from **Section 2.1**.

7.2. Using package Y12M in the solution of linear least squares problems by augmentation

After the transformation of the linear least squares problem (7.1) to the system of linear algebraic equation (7.2) by the use of (7.3) any sparse solver can be applied to solve (7.2), and it is clear that the last n components in the solution y of (7.2) form the solution x of the original problem. Some numerical experiments obtained with **Y12M** will be discussed in this section.

Test matrices of class $F2(1500,n,125,6,100.0)$, $n=500(100)1400$, were used in an experiment. The matrices $F2(m,n,c,r,\alpha)$ were introduced in **Section 3.7**. By the choice made here the number of equations in system (7.2) is varied ($p=m+n=2000(100)2900$), but the number of non-zero elements is fixed ($NZ=2(rm+110)+m=19720$); note that this is the number of non-zero elements in matrix B.

The **DS** option of **Y12M** (see **Chapter 5**) was used with a drop-tolerance $T=10^{-12}$. Both results obtained by keeping matrix L and results obtained by removing matrix L are given in **Table 7.1 - Table 7.3**.

The **IR** option of **Y12M** was used with five different values of the drop-tolerance: $T=10^{-k}$, $k=0(1)4$. The results obtained by the **IR** option are also given in **Table 7.1 - Table 7.3**.

The right-hand side vectors were chosen so that the residual vectors are equal to zero for all ten problems and, moreover, so that all components of the solution vectors are equal to one. There are two reasons for such a choice: (i) it is very easy to calculate the right-hand side vectors and, what is even more important, (ii) it is very easy to check the accuracy of the solutions calculated by the code.

The parameter α from (7.2) (this parameter is not the same as the parameter α in the test matrices of class F2) was fixed: $\alpha=1.0$ in all examples in this section.

The results given in **Table 7.1** show clearly that storage can be reduced considerably when the non-zero elements of matrix L are removed. This is very important for some machines like **CDC Cyber**, where the storage requirements may cause difficulties. It is also seen from **Table 7.1** that the really important savings are achieved when the **IR** option with large drop-tolerances is used. For these matrices savings are achieved when the drop-tolerance is varied in a quite large interval.

The results in **Table 7.2** show that the computing time for the **DS** option is fairly insensitive to whether the elements of matrix L are kept or not, while the use of the **IR** option with large drop-tolerances leads to a significant improvement. Savings are achieved when the drop-tolerance is varied in a quite large interval.

The results in **Table 7.3** show that the accuracy is reasonably good when the **DS** option is used, but is normally much better with the **IR** option; often close to the machine precision in the latter case.

	DS option		IR option				
n	L kept	L removed	T=0.0001	T=0.001	T=0.01	T=0.1	T=1.0
500	255889	111192	101610	79581	64110	55412	32129
600	294191	136040	88077	68227	54746	47634	26915
700	300305	140801	69227	57919	47160	39758	23930
800	270602	136181	66526	51133	44452	37907	23432
900	260079	138383	63872	51146	44535	36541	19771
1000	239894	126535	62712	49882	42650	35036	19782
1100	204261	96729	63175	50419	42972	35963	19812
1200	217605	106094	64716	52070	44533	35888	19910
1300	181777	84174	65606	52073	44971	34946	20585
1400	175433	79005	64180	53263	46794	36833	21529

Table 7.1
Values of parameter COUNT obtained in the solution of linear least
squares problems with matrices A = F2(1500,n,125,6,100.0).

	DS option		IR option				
n	L kept	L removed	T=0.0001	T=0.001	T=0.01	T=0.1	T=1.0
500	130.15	151.71	18.74(4)	12.68(4)	9.49(5)	8.11(9)	3.3(3)
600	178.38	188.47	16.48(4)	11.12(4)	8.00(5)	6.87(9)	2.6(3)
700	182.78	165.66	12.10(3)	9.03(4)	6.78(5)	5.63(9)	2.7(6)
800	137.86	140.22	11.16(3)	7.67(4)	6.29(5)	5.27(8)	4.7(24)
900	127.86	133.30	10.41(3)	7.49(4)	6.24(5)	5.13(8)	5.2(37)
1000	108.53	113.93	9.55(3)	7.06(4)	6.06(5)	4.99(9)	2.9(14)
1100	75.88	72.44	9.48(3)	6.99(4)	5.93(5)	5.13(9)	3.1(15)
1200	82.68	80.25	9.58(3)	7.20(4)	6.06(5)	4.98(7)	7.3(50)
1300	57.55	56.45	9.91(4)	7.07(4)	6.14(5)	4.91(8)	5.5(33)
1400	51.39	48.62	9.25(3)	7.17(4)	6.46(5)	5.17(8)	4.2(21)

Table 7.2
Computing times obtained in the solution of linear least
squares problems with matrices A = F2(1500,n,125,6,100.0).
The times are given in seconds, the computer is an IBM 3033.
The number of iterations are given in bracckets.

An attempt to solve the same problems with the old version of **MA28** was carried out. The results obtained with **A=F2(1500,500,125,6,100.0)** are presented in **Table 7.4**. One hour CPU time on the **IBM 3033** computer was not sufficient to solve the second problems (with n=600), and no attempt to solve the other eight problems was made. The reasons for the very bad performance of **MA28** for this type of problems were studied in **Section 4.4** (see Example 4.1 and Example 4.2).

	DS option		IR option				
n	L kept	L removed	T=0.0001	T=0.001	T=0.01	T=0.1	T=1.0
500	3.5E-3	2.6E-3	1.0E-6	1.0E-6	1.0E-6	1.0E-6	2.3E-4
600	3.0E-3	2.6E-3	1.3E-6	1.3E-6	1.3E-6	1.3E-6	5.9E-5
700	2.5E-3	3.4E-3	4.3E-6	4.3E-6	4.3E-6	4.3E-6	3.1E-4
800	2.0E-3	3.1E-3	1.2E-7	1.2E-7	1.2E-7	1.2E-7	1.4E-5
900	2.6E-3	3.2E-3	6.0E-8	6.0E-8	6.0E-8	6.0E-8	6.0E-8
1000	2.1E-3	8.7E-4	6.0E-8	6.0E-8	6.0E-8	6.0E-8	3.5E-4
1100	4.6E-4	1.5E-3	6.0E-8	6.0E-8	6.0E-8	6.0E-8	7.1E-4
1200	3.5E-3	1.3E-3	6.0E-8	6.0E-8	6.0E-8	6.0E-8	6.0E-8
1300	1.3E-3	8.3E-4	6.0E-8	6.0E-8	6.0E-8	6.0E-8	6.0E-8
1400	1.3E-3	1.7E-3	6.0E-8	6.0E-8	6.0E-8	6.0E-8	6.0E-8

Table 7.3
The accuracy obtained in the solution of linear least squares problems with matrices A = F2(1500,n,125,6,100.0). The residual vectors are calculated in double precision and then rounded to single precision; therefore the accuracy achieved is comparable to the machine precision when IR converges.

Characteristics compared	MA28	DS - Y12M (L kept)	IR - Y12M (T=0.01)
Accuracy	1.7E-2	3.5E-3	1.0E-6
Computing time	1466.0	130.15	9.49
Storage	271356	255889	64110

Table 7.4
Results obtained in the solution of a linear least squares problem with A=F2(1500,500,125,6,100.0) by using the old version of MA28 and two options of Y12M (an IBM 3033 is used in these runs).

7.3. The choice of α in the solution of linear least squares problems by the method of augmentation

A parameter α is introduced in (7.3). This parameter can be used to reduce the condition number of matrix B in the augmented system of linear algebraic equations. In [21,22] it is shown that

(7.4) $\kappa(B) \approx \sqrt{2}\kappa(A)$ when $\alpha \approx \sigma_n/\sqrt{2}$,

where $\kappa(A)$ is the spectral condition number of A defined by

(7.5) $\kappa(A) = \|A\|_2\|A^\dagger\|_2 = \sigma_1/\sigma_n$,

and

(7.6) $\sigma_1 \geq \sigma_2 \geq \ldots \geq \sigma_n > 0$

are the singular values of matrix A   (the square roots of the eigenvalues of matrix $A^T A$).

Moreover, in [21,22] is also shown that

(7.7) $\kappa(B) \geq \kappa^2(A)$ for $\alpha \geq \sigma_1\sqrt{2}$

and that $\kappa(B)$ increases again for $\alpha < \sigma_n/\sqrt{2}$ (this effect being also discussed in [23]). If the well-known system of normal equations (introduced in **Example 3.2**),

(7.8) $A^T A x = A^T b$,

is used, then the spectral condition number of $A^T A$ is

(7.9) $\kappa(A^T A) = \kappa^2(A)$.

From (7.4) and (7.9) it follows that if A is ill-conditioned and if $\alpha \approx \sigma_n/\sqrt{2}$, then the method of augmentation will perform better than the method based on forming the system of normal equations. If σ_n^* is an estimate of σ_n, then $\alpha \approx \sigma_n^*/\sqrt{2}$ will normally be a good choice of α when the method of augmentation is used. Unfortunately, σ_n is in general not known and it may be very difficult to obtain a good estimate of this singular value. Therefore the method of augmentation is usually applied with $\alpha = 1$ (see, for example, [75]. However, sometimes this choice causes difficulties. In this section it will be explained why the difficulties arise and what can be done to improve the results.

Assume temporarily that $\alpha = 1$. Assume also that the diagonal elements of matrix I are chosen as pivots during the first m stages of the GE. Then the active part of matrix B from (7.3) at stage m+1 is $A^T A$ and it is apparent that the method of augmentation has no advantage, compared to the method based on (7.8), when this happens. Therefore one should attempt to avoid the use of pivots among the diagonal elements of I during the first m stages of the GE. When matrix A is dense and when partial pivoting is used, the stability requirements often prevent the selection of pivots among the diagonal elements of I during the first m stages. However, the situation may change when matrix A is sparse and sparse technique is used. In the latter case **the stability requirements are normally relaxed** (in an attempt to preserve sparsity). Moreover, the diagonal elements of I have very often best Markowitz cost and are candidates for pivots (because of the symmetry of B; see **Example 4.1** and **Example 4.2**). Therefore there is a real danger that many pivots during the first m stages of the GE will be diagonal elements of I. If this happens, then the active part of B at stage m+1 will be close to $A^T A$.

It is clear from the above discussion that one should avoid the choice of diagonal elements of I as pivots. If the **DS** option with T=0.0 (or a very small T) is used, this could easily be done by **(i)** taking a sufficiently small value of α, and **(ii)** using sufficiently large values of the stability factor u (see **Section 4.4**). Then the diagonal elements of αI will not belong to the stability sets B_s (s=1,2,...,m), see (4.11) and (4.24), and, therefore, will not be chosen as pivots.

The situation becomes more complicated when an IR option with large drop-tolerance T is used. A very small α will lead now to removing diagonal elements of αI (which is equivalent to the choice $\alpha=0$). Therefore the choice

(7.10) $\alpha \geq max(5.0*T, 10^{-5})$

seems to be a good one. In many case it prevents removing diagonal elements of αI, while the diagonal elements are still not belonging to the stability sets B_s.

One may try to improve (7.10) by replacing the constants 5.0 and 10^{-5} by some quantities that are problem dependent. Let

$$(7.11) \quad c = \min_{\substack{1 \leq i \leq p \\ 1 \leq j \leq p}} [\max (|b_{ij}|)], \qquad b_{ij} \in B .$$

Then α can be chosen by using the following criterion:

(7.12) $\alpha \geq max(\gamma_1 T, \gamma_2 c),$

where $\gamma_1 > 1$ and $0 < \gamma_2 < 1$ are suitably chosen constants.

Remark 7.1. The choice of α is made not in order to reduce the condition number of matrix B, but to prevent the selection of pivots among the diagonal elements of αI and to prevent the removal of these elements when a large drop-tolerance is selected. This choice will not guarantee that the above two purposes are achieved in all cases (but note that it is rather difficult to guarantee any statement in which rounding errors are involved: even partial pivoting gives no guarantee for a stable GE process). Nevertheless, many experiments indicate that a choice based on these ideas can efficiently be used with the method of augmentation. Some numerical results will be given in the following section. ∎

Remark 7.2. Very often $\alpha = 1$ is used. Moreover, very often this choice gives good results. However, when the computer-oriented manner of exploiting sparsity is used, it may be worthwhile to choose some small, but not too small, value of α. Some examples that illustrate this will be given in the next section. ∎

7.4. Using small values of α in the solution of linear least squares problems by augmentation.

The value of α used in **Section 7.2** is one. The results obtained with this value of α are good for these matrices because:

(i) the diagonal elements of matrix I are not the largest (in absolute value) elements of matrix B from (7.3),
(ii) the matrices that used in Section 7.2 are not very ill-conditioned.

However, in general these two conditions will not be satisfied. To illustrate that **sometimes** one should be careful with the choice of α, an experiment with ten matrices that have been used in many other studies (see, for example, [112-113,166-167,322]) will be described in this section. This set of ten matrices will be called here the **Waterloo set of sparse test matrices** (some of the matrices from the Waterloo set are included in the Harwell-Boeing set, [71,72]).

Some characteristics concerning the matrices from the Waterloo set are given in **Table 7.5**. This set is rather representative. The numbers of rows vary from 313 to 1850, the numbers of columns vary from 176 to 712, and the numbers of non-zeros in matrix A vary from 1557 to 8755. The numbers of non-zeros in matrix B are much more important when augmented systems are solved: if the number of non-zeros in A is NZ, then the number of non-zeros in B is 2NZ+m.

Problem	m	n	NZ	Short characteristics of the problem
1	313	176	1557	Sudan survey data
2	1033	320	4732	Analysis of gravity-meter observations
3	1033	320	4719	Another analysis of gravity-meter observations
4	1850	712	8755	Similar to Problem 2 but larger
5	1850	712	8638	Similar to Problem 3 but larger
6	784	225	3136	15x15 grid-problem
7	1444	400	5776	20x20 grid-problem
8	1512	402	7152	3x3 geodetic network with 2 observations per node
9	1488	784	7040	4x4 geodetic network with 1 observation per node
10	900	269	4208	Geodetic problem provided by the US National Geodetic Survey

Table 7.5
Some characteristics of the test-matrices used in this section.

The use of $\alpha=1$ causes difficulties in the solution of some of these systems (at least, when single precision on IBM-like machines is selected). The solution of the third system is very inaccurate when the **DS** option with T=0.0 (the classical approach of exploiting sparsity) is applied. The attempt to improve it by using the **IR** option is not successful even with T=0.0. However, the use of the simple criterion (7.10) to determine α is very useful in this situation; see the results in **Table 7.6**.

Scaling (column equilibration) was used to obtain these results. The magnitude of the elements of the matrices from **Table 7.5** is not known to the author (a huge file must be examined to get some information about the non-zeros). Therefore it is difficult to determine a good value of the drop-tolerance T when (1.9) is used. Scaling facilitates the choice of the drop-tolerance. The largest (in absolute value) elements in the columns become equal to one, and T<1.0 has to be chosen.

The first trial is always with T=0.1. If the run is not successful because the **IR** does not converge, then the drop-tolerance is reduced by a factor of 10 and another run is performed. This can happen several times. If the run is not successful because the factorization is not completed (the drop-tolerance is too large and all elements in a row or column are removed), then the drop-tolerance is reduced by a factor of 100 and another run is performed. This can also happen several times. If several trials have been carried out until the solution is computed, then the computing time given is the total time needed to obtain the solution (the sum of the computing times for all trials). The final tolerance used is given in **Table 7.6**. It should be noted that this strategy of varying the drop-tolerance often gives quite good results even if one system only is to be solved, but it is especially efficient when a sequence of many systems has to be treated; this problem will be discussed in the next chapter.

Problem	p=m+n	2NZ+m	DS option with T=0			IR option with starting T=0.1			
			COUNT	Time	Accuracy	T	COUNT	Time	Accuracy
1	487	3427	13585	2.03	9.9E-6	1.0E-1	5785	0.98(20)	8.9E-12
2	1353	10497	23957	4.11	7.5E-6	1.0E-2	14600	5.44(17)	3.4E-12
3	1353	10471	22766	3.85	2.0E-4	1.0E-3	15027	3.49(16)	4.8E-12
4	2562	19360	79930	20.35	4.2E-6	1.0E-2	35806	13.25(14)	1.2E-12
5	2562	19122	69156	17.25	1.9E-5	1.0E-3	38716	11.17(23)	1.7E-12
6	1009	7056	44207	9.41	4.9E-6	1.0E-1	8564	1.25(14)	8.9E-12
7	1844	12996	135183	57.94	4.1E-6	1.0E-1	16075	2.57(17)	9.5E-12
8	1914	15816	80158	15.51	6.1E-6	1.0E-1	19252	3.44(20)	3.4E-12
9	2272	15568	94311	22.01	1.0E-5	1.0E-2	53223	12.47(20)	2.6E-12
10	1169	9316	46713	10.72	1.8E-2	1.0E-5	34519	11.99(48)	1.3E-11

Table 7.6
Numerical results obtained in the solution of problems whose matrices are described in Table 7.5 by using augmentation with a small α.
The number of iterations are given in brackets.
The computing times are in seconds on IBM 3081 (single precision is used).

The version of **Y12M** in which the Scandinavian way of using the **IR** is implemented (see Section 5.5) is used in these runs. Therefore, if the **IR** converges, then the accuracy achieved is about twice greater than the machine accuracy in single precision.

Remark 7.3. The use of criterion (7.10) to select a value of α leads to choosing off-diagonal elements of matrix B as pivots. This often causes a poor preservation of sparsity. This phenomenon has already been observed for matrices of class $E(n,c)$; see the comments after **Table 4.4**. The use of $\alpha=1$, which allows diagonal elements as pivots, sometimes gives much better results (with regard to storage and computing time). However, the solution of some problems can be very inaccurate and, as mentioned above, precisely this is the case for the third problem in **Table 7.5**. The conclusion is that, while it may be better to use $\alpha=1$ in connection with the preservation of sparsity, in general it is safer to select a value of α by the criteria given in Section 7.3.

7.5. Concluding remarks and references concerning the solution of linear least squares problems by augmentation

The use of the augmented system (7.3) in the solution of linear least squares problems has been discussed in [21-22,25]. This system is often solved by orthogonalization methods [21-22,30]. The orthogonalization methods work well when linear least squares are treated directly (see Chapter 12 - Chapter 16), but these methods are not very suitable for **sparse** systems of linear algebraic equations. **Hachtel** [138] was probably the first to suggest the use of GE in the solution of the augmented system of linear algebraic equations (7.3) in the case where B is sparse (see also [23,75]). The method of augmentation (with the use of the GE in the solution of the augmented system) is one of the methods recommended in [75], where several methods for solving linear least squares problems are compared. In recent surveys of methods for solving linear least squares problems, methods based on the Givens orthogonalization are normally recommended (as, for example, in [143,153,306]). Some comparisons of results obtained by the method of augmentation with results obtained by a method based on the Givens orthogonalization could be made by using the results given in **Chapter 16**.

The simplicity of the code for creating the structure of the augmented system (7.2) in the case where the input storage scheme is used is remarkable. To the author's knowledge this fact was first emphasized in [306].

The results given in **Table 7.4** show once again that the selection of a good pivotal strategy is very important (for some classes of problems at least) and that the results can be improved very significantly when this selection is carefully made.

Until now the parameter α has been used mainly in an attempt to decrease the condition number of matrix B. For sparse matrix techniques it is even more important to select this parameter so that off-diagonal elements are chosen as pivots during the first m stages. Some criteria, by which this can be achieved when a general sparse matrix package for solving systems of linear algebraic equations is used, are given in **Section 7.3** and tested in **Section 7.4**. It might be even better if a special code for solving augmented systems, in which the use of diagonal elements of the augmented matrix as pivots is avoided as long as possible, were developed. Some ideas proposed in **Arioli et al.** [6,7,9] could be applied in such a code.

Matrix B in (7.2) is symmetric but indefinite. The symmetry can be exploited by using the so-called (2x2)-pivots. A code, in which this has been done, is described in [175]. This code has been compared with **Y12M** and the latter code, in which symmetry is not exploited, performed better (with regard to computing time). The reason for this is probably the fact that **the use of (2x2)-pivots in a sparse code is rather expensive**. However, more experiments are needed in order to draw a definite conclusion about whether it is profitable to exploit the symmetry of B.

SPARSE MATRIX TECHNIQUE

FOR ORDINARY DIFFERENTIAL EQUATIONS

Large systems of ordinary differential equations (ODE's) appear in a natural way in many fields of science and engineering. The application of sparse matrix technique is a very useful option in a package for solving such systems numerically. Such an option, the code SPAR1, is described in this chapter. SPAR1 is written for systems of linear ODE's, but the same ideas can be applied to systems of non-linear ODE's.

Several criteria for comparing SPAR1 with a corresponding code, DENS1, in which the same integration method is used but sparsity is not exploited, are formulated.

Code SPAR1 can be used in several modes, but experiments indicate that it is efficient to apply the computer-oriented manner of exploiting sparsity (see Chapter 1); i.e. to introduce a positive drop-tolerance and to remove all elements that become "small" in some sense during the computations.

The discretization of a system of linear ODE's leads to a long sequence of systems of linear algebraic equations. This fact can be exploited to adjust the drop-tolerance at the beginning of the computation and then to carry out the remaining computation with a nearly optimal drop-tolerance. The automatic variation of the drop-tolerance is an important tool for increasing the efficiency of the computational process when long sequences of systems of algebraic equations are to be solved. It should be emphasized here that solving systems of ODE's is used in this chapter only in order to facilitate the exposition of the results. The main idea, an automatic adjustment of the drop-tolerance during the computation (by the code itself), is fairly general and can be used (perhaps with slight modifications) any time a long sequence of systems of linear algebraic equations is to be solved.

8.1. Statement of the problem

Consider a system of s linear ordinary differential equations (ODE's):

$$(8.1) \quad y' = A(t)y + b(t), \qquad t \in [0,\tau] \subset \mathbf{R}, \qquad y \in \Omega \subset \mathbf{R}^s,$$

where

(i) for any $t \in [0,\tau]$ the values of $A(t) \in \mathbf{R}^{s \times s}$ and $b(t) \in \mathbf{R}^{s \times 1}$ are either given or some sufficiently accurate approximations to these values can be calculated,

(ii) an initial value $y(0) = \eta$ is either given or a sufficiently accurate approximation y_0 to η can be obtained in some way.

If system (8.1) is stiff (roughly speaking, the Jacobian matrix $A(t)$ has some very large eigenvalues; in absolute value), then approximations to the values of its exact solution at the points of the grid:

(8.2) $G_N = \{ t_n \ / \ t_0=0, \ t_{n+1}=t_n+h_{n+1}, \ h_{n+1}>0, \ n=0(1)N-1, \ t_N=\tau, \ N\in\mathbb{N} \}$

are usually calculated using **implicit** numerical methods with infinite regions of absolute stability (regions in which the roots of certain polynomials induced when the numerical method chosen is applied to the special problem $y'=\lambda y$, $\lambda \in \mathbb{C}$, are inside the unit circle $|z| \le 1$ for any $h\lambda \in \mathbb{C}$, $h > 0$). The application of any implicit method in the calculation of an approximation y_n to the exact solution $y(t_n)$ for $t_n \in G_N$ leads to the solution of one or several systems of linear algebraic equations. Two examples are given below in order to illustrate this statement.

Example 8.1. Assume that a **BDF** (a backward differentiation formula):

(8.3) $y_n = \sum\limits_{i=1}^{k} \alpha_i(\overline{h}_n)y_{n-i} + h_n\beta[A(t_n)y_n+b(t_n)]$,

where the coefficients $\alpha_i(\overline{h}_n)$ depend on

(8.4) $\overline{h}_n = (h_{n-1}/h_n, h_{n-2}/h_n, \ldots, h_{n-k+1}/h_n)$,

is used in the solution of (8.1) at step n. Then **one** system of linear algebraic equations

(8.5) $[h_n^*I+A(t_n)]y_n = h_n^* \sum\limits_{i=1}^{k} \alpha_i(\overline{h}_n)y_n - b(t_n)$, $h_n^* \overset{def}{=} -(h_n\beta)^{-1}$,

must be solved (at each time step n) in order to calculate y_n .   ∎

Example 8.2. Assume that a k-stage diagonally implicit Runge-Kutta method (a **DIRK** method; see [2]),

(8.6) $k_i(h_n) = A(t_n+\alpha_ih_n)\{y_{n-1}+h_n[\sum\limits_{j=1}^{i-1} \beta_{ij}k_j(h_n)+\gamma k_i(h_n)]\} + b(t_n+\alpha_ih_n)$,

$i = 1,2,\ldots,k$,

(8.7) $y_n = y_{n-1} + h_n \sum\limits_{i=1}^{k} p_ik_i(h_n)$,

is used at step n in the numerical treatment of (8.1). Then **k** systems of linear algebraic equations

(8.8) $[h_n^*I+A(t_n+\alpha_ih_n)]k_i(h_n) = h_n^* \{A(t_n+\alpha_ih_n)[y_{n-1}+h_n \sum\limits_{j=1}^{i-1} \beta_{ij}k_jh_n)]$

$+ b(t_n+\alpha_ih_n))$, $i=1,2,\ldots,k$, $h_n^* \overset{def}{=} -(h_n\gamma)^{-1}$,

must be solved in order to determine y_n from (8.7).   ∎

Consider (8.5) and (8.8). Denote the unknown vectors by z, the right-hand side vectors by c, and the coefficient matrices by B. It is apparent that the discretization process for systems of **linear ODE's** leads to the solution of long sequences of systems of linear algebraic equations

(8.9) $Bz = c$

both in the case where **BDF's** are used and in the case where methods of Runge-Kutta type are applied. The solution of systems (8.9) is a very considerable part of the computational work in the numerical treatment of (8.1) when s is large. If matrix $A(t)$ from (8.1) is sparse, then matrix B is also sparse, and the sparsity should be exploited to reduce both the storage and the computing time. The implementation of sparse matrix technique in the solution of stiff systems of **ODE's** will be discussed in this chapter.

The main ideas used in this chapter are fairly general and can be applied to many numerical methods for solving **ODE's**. However, in order to facilitate the description and in order to define some criteria for comparison with the case where matrix B is **dense** (or sparse, but the sparsity is not exploited), it is more convenient to consider the particular integration method implemented in a dense code **DENS1** and in a sparse code **SPAR1**; [333-337]. The integration method will be sketched in the following section and after that different options of package **Y12M**, which is a part of the sparse code **SPAR1**, will be discussed.

Remark 8.1. It is assumed that the stepsize $h_n = t_n - t_{n-1}$ can be varied during the numerical integration of (8.1). It is also assumed that the variation of the stepsize does not affect the fundamental properties (consistency, zero-stability and convergence) of the integration method used. No attempt to justify this assumption will be made here. However, some ideas from [290,293,297,298,300,302,304,305,307,308,313,314] can be applied to show that the assumption made is reasonable (in fact such an assumption is made in any code for solving **ODE's**, but not always is this justified for the particular integration methods used). ∎

Remark 8.2. It is assumed that matrix $A(t)$ is **a general sparse matrix**. If this is not so, then matrix $A(t)$ has some special property and/or some special pattern. It may be more efficient to exploit the special property and/or the special pattern of $A(t)$. For example, if the matrix is such that purely iterative methods are convergent for the systems of type (8.9), then it may be more profitable to apply some of the subroutines in package **ITPACK** or in package **NSPCG** [183,286,287]. Another example is the case of band matrices with narrow bandwidth. It should be mentioned that a code **BAND1** [339] has been developed for systems of **linear ODE's** whose Jacobian is a band matrix with a narrow bandwidth. ∎

8.2. The integration algorithm used in the software

Assume that y_0 is a sufficiently accurate approximation to η and let $t_n \in G_N$ for n=0,1,...,N. If all approximations y_j to $y(t_j)$ for j=1,2,...,n are already computed, then the next approximation y_{n+1} is obtained by

(8.10) $y_{n+1} = y_n + 0.5h_n[k_1(h_n) + k_2(h_n)]$,

where

(8.11) $\bar{A}_n k_1(h_n) = \gamma_n[A(t_n+0.5h_n)y_n + b(t_n+0.5h_n)]$,

(8.12) $\bar{A}_n k_2(h_n) = \gamma_n\{A(t_n+0.5h_n)[y_n+h_n\beta k_1(h_n)] + b(t_n+0.5h_n)\}$,

(8.13) $\bar{A}_n \overset{def}{=} \gamma_n I + A(t_n+0.5h_n)$, $I \in R^{sxs}$ *being the identity matrix*,

(8.14) $\beta = \sqrt{2} - 1$, $\gamma = 1 - 0.5\sqrt{2}$, $\gamma_n \overset{def}{=} -(h_n\gamma)^{-1}$.

The approximation y_{n+1} found by (8.10)-(8.14) is of order **two**. This means that $y_{n+1} - y(t_{n+1}) = O(h_n^3)$. By applying **an embedding technique** ([294,333]) a third-order approximation is calculated:

(8.15) $\bar{y}_{n+1} = y_n + (h_n/6)\{2[k_1(h_n)+k_2(h_n)] + k_3(h_n) + k_4(h_n)\}$,

where $k_1(h_n)$ and $k_2(h_n)$ are the same as in (8.10) (because of the embedding technique used), while $k_3(h_n)$ and $k_4(h_n)$ are computed by

(8.16) $k_3(h_n) = A(t_n)y_n + b(t_n)$,

(8.17) $k_4(h_n) = A(t_n+h_n)\{y_n+h_n[\beta k_1(h_n)-\beta k_2(h_n)+k_3(h_n)]\} + b(t_n+h_n)$.

A non-negative **acceptability** function

(8.18) $T(t_n) = c(\|y_n-\bar{y}_n\|) = (ch_{n-1}/6)(\|k_1(h_n)+k_2(h_n)+k_3(h_n)+k_4(h_n)\|)$

is defined for $\forall t_n \in G_N$ ($c \geq 1$ being a constant) and the approximation y_{n+1} is **declared** as acceptable when

(8.19) $T(t_{n+1}) \leq \epsilon(t_{n+1})$,

where $\epsilon(t_n)$ is some **error-tolerance** function defined for $\forall t_n \in G_N$.

If the acceptability check (8.19) fails, then the step is not always rejected immediately. There are options where extrapolation techniques are used in a **second acceptability check**. The main principles used in these options can be described as follows. Calculate, by using (8.10)-8.14), two approximations y_{n+2} and y_{n+2}^* starting with y_{n+1} (and applying $h_{n+1} = h_{n+1}$) and with y_n (and applying a stepsize $2h_n$) respectively. Define a **second acceptability function**

(8.20) $T^*(t_n) = (c^*/3)(\|y_n-y_n^*\|)$ ($c^* \geq 1$ being a constant)

on the set of $\forall t_n \in G_N$ for which (8.19) is **not** satisfied. If

(8.21) $T^*(t_{n+2}) \leq \epsilon(t_{n+2})$,

then both y_{n+1} and y_{n+2} are declared acceptable. If (8.21) is not satisfied, then both y_{n+1} and y_{n+2} are rejected and the calculations are restarted by computing a new approximation y_{n+1} with a reduced stepsize.

The algorithm defined by (8.10)-(8.21) requires the solution of two systems of linear algebraic equations per successful step which is accepted by the first acceptability check (8.19). If the second acceptability check is to be applied (let us repeat that this happens only when (8.19) indicates failure and when the user has specified an option in which the second acceptability check (8.21) is used), then the code carries out two small steps (with stepsize h_n and h_{n+1} with $h_n=h_{n+1}$) and one large step (with stepsize $2h_n$). If only the small stepsizes are counted, then the code solves three systems of linear algebraic equations per successful step when (8.21) is used.

The above discussion shows that the solution of linear algebraic equations is a substantial part of the integration process. This part of the computations is discussed below. Let

$$(8.22) \quad b_1^*(t_n+0.5h_n) = \gamma_n[A(t_n+0.5h_n)y_n + b(t_n+0.5h_n)],$$

$$(8.23) \quad b_2^*(t_n+0.5h_n) = \gamma_n\{A(t_n+0.5h_n)[y_n+h_n\beta k_1(h_n)] + b(t_n+0.5h_n)\},$$

and assume that the decomposition

$$(8.24) \quad L_nU_n = P_n\overline{A}_nQ_n$$

is calculated at step n by the use of GE (but not necessarily by the use of package Y12M). The systems (8.11) and (8.12) can then be rewritten as

$$(8.25) \quad P_n^TL_nU_nQ_n^Tk_m(h_n) = b_m^*(t_n+0.5h_n) \qquad (\ m=1,2\ )\ .$$

If systems (8.25) are large (and this is the case when the Jacobian matrix $A(t)$ is sparse), then it is not efficient to calculate the factorization (8.24) at each integration step. It is normally better to keep an old factorization L_jU_j with $j<n$ and to apply the iterative refinement processes

$$(8.26) \quad r_m^{[i]}(h_n) = b_m^*(t_n+0.5h_n) - \overline{A}_nk_m^{[i]}(h_n), \qquad i=0,1,\ldots,p_m-1, \quad m=1,2,$$

$$(8.27) \quad d_m^{[i]}(h_n) = Q_j(L_jU_j)^{-1}P_jr_m^{[i]}(h_n), \qquad i=0,1,\ldots,p_m-1, \quad m=1,2,$$

$$(8.28) \quad k_m^{[i+1]}(h_n) = k_m^{[i]}(h_n) + d_m^{[i]}(h_n), \qquad i=0,1,\ldots,p_m-1, \quad m=1,2,$$

in the calculation first of $k_1(h_n)$ and then of $k_2(h_n)$. In principle, this is the same iterative process as that defined by (5.4)-(5.6). The notation is more complicated because the IR process is now a part of a larger problem. Therefore it is necessary to use a special notation for the step numbers relevant for the matrices and vectors involved in the computations (the indices n and j being used for this purpose). It is also necessary to distinguish between the first and the second system (index m is used for this purpose). Finally, it is necessary to show the number of the current iteration (index i in square brackets being used for this purpose). The IR processes (8.26)-(8.28) may fail to converge not only because the GE is performed (on the computer used) with rounding errors, but also because the decomposition L_jU_j applied at step n is too different from the decomposition L_nU_n when $n>j$.

Starting approximations for the two iterative processes can be obtained either by using $k_m^{[0]}(h_n) = k_m(h_{n-1})$ or by solving

$$(8.29) \quad k_m^{[0]}(h_n) = Q_j(L_jU_j)^{-1}P_jb_m^*(t_n+0.5h_n) \qquad (m=1,2) .$$

The **IR** processes (8.26)-(8.28) are terminated if for some i one of the following stopping criteria is satisfied:

$$(8.30) \quad \text{RATE(i)}\geq 1 \quad \wedge \quad i>1 \quad (\text{RATE(i)} \overset{def}{=} \|d_m^{[i]}(h_n)\|/\|d_m^{[i-1]}(h_n)\|),$$

$$(8.31) \quad i=\text{MAXIT} \qquad (\text{ the default value being } \text{MAXIT}=3),$$

$$(8.32) \quad \|d_m^{[i]}(h_n)\|\leq\{\delta[1-\text{RATE(i)}]\epsilon(t_n)/(0.5h_n) \quad \wedge \quad \text{RATE(i)}<1 \quad (\delta=0.1).$$

The codes **DENS1** and **SPAR1** that will be compared in the following sections allow also termination of the iteration when the norm of the residual vector becomes small. This stopping criterion is the same as (8.32), but $\delta=0.1$ is replaced by $\delta=0.01$.

If the **IR** is stopped either by (8.30) or by (8.31), then the **acceptability criterion with regard to the system of linear algebraic equations (8.32)** is normally not satisfied and the result should not be used in the further calculations. One of the following three actions has to be performed when this happens. A new factorization L_nU_n has to be computed if $n>j$. If $n=j$ and if dropping is used (T>0.0), then the drop-tolerance should be reduced and L_nU_n should be recalculated. If $n=j$ and if no dropping is used (T=0.0), then the stepsize h_n should be reduced and L_nU_n should be recalculated again.

If (8.32) is satisfied, then the result is acceptable. The code sets $k_m(h_n) = k_m^{[i+1]}(h_n)$ and continues the calculations. The division with $0.5h_n$ in the acceptability criterion (8.32) can be explained as follows. Look at (8.10) and assume that $\epsilon(t_{n-1})$ does not differ too much from $\epsilon(t_n)$. It should be expected that y_{n-1} has been calculated so that

$$(8.33) \quad \|y(t_{n-1})-y_{n-1}\| \approx \epsilon(t_{n-1})$$

Therefore it is natural to require that

$$(8.34) \quad 0.5h_n\|d_m^{[i]}(h_n)\| \approx \epsilon(t_{n-1})$$

when **IR** is terminated. It is expected that for the value of i for which **IR** is stopped by the third stopping criterion the following relation holds:

$$(8.35) \quad \|d_m^{[i]}(h_n)\| \approx \|k_m(h_n) - k_m^{[i]}(h_n)\| .$$

This means that the code attempts to stop the **IR** process when the accuracy obtained in the calculation of an approximation to $k_m(h_n)$ is judged to be sufficient to achieve the accuracy required at the integration step n. The factor $\delta=0.1$ is used in an attempt to avoid the influence of errors due to the solution of the system of linear algebraic equations on the accuracy of y_n. The factor $1-\text{RATE(i)}$ is useful when the **IR** process is slowly convergent. If this is so, then $\|d_m^{[i]}(h_n)\|$ could be small even when $k_m^{[i]}(h_n)$ is not sufficiently close to $k_m(h_n)$.

The use of (8.32) with the correction vector $d_m^{[i]}(h_n)$ replaced by the residual vector $r_m^{[i]}(h_n)$ (and with $\delta=0.01$) can be justified in a similar way.

Assume that (8.30)-(8.32) are applied in the solution of the first system of linear algebraic equations (i.e. $m=1$). Assume also that the acceptability criterion (8.32) is not satisfied and a new decomposition L_nU_n is calculated. If no dropping is used, then the code sets $k_1(h_n)$ and $k_2(h_n)$ equal to $Q_n(L_nU_n)^{-1}P_nb_1^*(t_n+0.5h_n)$ and $Q_n(L_nU_n)^{-1}P_nb_2^*(t_n+0.5h_n)$ respectively. This means that if a new decomposition has been calculated at integration step n, then the codes solve the systems of linear algebraic equations directly (no IR process is carried out) when no "small" elements are removed. For the dense code this is always true; "small" elements are never dropped. For the sparse code this means that the above simplification is only possible when the classical way of exploiting sparsity is used.

Assume now that the acceptability criterion (8.32) is not satisfied in the solution of the second system of linear algebraic equations ($m=2$). Then again a new decomposition L_nU_n is calculated. The codes set $k_2(h_n)$ equal to $Q_n(L_nU_n)^{-1}P_nb_2^*(t_n+0.5h_n)$ when no dropping is used; i.e. only the second system is solved directly in this situation.

If sparse matrix technique is used and if "small" elements are removed, then IR is to be used even at steps where a new decomposition is calculated.

The practical implementation of the algorithms sketched above in two codes, DENS1 and SPAR1, will be discussed in the following sections. However, several remarks are needed before the discussion.

Remark 8.3. The integration algorithm (8.10)-(8.14) is a modified diagonally implicit Runge-Kutta method (an MDIRKM, [294]). Integration methods of this type have good stability properties (see more details about stability of numerical methods for systems of ODE's and some other related concepts in [36-38,106,108,140,162,163,217,230]). The algorithm that is defined by (8.10)-(8.14) is AN-stable, [294]. However, the auxiliary integration method, defined by (8.15)-(10-17), does not have good stability properties; it is not even A-stable. This means that if the system of ODE's is very stiff, then the first acceptability check may indicate a failure also when the approximation y_{n+1} obtained by (8.10) is acceptable. Such cases have been observed in some experiments, [333]. The introduction of the second acceptability check is very efficient in this situation; by using this check the false failure is normally detected. It should be emphasized that the second check is only activated when the first one fails. If the system of ODE's is not extremely stiff, then as a rule the number of steps where only the first check is used is large. This increases the efficiency, because the first check is much cheaper; no systems of linear algebraic equations are solved in (8.16) and (8.17). In the discussion in the following sections it is assumed that the second acceptability check is used only occasionally. This assumption is based on many experiments. However, as mentioned above, examples of very stiff ODE's, where the second check has to be activated many times, can be constructed. ∎

Remark 8.4. The fact that the system (8.1) is **linear** is exploited not only to represent the integration algorithms discussed in this chapter in a more transparent form, but also in the derivation of some of the integration

algorithms. The approximation calculated by (8.15) is of order three only
for linear systems of ODE's (and, thus, the first acceptability check works
only for linear systems). A very efficient integration algorithm has been
designed by exploiting the linearity; see [333-337]. To illustrate this we
mention here that: (i) only two function evaluations are needed per successful
step when the first acceptability check is used, and (ii) if the step is
rejected then there is no need to recalculate $k_3(h_n)$, assuming that $k_3(h_n)$
is calculated before the calculation of $k_1(h_n)$ and $k_2(h_n)$. ∎

   Remark 8.5. A careful study of **Example 8.1** and **Example 8.2** shows
that the use of BDF's is in general more profitable than the use of formulae
of Runge-Kutta type when the computing time and the storage are counted.
However, the latter formulae may become less time consuming for special
problems with Jacobian matrices $A(t)$ that have eigenvalues close to the
imaginary axis. This is the case for problems arising in nuclear magnetic
resonance spectroscopy (illustrated in **Fig.1** in [211]). Some experiments
carried out with different codes for solving ODE's [36,37,149], including
a code based on BDF's, show that the codes DENS1 and SPAR1, which are
based on the algorithm sketched in this section, perform best for spectroscopic problems. However, the code based on BDF's performed best when
problems whose Jacobian matrices have only real eigenvalues were solved. ∎

   Remark 8.6. It must be emphasized that the two systems of linear
algebraic equations (8.11) and (8.12) have the **same** coefficient matrix
$\bar{A}_n$ $(n=1,2,\ldots,N)$. The application of other two-stage Runge-Kutta methods in
the discretization of (8.1) leads to the solution of two systems of linear
algebraic equations with **different** matrices. This means that the MDIRKM's
are clearly better than the other Runge-Kutta methods when small problems
(8.1) are solved. The application of simple GE at each integration step is
the best choice for such problems; [333]. The MDIRKM's require only one
decomposition per successful integration step with the first acceptability
check, while the number is **two** when any other two-stage Runge-Kutta method
is used. If the system (8.1) is large and if IR is used, then the advantage
of the MDIRKM's is not very clear. Nevertheless, the fact that the two
systems (8.11) and (8.12) have the same coefficient matrix indicates that
if the IR process is convergent in the solution of (8.11), then it will
also be convergent in the solution of (8.12), because convergence depends
only on matrix $\bar{A}_n - \bar{A}_j$, where $\bar{A}_j$ with $j < n$ is the matrix whose
decomposition is used at step n. In order to enhance the possibility of
terminating the second IR process with (8.32), larger values of MAXIT
might be used in the stopping criteria when (8.12) is solved. The successful
termination of the second IR process is desirable, because if the result
produced by these calculations is rejected, then a new decomposition is needed
and the calculations for finding $k_2(h_n)$ are to be repeated. ∎

   Remark 8.7. It might seem to be necessary to keep a copy of matrix $\bar{A}_n$
when IR is used; see (8.26). However, this is not true; matrix $\bar{A}_n$ can be
calculated by (8.13) using matrix $A(t_n+0.5h_n)$, which is needed after the
calculation of $k_1(h_n)$, and therefore must be kept in any case. The
calculation of $\bar{A}_n$ by (8.26) is very cheap; only $O(s)$ extra arithmetic
operations are needed. ∎

8.3. Transition from a dense code to a sparse code

In the code **DENS1**, by which problems (8.1) with dense matrices $A(t)$ are solved, two two-dimensional **REAL** arrays, AORIG and ALU, are used. At step n the elements of matrix $A(t_n + 0.5h_n)$ are held in AORIG, while the factors L_j and U_j $(j \leq n)$ are held in ALU. The LU-factorization is performed by using partial pivoting and information about the pivotal interchanges is kept in a one-dimensional **INTEGER** array IPVT. Several extra one-dimensional **REAL** arrays are needed for the vectors $k_1(h_n)$, $k_2(h_n)$, $k_3(h_n)$, $k_4(h_n)$, y_n and y_{n-1}, as well as for some intermediate results used in **IR**.

The main principles used in the transition from the dense code **DENS1** to the sparse code **SPAR1** are very simple and can be formulated as follows:

(i)	keep all one-dimensional arrays, excepting IPVT, the same as in code DENS1,
(ii)	use the arrays AORIG, CNORIG and RNORIG from the input storage scheme defined in Section 2.1 instead of the two-dimensional array AORIG in code DENS1,
(iii)	use the arrays ALU, CNLU, RNLU, HA and PIVOT from the dynamic storage scheme defined in Section 2.4 instead of the two-dimensional array ALU and the array IPVT in code DENS1.

The arrays used in code **SPAR1** instead of the arrays AORIG, ALU and IPVT from code **DENS1** are given in **Table 8.1**.

Name of the array	Type of the array	Length of the array
AORIG	REAL	NZ
CNORIG	INTEGER	NZ
RNORIG	INTEGER	NZ
ALU	REAL	NN
CNLU	INTEGER	NN
RNLU	INTEGER	NN1
HA	INTEGER	11*NS
PIVOT	REAL	NS

Table 8.1
The arrays used in code SPAR1 instead of the main arrays in code DENS1
(the two two-dimensional arrays and the array in which information
about the pivotal interchanges is stored). NS is used for the number
s of equations in system (8.1).

8.4. Comparing the storage requirements of the two codes

The same integration algorithm (the **MDIRKM** described in **Section 8.2**) with the same rules for accepting or rejecting the approximation calculated in a given step has been implemented in codes **DENS1** and **SPAR1**. A dense matrix technique (**DMT**) is used in **DENS1**, while a sparse matrix technique (**SMT**) is used in **SPAR1**. This is the only difference between the two codes. Therefore a comparison of the two codes will show in a rather clear way when one of them should be preferred. In this section a comparison with regard to the storage requirements is carried out.

Assume that the real numbers are declared in single precision and that an integer occupies the same storage in the computer memory as a real number. Then

(8.36) $S = 3NZ + 2NN + NN1 + 12NS$

locations in the computer memory are used in **SPAR1**, while

(8.37) $D = 2(NS)^2 + NS$

locations are used in **DENS1**. As in **Section 5.3** let:

(8.38) $NN = \nu NZ$ and $NN1 = 0.6\nu NZ$.

It is easily seen that

(8.39) $S \leq D$ $\Leftrightarrow$ $NZ/(NS)^2 \leq g(\nu,NS)$,

where

$$(8.40) \quad g(\nu,NS) \overset{\text{def}}{=} (2-11/NS)/(3+2.6\nu) .$$

Since $\nu=2$ is the minimal value of ν (see **Remark 2.1**) and since

(8.41) $g(\nu,NS) < g(2,NS) < \underset{NS\to\infty}{lim} [g(2,NS)] < 0.244$,

it is obvious that the use of **DENS1** is more efficient (with regard to the storage requirements) than the use of **SPAR1** when

(8.42) $NZ/(NS)^2 > 0.25$.

The converse, however, is not true. The efficiency of **SPAR1** depends on NZ and NS (which are entirely determined by the particular problem solved) as well as on ν (which depends also on the expectation for fill-ins and, thus, on the way in which storage is reserved for the run under considera-tion). When the choice of ν is made, one can compute the value of $g(\nu,NS)$ and use (8.39) to decide whether **DENS1** and **SPAR1** should be preferred (if the storage requirements only are taken into account). Some values of function $g(\nu,NS)$ are given in **Table 8.2**. It is seen that this function is slowly varying in NS, while the variation in ν is considerable. It is also seen that even when the number of reserved locations is rather high ($\nu=6$), **SPAR1** is more efficient than **DENS1** if the non-zero elements are no more than 10%.

This is an excellent result, because $NZ/(NS)^2 < 0.1$ will nearly always be satisfied for sufficiently large matrices. For example, if $NS=1000$ and if $NZ/(NS)^2 > 0.1$, then the matrix will have more than 100 non-zeros per row; i.e. the matrix is not very sparse.

Remark 8.8. It must be emphasized that function $g(\nu,NS)$ is machine dependent. Two examples are given below to illustrate this. If INTEGER*2 statements are available on the computer used, then $S=2NZ+1.5NN+0.5NN1+6.5NS$, $g(\nu,NS)=(2-6/NS)/(2+1.8\nu)$ and $lim[g(2,NS)]=0.375$ as $NS \to \infty$. If both double precision and INTEGER*2 are used, then $S=1.5NZ+1.25NN+0.25NN1+3.75NS$, $g(\nu,NS)=(2-3.5/NS/(1.5-1.4\nu)$ and $lim[d(2,NS)]=0.465$ as $NS \to \infty$. Again, if the choice of ν is made, then one can compute $g(\nu,NS)$ and use (8.39) in order to decide which code is to be chosen. ∎

NS	$\nu=2$	$\nu=3$	$\nu=4$	$\nu=5$	$\nu=6$
100	0.230	0.175	0.141	0.118	0.102
250	0.239	0.181	0.146	0.122	0.105
1000	0.243	0.184	0.148	0.124	0.107
∞	0.244	0.185	0.149	0.125	0.108

Table 8.2
Values of function $g(\nu,NS)$ for different ν and NS .

Remark 8.9. The analysis made in this section is similar to that in Section 5.3. However, there are two important facts that should be pointed out.

If the system of ODE's is large, then IR has to be used both in the dense code and in the sparse code. In both cases one must try to use the same decomposition several times, because the decomposition is very expensive process. If the DS option is used (if one accepts to decompose the matrix at each integration step), this will not give any considerable reduction of the storage requirements; two large arrays, one for the LU factors and the other for the Jacobian matrix (the Jacobian matrix is used in (8.12) after the solution of (8.11)), are needed both in the IR option and in the DS option.

The second important fact that must be mentioned here is the following. The comparison is carried out by ignoring the arrays that are used both in DENS1 and in SPAR1, because the main purpose is not to show how many locations are used in each of these codes, but to show when the storage needed in one of them is greater than the storage needed in the other. ∎

8.5. Comparing the computing time for the two codes

Assume that an old decomposition $L_j U_j$ is used at step n (i.e. $j<n$). Assume also that the step is accepted by the first acceptability check (8.19). Then

$$(8.43) \quad N_{DENS1} = [2(p_1+p_2)+6]s^2 + O(s)$$

multiplications are needed at step n when **DENS1** is used. The number of additions is approximately the same. The numbers of iterations performed in the solution of the two systems of linear algebraic equations (8.11) and (8.12) are denoted by p_1 and p_2 respectively.

Assume that the number of non-zero elements in the factors L_j and U_j is NZ+NF (NF being the number of fill-ins created during the GE when the classical manner of exploiting sparsity is used). Write NZ+NF=μNZ ($\mu \geq 1$). Assume also that **SPAR1** is applied. Then

(8.44) $N_{SPAR1} = [(1+\mu)(p_1+p_2)+2\mu+4]NZ + O(s)$

multiplications are needed at a successful step n when an old decomposition is used (j<n) and when the step is accepted by (8.19). The number of additions is approximately the same.

Let the number of equations s be sufficiently large. Then it is obvious that:

(8.45) $N_{SPAR1} < N_{DENS1}$ $\Leftrightarrow$ $NZ/s^2 < [2(p_1+p_2)+6]/[(1+\mu)(p_1+p_2)+2\mu+4]$.

Assume that:

(i)   the number N of integration steps is large,
(ii)   only a few decompositions are computed,
(iii)   the steps are accepted mainly by (8.19).

Let p be the average number of iterations per integration step. The relationship (8.45) indicates that if s is sufficiently large and if the above assumptions are satisfied, then **SPAR1** will be more efficient than the code **DENS1** (with regard to the arithmetic operations neeeded) when

(8.46) $NZ/s^2 < g^*(\mu,p)$,

where

(8.47) $g^*(\mu,p) \overset{\text{def}}{=} (2p+6)/[(1+\mu)p+2\mu+4]$.

μ	p=4	p=6	p=8	p=10
1.5	0.82	0.82	0.81	0.81
2.0	0.70	0.69	0.69	0.68
2.5	0.61	0.60	0.60	0.59
3.0	0.54	0.53	0.52	0.52
3.5	0.48	0.47	0.47	0.46
4.0	0.44	0.43	0.42	0.42

Table 8.3
Values of function $g^*(\mu,p)$ for some values of μ and p .

Some values of function $g^*(\mu,p)$ are given in **Table 8.3**. This function is very slowly varying in p, while the variation in μ is considerable. It is clear that at steps where an old decomposition is used **SPAR1** will perform more efficiently than **DENS1** (on sequential computers at least) even if the matrix is not very sparse. However, the computational cost at steps where a decomposition is to be calculated is rather large and, therefore, a careful analysis of this situation is necessary.

Assume that a decomposition is carried out at step n and that the approximation calculated by (8.10) is accepted by (8.19). Then

$$(8.48) \qquad N^*_{DENS1} = s^3/3 + 0(s)$$

multiplications are needed in the decompostion at step n when **DENS1** is used. The number of additions is approximately the same.

Assume that $NZ+NF=\mu NZ$ is the number of non-zero elements in L_n and U_n ($\mu \geq 1$; it is again assumed that the classical manner of exploiting sparsity is used). A very crude estimation of the number corresponding to that given by (8.48), which could be used in connection with **SPAR1**, can be writtens as

$$(8.49) \qquad N^*_{SPARS1} = 2\mu^2(NZ)^2/s + 0(s) \ .$$

This number, N^*_{SPARS1}, is found by using the following assumptions:

(i)	**the distribution of the non-zeros in active submatrices is fairly uniform,**
(ii)	**the average number of rows involved in a stage of the GE is $\mu NZ/s$, and the average number of non-zeros per row is also $\mu NZ/s$.**

If these assumptions are satisfied, then the average number of multiplications per stage (in the GE process) is $\mu^2(NZ)^2/s^2$. It is clear after this observation that (8.49) holds if **(i)-(ii)** hold and, moreover, if an additional factor 2 is introduced. By this factor an attempt to take into account the considerable overhead (e.g. for the use of indirect indices, the merging of rows with different sparsity patterns, the pivotal search, etc.) is made.

If (8.49) holds and if s is sufficiently large, then

$$(8.50) \qquad N^*_{DENS1} < N^*_{SARS1} \qquad \Leftrightarrow \qquad NZ/s^2 > \overset{\text{def}}{\bar{g}(\mu)} = (\mu\sqrt{6})^{-1} \ .$$

It should be noted that while (8.45) is rather reliable, (8.50) is derived by imposing severe restrictions on the non-zero elements and their distribution. Nevertheless, (8.50) indicates clearly that the values of NZ/s^2 for which **SPAR1** becomes more efficient than **DENS1** (on sequential machines at least) are much smaller for steps where decomposition is to be carried out (comparing with steps where an old decomposition is used). As an illustration only, let $\mu=2.5$ and p=10. Then the conditions in (8.45) and (8.50) are satisfied for $NZ/s^2 < 0.59$ and $NZ/s^2 < 0.16$ respectively. This shows that one should attempt **either** to reduce the number of steps at which a decomposition is carried out (by restricting the variation of the stepsize and/or by using a large MAXIT in the **IR** processes) **or** to reduce μ. The experiments carried out in an attempt to design the stepsize selection strategy so that the number of decompositions is minimized were successful for some particular problems, but in many cases the opposite effect was observed (the number of steps was increased considerably without a significant change of the number of decompositions). Therefore, attempts to design a device for reducing μ were carried out. These attempts are based on different implementations of the computer-oriented manner of exploiting sparsity and will be discussed in the following sections. However, first some results obtained when the classical manner of exploiting sparsity is applied in **SPAR1** will be presented and compared with results obtained by **DENS1**. These results indicate that a reduction of μ is highly desirable; at least when certain problems modeling real physical phenomena are treated numerically.

8.6. Numerical experiments with DENS1 and SPAR1

Many numerical experiments with **DENS1** and **SPAR1** have been carried out in order to compare the performance of the two codes in different situations. Two spectroscopic problems will be used as working examples in this chapter, but many of the same conclusions are drawn from results obtained in other experiments. The first example is rather small; only 63 equations. Moreover it is not very sparse; $NZ/s^2=0.21$. Therefore, for this example, the performance of **DENS1** should be better than that of **SPAR1** according to the criteria derived in the previous two sections. The second example is larger; s=255. The density of the non-zero elements for the second example is $NZ/s^2=0.05$. Therefore **SPAR1** should perform better than **DENS1** for this example (again according to the criteria from the previous two sections).

The results for s=63 are given in **Table 8.4**. Since $\mu=2.35$ and $(\mu\sqrt{6})^{-1}=0.17$, **DENS1** is more efficient than **SPAR1** when a new decomposition is to be calculated (this happens at about 10% of the integration steps). The average number of substitutions per integration step is p=6.37. Thus, $g^*(2.35,6.37)=0.62$, which indicates that **SPAR1** is better than **DENS1** at steps where an old decomposition can be used (this happens at about 90% of the steps). It is seen that, as predicted, **DENS1** performs better, but the difference is not very great (many steps are carried out by using an old decomposition).

Compared characteristics	DENS1	SPAR1
Number of function evaluations	1571	1597
Number of successful integration steps	763	774
Number of decompositions	67	82
Number of substitutions	4554	5038
Non-zeros at the beginning of the GE (NZ)	839	839
Maximal number of occupied locations in ALU (COUNT)	3969	1972
Computing time (in seconds on an IBM 3081D computer)	60.98	63.09

Table 8.4
Results obtained in a run with a small spectroscopic problem (63 equations) with an error-tolerance $\epsilon = 10^{-3}$.

It is seen from **Table 8.4** that **SPAR1** uses more decompositions and substitutions than **DENS1**. The increase of the number of decompositions can easily be explained. The pivotal strategies in a sparse code are normally relaxed in comparison with the classical partial pivotal strategy used in a dense code (to preserve better the sparsity; see **Chapter 4**). Therefore the factors L and U computed by **SPAR1** are normally more inaccurate than those computed by **DENS1**. This means that these factors should be updated more times for the sparse code. Since the factors computed by **SPAR1** are more inaccurate than those computed by **DENS1**, also the rate of convergence of the **IR** processes tend to be slower when **SPAR1** is used. This explains the increase of the number of substitutions (a substitution consists of a forward and back susbstitution). Finally, the numbers of integration steps and function evaluations is also slightly increased (by 1% - 2%). It should be pointed out, however, that the increase of these characteristics for the sparse code does not lead to a considerable increase of the computing time. The slight increase of the computing time for **SPAR1** is mainly due to the fact that the steps where a decomposition is to be calculated are rather expensive.

If single precision is used, then **DENS1** is clearly better than **SPAR1** with regard to the storage requirements. Indeed, $\nu=3$ has to be used when the lengths of the main arrays of the dynamic storage scheme are to be determined; ν must be larger than μ because some space, "elbow room" (see **Section 2.7**), should be reserved for making copies of rows and columns at the end of the lists. The value of $g(\nu,NS)$ for $\nu=3$ and $NS=63$ is 0.17, which is less than $NZ/s^2=0.21$. If **INTEGER*2** statements are available and if double precision is used, then $g(3,63)=0.34$ (which is greater than $NZ/s^2=0.17$) and **SPAR1** becomes more efficient than **DENS1** with regard to the storage requirements.

Compared characteristics	DENS1	SPAR1
Number of function evaluations	973	979
Number of successful integration steps	470	473
Number of decompositions	59	67
Number of substitutions	2763	3015
Non-zeros at the beginning of the GE (NZ)	3323	3323
Maximal number of occupied locations in ALU (COUNT)	65025	20170
Computing time (in seconds on an IBM 3081D computer)	805.08	606.52

Table 8.5
Results obtained in a run with a medium spectroscopic problem
(255 equations) with an error-tolertance $\epsilon = 10^{-2}$.

The results for s=255 are given in **Table 8.5**. For this problem $\mu=6.07$ and $(\mu\sqrt{6})^{-1}=0.07$. This means that **SPAR1** is the better choice at steps where a decomposition is to be calculated because $NZ/s^2=0.05$. The avarage number of substitution per integration step is p=6.37 and, therefore, $g^*(\mu,p)=g^*(6.07,6.37)=0.31$. This shows that **SPAR1** is more efficient than **DENS1** also at steps where an old decomposition is used. Again, as for the small example, both the numbers of decompositions and substitutions are smaller for **DENS1**. Nevertheless, the computing time for the sparse code is considerably smaller than that for the dense code.

For this example **SPAR1** is better than **DENS1** also when the storage requirements are taken into account. It is reasonable to use $\nu=7$ for this problem (though $\nu=6.5$ will be sufficient) and $g(\nu,NS)=g(7,255)=0.09$ is larger than $NZ/s^2=0.05$.

The results given in this sections as well as the results in the previous two sections show that it is worthwhile to try to reduce the parameter ν. This can be done by using the computer-oriented manner of exploiting sparsity, which will be discussed in the next section.

8.7. Dropping small elements during the GE process

It has already been demonstrated that the use of the computer-oriented manner of exploiting sparsity gives excellent results for some problems. In this and the following sections the use of the computer-oriented manner will be discussed in more detail in connection with long sequences of systems of linear algebraic equations arising from the discretization of **ODE's**. The absolute drop-tolerance defined in (1.9) is used to drop "small" elements, but the use of a relative drop-tolerance could be even more profitable. Numerical results, obtained in the solution of the same problems as in the previous section, are given in **Table 8.6** and in **Table 8.7**. The results for T=0.0 are the same as the results for code **SPAR1** in **Table 8.4** and **Table 8.5**. These results are given only in order to facilitate the comparison.

Compared characteristics	T=0.0	T=2.0
Number of function evaluations	1597	1547
Number of successful integration steps	774	749
Number of decompositions	82	85
Number of substitutions	5038	5364
Non-zeros at the beginning of the GE (NZ)	839	839
Maximal number of occupied locations in ALU (COUNT)	1972	896
Computing time (in seconds on an IBM 3081D computer)	63.09	39.21

Table 8.6
Results obtained in a run with a small spectroscopic problem (63 equations) with an error-tolertance $\epsilon = 10^{-3}$ and with two values of the drop-tolerance T (the classical manner of exploiting sparsity is used when T=0.0 and the results for T=0.0 are the same as the results for code SPAR1 in Table 8.4).

The first three of the quantities (the function evaluations, the integration steps and the decompositions) that are compared in **Table 8.6** and **Table 8.7** are nearly the same for T=0.0 and for T>0.0 (slightly smaller when T=0.0 is applied). The number of substitutions is increased when T>0.0 is used, but, nevertheless, the total computing times are decreased for both problems. The decrease of the computing time is greater for the large problem where such a decrease is much more desirable.

The values of μ are 1.07 (when s=63) and 2.32 (when s=255). This is a very considerable reduction compared with the case where the classical manner of exploiting sparsity is used (T=0.0), where the corresponding numbers are 2.35 and 6.07; see the previous section. SPAR1, when used together with the computer-oriented manner of exploiting sparsity, performs better than DENS1 for both problems with respect to both computing time and storage.

Compared characteristics	T=0.0	T=1.0
Number of function evaluations	979	985
Number of successful integration steps	473	476
Number of decompositions	67	68
Number of substitutions	3015	3499
Non-zeros at the beginning of the GE (NZ)	3323	3323
Maximal number of occupied locations in ALU (COUNT)	20170	7706
Computing time (in seconds on an IBM 3081D computer)	606.52	161.37

Table 8.7
Results obtained in a run with a medium spectroscopic problem
(255 equations) with an error-tolertance $\epsilon = 10^{-3}$
and with two different values of the drop-tolerance T
(the classical manner of exploiting sparsity is used when T=0.0
and the results for T=0.0 are the same as the results for
code SPAR1 in Table 8.5).

8.8. Dropping small elements before the start of the GE process

"Small" elements are dropped only during the GE transformations in the device described in the previous section. However, in many cases some of the elements of matrix $\bar{A}_n$ from (8.13) are also "small"; or, in other words, (1.9) is satisfied for these elements. Thus it is worthwhile to scan the elements of this matrix before the beginning of the GE and to remove all elements that satisfy (1.9). This possibility was mentioned in Remark 2.2, but the effect of such a previous scan is first demonstrated in this section. The two working examples were run with an option in SPAR1 where all "small" elements were removed before the beginning of the GE. The results are given in Table 8.8 and in Table 8.9. The best results obtained until now, the results from the previous section, are also given in these two tables (in order to facilitate the comparison).

It is seen that NZ1, the number of elements after the scan, may be considerably smaller than NZ. Moreover, even COUNT, the maximal number of occupied locations in the arrays of the dynamic storage scheme, could be smaller than NZ. This means that if a scan of the elements of the matrix is carried out before the start of the GE, then it is not necessarily true that $\mu \geq 1$ ($\mu = (NZ1+NF)/NZ=0.62$ for the small problem with a scan). This shows that the device may be very efficient if there are many "small" elements (as for s=63). However, even when there are not many elements that are removed during the scan (as in the second example), savings both in storage and in computing time may still be achieved; see Table 8.9.

Compared characteristics	No scan	Scan
Number of function evaluations	1547	1583
Number of successful integration steps	749	768
Number of decompositions	85	84
Number of substitutions	5364	5407
Non-zeros at the beginning of the GE (NZ1)	839	336
Maximal number of occupied locations in ALU (COUNT)	896	518
Computing time (in seconds on an IBM 3081D computer)	39.21	29.24

Table 8.8
Results obtained in a run with a small spectroscopic problem (63 equations)
with an error-tolertance $\epsilon = 10^{-3}$ and with drop-tolerance T=2.0.
Two options are used (without a previous scan and with a previous scan).
NZ1=NZ when no scan is applied. The results witout a previous scan are
the same as the results with T=2.0 in Table 8.6.

Compared characteristics	No scan	Scan
Number of function evaluations	985	985
Number of successful integration steps	476	476
Number of decompositions	68	68
Number of substitutions	3499	3588
Non-zeros at the beginning of the GE (NZ1)	3323	2171
Maximal number of occupied locations in ALU (COUNT)	7707	6523
Computing time (in seconds on an IBM 3081D computer)	161.37	144.21

Table 8.9
Results obtained in a run with a medium spectroscopic problem
(255 equations) with an error-tolertance $\epsilon = 10^{-3}$ and with T=1.0.
Two options are used (without a previous scan and with a previous scan).
NZ1=NZ when no scan is applied. The results witout a previous scan are
the same as the results with T=1.0 in Table 8.7).

It should be mentioned that the elements of matrix $\bar{A}_n$ are recalculated
by using (8.13) at every integration step n because this matrix is needed
in the calculation of $r_m^{[i]}(h_n)$; see Remark 8.7 and (8.26). No element is
removed from $\bar{A}_n$ when this matrix is used in the calculation of the residual
vectors $r_m^{[i]}(h_n)$. In fact, $\bar{A}_n$ is never explicitly formed when the residual
vectors are calculated; matrix $A(t_n+0.5h_n)$, kept in the input storage scheme,
is actually used. The scan described in this section is carried out only
before GE and is used only in the calculatuons of the factors L_n and U_n.

It should also be mentioned that NZ1 can vary from one integration step
to another, because the elements of the Jacobian matrix are in general time-
dependent. The values of NZ1 given in the tables are the largest values
found during the whole computational process.

8.9. Automatic determination of the drop-tolerance

The values of the drop-tolerance T used in the previous two sections
have been found by several trials. Assume that only one system of linear

algebraic equations is to be solved. Then, as a rule, it is not very difficult
to determine a good value of T (i.e. a T > 0.0 by which results that are
better than the results obtained with T = 0.0 can be achieved). However, it
is rather difficult to find the optimal value of T (or a value that is
nearly optimal) when one system only is to be solved.

Consider a long sequence of systems of linear algebraic equations. For
such a sequence it is possible to apply a device by which the code will
attempt to determine values of T that are nearly optimal. Moreover, the
attempt is carried out automatically, by the code, in the course of the
solution process. Let $a=max(|\bar{a}_{ij}|)$, where $\bar{a}_{ij} \in \bar{A}_n$ (i=1,2,...,s, j=1,2,...s).
Set $T_{initial}=0.1a$. Begin with $T=T_{initial}$ and set $T:=0.5T$ every time when
the the **IR** process fails to converge at steps where a new LU-decomposition
has been computed (i.e. at steps n with $n \in \{1,2,...,N\}$ at which L_n and
U_n are used in the **IR** process). If T has been reduced J times (J=10
in **SPAR1**), then T=0 is used in the further computations. The device
described above is very simple. The main idea is: **carry out some extra work
in the starting phase of the computations and then solve efficiently the
remaining systems using the nearly optimal drop-tolerance found during the
starting phase.**

Note that it is not necessary to assume that the sequence of systems of
linear algebraic equations appears after the discretization of a system of
ODE's. However, if this is so, then an additional device, by which the drop-
tolerance could be increased, may be incorporated in the algorithm described
in the previous paragraph (where T can only be decreased). Assume that an
integration step, say n, is rejected several times. After every rejection the
stepsize is decreased by the code (in the algorithm implemented in **SPAR1** the
stepsize is decreased by a factor of two after every rejection). This means
that the diagonal elements of $\bar{A}_n$ are normally increased by a rather
considerable factor after each rejection; see (8.13)-(8.14). One should
expect that the matrix $\bar{A}_n$ becomes diagonally dominant (or at least closer
to a diagonally dominant) after several rejections of the stepsize, and if
this is so, then the **IR** process will converge even if T is large. This
observation is used to formulate a criterion for increasing the drop-
tolerance: **if the stepsize is rejected twice at some step n, then the code
sets T:=2T.**

The results obtained by using the ideas described in this section for the
two working examples are given in **Table 8.10** and in **Table 8.11**. It is seen
that the code is able to determine a nearly optimal drop-tolerance. The
computing times are slightly increased (compared with the times in the tables
from the previous section). This should be expected because of the extra work
needed to determine the final drop-tolerance. The differences, however, are
very small. This is true even when the final drop-tolerance found by the code
is smaller than the optimal one (because also in this case the code performs
a part of the computations with a drop-tolerance which is larger than the
final one). Taking into account that the user will in general not be able to
find the optimal value of T (this being especially true when the problem is
large), the algorithm for an automatic determination of nearly optimal values
for the drop-tolerance must be considered as very efficient.

Compared characteristics	Constant T	Variable T
Number of function evaluations	1583	1595
Number of successful integration steps	768	774
Number of decompositions	84	92
Number of substitutions	5407	5884
Non-zeros at the beginning of the GE (NZ)	839	839
Non-zeros in ALU after the scan (NZ1)	336	315
Maximal number of occupied locations in ALU	518	475
Initial value of the drop-tolerance T	2.0	94.40
Final value of the drop-tolerance T	2.0	2.17
Computing time (in seconds on IBM 3081D)	29.24	30.12

Table 8.10
Results obtained in a run with a small spectroscopic problem
(63 equations) with an error-tolertance $\epsilon = 10^{-3}$ and with T=2.0.
Two options are used: with a constant drop-tolerance (the results are
the same as the results in Table 8.8) and with a variable drop-tolerance.
NZ1 is the largest number of non-zeros in ALU after a scan; this number is
obtained with the smallest T used in the computations when the option
with a variable drop-tolerance is applied).

Compared characteristics	Constant T	Variable T
Number of function evaluations	985	1003
Number of successful integration steps	476	484
Number of decompositions	68	81
Number of substitutions	3588	3852
Non-zeros at the beginning of the GE (NZ)	3323	3323
Non-zeros in ALU after the scan (NZ1)	2171	2283
Maximal number of occupied locations in ALU	6523	6634
Initial value of the drop-tolerance T	1.0	55.54
Final value of the drop-tolerance T	1.0	0.87
Computing time (in seconds on IBM 3081D)	144.21	148.60

Table 8.11
Results obtained in a run with a medium spectroscopic problem
(255 equations) with an error-tolertance $\epsilon = 10^{-2}$ and with T=1.0.
Two options are used: with a constant drop-tolerance (the results are
the same as the results in Table 8.9) and with a variable drop-tolerance.
NZ1 is the largest number of non-zeros in ALU after a scan; this number is
obtained with the smallest T used in the computations when the option
with a variable drop-tolerance is applied).

8.10. Application to difficult problems

The algorithms described in Section 8.7 - Section 8.9 are very efficient
when the problem solved is stringent in the sense that the straightforward
application of the classical manner of exploiting sparsity (where small

elements are not removed) leads to a great number of fill-ins. If the problem
is large, then many fill-ins are often created during the decomposition. An
example will be given in this section.

Consider a spectroscopic problem with 1023   equations. The number of
non-zero elements in matrix A(t) is normally very large for such problems.
In the particular example that will be discussed here the number of non-zero
elements is NZ=64517 (i.e. the average number of non-zero elements per row
is 64). The results obtained by the code SPAR1 with an automatic
determination of the drop-tolerance and with a previous scan for removing
small elements are given in Table 8.12.

An attempt to perform the corresponding computations by using the
classical manner of exploiting sparsity (i.e. without dropping small elements)
has also been carried out. The subroutine spent about one hour CPU time on an
IBM 3081D to integrate (8.1) over only about 1.5% of the desired (by the
physisists) time interval: the system of ODE's was integrated over the
interval [0,1.55], while it was required to integrate it over [0,100.0]. Four
decompositions were performed in this run. The number of occupied locations
in ALU was 352470, and this shows why the computational process is so
expensive; see also the results given in Table 2.3.

Compared characteristics	Variable T
Number of function evaluations	1615
Number of successful integration steps	788
Number of decompositions	76
Number of substitutions	4701
Non-zeros at the beginning of the GE (NZ)	64517
Non-zeros in ALU after the scan (NZ1)	13571
Maximal number of occupied locations in ALU	75013
Initial value of the drop-tolerance T	146.00
Final value of the drop-tolerance T	1.14
Computing time (in seconds on IBM 3081D)	2124.00

Table 8.12
Results obtained in a run with a stringent spectroscopic problem
(1023 equations) with an error-tolertance $\epsilon = 10^{-2}$.
NZ1 is the largest number of non-zeros in ALU after a scan; this number
is obtained with the smallest T used in the computations when the option
with a variable drop-tolerance is applied).

The example illustrates the great efficiency of the algorithms studied
in this chapter for some classes of problems. However, it should be pointed
out (once again!) that there are problems where only a few elements are
dropped. For such problems the algorithms discussed here will still work, but
will not be very efficient, and it will be better to use the classical manner
of exploiting sparsity (where no attempt to drop small non-zero elements is
carried out). This is the reason for keeping options based on both manners in
SPAR1. Nevertheless, it is important to emphasize that even if an option
based on the algorithms from Section 8.7 - Section 8.9 for a problem where
no element is dropped, the penalty (in terms of computing time) will not be
large, while the above example shows that an option in which no dropping of
small elements is allowed should be used very carefully. This is because if

such an option is applied to a stringent problem in the solution of which many
fill-ins appear, then the results may be catastrophic. Thus, an option based
on the computer-oriented manner of exploiting sparsity is absolutely necessary
in a software for solving ODE's, but it is also desirable to have options in
which the classical manner is applied.

Large spectroscopic problems are not the only class of problems which
cause great difficulties for the classical manner of exploiting sparsity. A
similar situation appears, for example, when models simulating certain biological patterns are treated numerically (see Table 1.3). These problems,
which seem to be very innocent (only about 7.5 elements per row; in average),
are extremely difficult when no small elements are removed during the GE;
the number of elements in L and U is about 283000. If small elements are
removed, the corresponding figure is only about 28000.

8.11. Some extensions of the results

Several assumptions were made in the beginning of this chapter. It was
stated there that these were needed only to facilitate the exposition of the
results. It is typical for many applications that the situation is much more
complicated than the case where a system of linear algebraic equations is to
be solved. This can be seen by comparing this chapter with the previous
chapters and recalling the simplifying assumptions made in this chapter. In
this section it will be shown how some of the assumptions made earlier can be
removed.

Remark 8.10.   SPAR1   is based on a particular integration method
(sketched in Section 8.2). However, the main ideas:

   (i) application of sparse matrix technique,
  (ii) introduction of a drop-tolerance,
 (iii) development of a device for an automatic determination
       of the drop-tolerance,

can be applied in connection with any other integration method (as, for
example, some BDF, see Example 8.1). The functions from (8.40), (8.47) and
(8.50) are dependent on the integration method used, but these functions can
easily be modified when the method is changed. It should also be noted that
these functions are not only dependent on the basic integration method, but
also on the device used in the error control.   ■

Remark 8.11.   The Jacobian matrices of the systems of ODE's are
sometimes band matrices (especially when these systems arise from some space
discretization of partial differential equations). If the bandwidth of the
matrix is narrow, then it is usually worthwhile to exploit the bandedness of
the Jacobian matrix. A code BAND1 for solving such systems has been developed
(subroutines from the well-known package LINPACK, [55], are attached to
BAND1). A comparison of BAND1 and SPAR1 will be carried out in the near
future. Such a comparison is needed because while it is clear that the use of
sparse matrix technique only is not competitive with the use of band matrix
technique, the result of the comparison is not very clear when the sparse
matrix technique is combined with the powerful application of positive values
of the drop-tolerance. Some numerical results, which indicate that such a
comparison is desirable, are given in [327,341]; see also Table 5.5.   ■

Remark 8.12.   **SPAR1** is designed for systems of **linear ODE's**. There are two reasons for this:

(1)   **linear ODE's** appear very often in applications,

(2)   linearity can be exploited because:

   (a)   the evaluation of the Jacobian matrix which causes problems for **non-linear ODE's** can easily be optimized,

   (b)   the integration formulae can be written in a more readable form when f(t,y)=A(t)y+b(t),

   (c)   special algorithms only for **linear ODE's** can be developed,

   (d)   the expression A(t)y+b(t) is computed in the code, and the developer has the responsibility for optimizing the calculation of A(t)y+b(t) (in the codes for **non-linear ODE's**, where the right-hand side f(t,y) is required at every step, this work has to performed by the user).

When sparse matrix technique is used, (d) is (or, at least may be) very important: the user is not forced to study the sparse matrix technique used in the software in order to find out how to calculate the right-hand side of the system of **ODE's** in an efficient way that is compatible with the storage scheme used in the software.

The above analysis shows clearly that if the system of **ODE's** is linear, then the linearity **should** be exploited. At the same time, it must also be emphasized that the main ideas applied in the development of **SPAR1** may be applied in the solution of non-linear systems of **ODE's**; y'=f(t,y). For such systems some quasi-Newton iterative process should be used instead of **IR**. The shifted Jacobian matrix $\gamma_n I + \partial f/\partial y$ has to be factorized. The computation of $\partial f/\partial y$ may be very expensive, but when this matrix is calculated, one can (in principle, at least) use the same ideas as in the previous sections. There are plans to develop a code to treat the air pollution model discussed in [302,307,313-317], where non-linear systems of **ODE's** are to be solved in the chemical part ([151]).   ∎

8.12. Concluding remarks and references concerning the solution of ODE's by using sparse matrix technique

The sparse matrix technique presented in this chapter was applied in the solution of some spectroscopic problems (discussed in [211]) that are described by systems of linear **ODE's**. The code used in the first application, **Y12N**, is studied in [212-213]. The new code, **SPAR1**, is in fact an improvement of **Y12N**. The code **SPAR1** is fairly efficient when **spectroscopic problems** are treated numerically, but it could be used also in some other situations if the Jacobian matrix A(t) has eigenvalues close to the imaginary axis; [337-338]. If this condition is not satisfied, then other methods (for example, **BDF's**, see **Section 8.1**) should be used.

The integration method described in **Section 8.2** is developed in [294]. The implementation of this method in software for solving linear **ODE's** is discussed in [335-338]. **SPAR1** is fully documented in [335].

It is important to emphasize again that: **one cannot save any storage when the DS option is used instead of the IR option** in the case where systems of linear **ODE's** are solved. Indeed, two systems of linear algebraic equations are to be solved at each integration step and the Jacobian matrix is also needed in the solution of the second system. Therefore, the Jacobian matrix cannot be overwritten by the factors L_n and U_n even if a new decomposition is to be calculated at every integration step. Thus the situation is very different from the case where only one system of linear algebraic equations is solved and where the original matrix can be overwritten by the factors L and U. In the latter case one can save some storage when the **DS** is used if only a few fill-ins appear during the decomposition.

The computer-oriented manner of exploiting sparsity is efficient when systems of linear **ODE's** are solved not only because the Jacobian matrix has always to be kept and cannot be overwritten by the factors L_n and U_n, but also because nearly optimal values of the drop-tolerance can be determined automatically. The illustration in **Section 8.10** shows that **the option with an automatic determination of a nearly optimal drop-tolerance is practically the only option by which some difficult problems can be handled numerically.**

It is probably better to apply some conjugate gradient-type method (see **Chapter 11**) instead of the simple **IR**. It will be possible to use larger values of the drop-tolerance when a conjugate gradient-type method is implemented in the software. However, the use of a very large drop-tolerance may increase dramatically the number of substitutions performed during the solution process. Nevertheless, it is worthwhile to try to develop a code where the same ideas are used, but the **IR** is replaced by some conjugate gradient-type method. It should be mentioned here that conjugate gradient-type methods are used in some codes; see, for example, [108].

CHAPTER 9

CONDITION NUMBER ESTIMATORS

IN A SPARSE MATRIX SOFTWARE

The stability of the computational process in the solution of systems of linear algebraic equations Ax=b by the GE depends on the condition number of matrix A. Reliable and efficient algorithms for calculating estimates of the condition number of a matrix are given in [43]. The application of these algorithms in sparse matrix software (the code actually used is package Y12M, [331,341], but the same ideas could be applied for other codes also) is discussed in this chapter. Three algorithms have been implemented in Y12M and tested on a very large set of problems. The influence of the stability factor u that is used in the pivotal strategy (see **Chapter 4**) and the drop-tolerance T (see **Chapter 5**) on the accuracy of the estimates of the condition number is also studied.

9.1. Influence of the condition number on the accuracy of the solution of a system of linear algebraic equations

Consider the system of linear algebraic equations

(9.1) $Ax=b$, $A \in R^{n \times n}$, $b \in R^{n \times 1}$, $x \in R^{n \times 1}$, $rank(A)=n$.

Assume that the GE is used to obtain a decomposition (or a factorization) of the coefficient matrix:

(9.2) $LU=A+E$, $L \in R^{n \times n}$, $U \in R^{n \times n}$, $E \in R^{n \times n}$,

where L is a unit triangular matrix, U is an upper triangular matrix and E is a perturbation matrix. The GE process is normally carried out by using some pivoting (see **Chapter 4**). Any pivoting can be considered as a multiplication of matrix A by permutation matrices. In this chapter (excepting **Section 9.5**) it is assumed that such a multiplication is performed before the beginning of the GE and the notation A is also used for the matrix obtained after the multiplication. It is clear that this is not a restriction.

Assume that $\bar{x}$ satisfies the system:

(9.3) $(A+E)\bar{x}=b$.

Then the following theorem can be proved (see Theorem 4.4.2 in [231]):

Theorem 9.1. Assume that:

(i) $x \neq 0$ satisfies (9.1),

(ii) $\bar{x}$ satisfies (9.3),

(iii) $\|A^{-1}\| \|E\| < 1$.

Then

(9.4) $(\|x-\bar{x}\|)/\|x\| \le \kappa(A)(\|E\|/\|A\|)/[1-\kappa(A)(\|E\|/\|A\|)]$,

where $\kappa(A) = \|A\|\|A^{-1}\|$ is the condition number of matrix A with regard to the matrix norm chosen. ■

If GE with partial pivoting is carried out on a computer, then

(9.5) $\|E\|_\infty \le \psi(n)\|A\|_\infty 10^{-t}$,

where $\psi(n)$ is a function of n that depends on the way the arithmetic operations are carried out (and, first and foremost, on the pivotal strategy), while t is the number of digits in the mantissa in the computer representation of the reals. Assume that

(9.6) $\kappa(A)\psi(n)10^{-t} < 1$, $\kappa(A)=\|A\|_\infty\|A^{-1}\|_\infty$.

Then the conditions of **Theorem 9.1** are satisfied and the assertion of the theorem can be rewritten as

(9.7) $(\|x-\bar{x}\|)/\|x\| \le \kappa(A)\psi(n)10^{-t}/[1-\kappa(A)\psi(n)10^{-t}]$.

Assume that:
    (i) $\psi(n)$ is not too large,
    (ii)   A is not too badly scaled,
    (iii) $\kappa(A) = 10^p$ with some p<t .

Then it is shown in [231], p. 196, that one one should expect that a solution computed in **t-digit** arithmetic to be accurate to about **t-p significant figures**. Thus, the accuracy of the solution depends strongly (or, at least, will often depend strongly) on the condition number of matrix A if (i)-(iii) are satisfied.

The above discussion is an illustration of the great influence of the condition number of the matrix on the solution when a computer arithmetic with rounding errors is used (see also [231,259,270-277]). Therefore it seems to be worthwhile to try to estimate the condition number of the coefficient matrix of the system to be solved.

9.2. Calculating estimates of the condition number of a matrix

Algorithms for calculating condition number estimates in a cheap and reliable way are discussed in [42-44,99,135,173,180]. Three algorithms described in [43,135] have been implemented in **Y12M** and will be discussed in this chapter. These algorithms are based on the following rules:

    (i)   compute $\|A\|$,
    (ii)   use GE to calculate L and U,
    (iii) solve $U^Tw=e$ and $L^Ty=w$ (e being a given vector),
    (iv)   solve Lv=y and Uz=v,
    (v)   calculate $A^*=\|z\|/\|y\|$ (and consider A^* as an approximation to $\|A^{-1}\|$),
    (vi)   set $k^*=\|A\|A^*$ and consider k^* as an estimate of the condition number of matrix A.

The approximation A^* is good when $\|y\|/\|e\|$ is as large as possible ([43], p. 371). Therefore an attempt to choose the components of vector e so that $\|y\|/\|e\|$ is large has to be carried out.

In the first algorithm from [43] (see also [99]) the components of e are determined as follows. Let e_1 be either 1 or -1. Let

$$(9.8) \quad u_{ii}w_i = e_i - \sum_{j=1}^{i-1} u_{ji}w_j \qquad (i=2,3,\ldots,n)$$

and set $e_i=1$ if the sum in (9.8) is negative and $e_i=-1$ otherwise. This algorithm is used in subroutine **DECOMP** in [99]. **DECOMP** is probably the first subroutine in which a device for estimating the condition number $\kappa(A)$ of a matrix is implemented.

The second algorithm in [43] is not so simple. Set again e_1 equal either to 1 or to -1. Consider the following relations:

$$(9.9) \quad t_i = \sum_{j=1}^{k-1} u_{ji}w_j \qquad (i=k,k+1,\ldots,n),$$

$$(9.10) \quad e_k^+ = sign(-t_k), \qquad e_k^- = -e_k^+ \qquad (k=2,3,\ldots,n),$$

$$(9.11) \quad w_k^+ = (e_k^+-t_k)/u_{kk}, \qquad w_k^- = (e_k^--t_k)/u_{kk} \qquad (k=2,3,\ldots,n),$$

$$(9.12) \quad t_i^+ = t_i + u_{ki}w_k^+, \qquad t_i^- = t_i + u_{ki}w_k^- \qquad (i=k+1,k+2,\ldots,n, \ k=2,3,\ldots,n),$$

$$(9.13) \quad T_k^+ = \sum_{i=k}^{n} t_i^+, \qquad T_k^- = \sum_{i=k}^{n} t_i^- \qquad (k=2,3,\ldots,n),$$

$$(9.14) \quad e_k = e_k^+ \ \text{if} \ T_k^+ > T_k^- \ \text{and} \ e_k = e_k^- \ \text{otherwise} \ (k=2,3,\ldots,n).$$

The algorithm defined by (9.9)-(9.14) is implemented in many **LINPACK** subroutines [55]). An attempt to avoid overflow and division by zero is carried out in the subroutines of **LINPACK** by scaling appropriate vectors during the calculations. For some matrices the number of scalings could be reduced by a simple modification of the **LINPACK** condition number estimator. The modified algorithm and its application are discussed in [135].

Three questions should be answered in connection with the implementation of the above algorithms in sparse matrix software:

(a) How can the sparsity scheme used in the software be applied in the calculations within the condition number estimator?

(b) What is the effect of the pivotal strategy used in the sparse matrix software on the accuracy of the results produced by the condition number estimator?

(c) What is the effect of dropping "small" elements during the GE on the accuracy of the results produced by the condition number estimator?

The first question concerns the efficiency of the sparse condition number estimators with regard to the computing time and the storage used. It must be stressed here that it **may** be very inefficient just to take a code for dense matrices and to modify it for sparse matrices. The fact that the matrix is sparse must be exploited and special techniques have to be applied.

The next two questions are related to the ability of the condition number estimator to calculate accurate estimates k*. It must be emphasized here **"that at best we can obtained** $\kappa(A+E)$**"** ([43], p. 373). Therefore in the calculation of k* by the use of the factors L and U an assumption that the GE is stable is made. If matrix A is **dense**, then it is commonly accepted that **"in practice Gaussian elimination with partial pivoting must be considered a stable algorithm"** ([231], p. 152), though examples where this is not true can be constructed ([271]). Thus, in the condition number estimators it is implicitly assumed that L and U are sufficiently accurate and that the norm of matrix E is small in some sense. Experience indicates that this is a realistic assumption for dense matrices. However, the situation may change when the matrix is sparse. The accuracy requirements during   GE are often relaxed (because the pivotal strategy attempts to preserve the sparsity better by using a large stability factor u and/or because some "small" elements are neglected when the computer-oriented manner of exploiting the sparsity is in use). Thus, **some difficulties connected with the accuracy of   GE   for sparse matrices may arise and a careful examination of the influence of the relaxation of the accuracy requirements during the factorization of sparse matrices on the accuracy of the results produced by the condition number estimators must be carried out.**

9.3. Application of the storage schemes used in package   Y12M

Let   ANORM1 = $\|A\|_1$. The main steps in the calculation of an estimate κ^* of the condition number of matrix A are the calculation of ANORM1 and the solution of $U^Tw=e$, $L^Ty=w$, $Lv=y$, $Uz=v$. The application of the storage scheme used in **Y12M** in the performance of these steps will be described in this section. The codes by which the calculations are carried out will also be given.

The input storage scheme is described in **Section 2.1**. This scheme can be outlined as follows. Let N and NZ be the order and the number of non-zeros of matrix A. The non-zeros are stored, in an arbitrary order, in the first NZ locations of array AORIG. If $a_{ij}\neq0$ is stored in AORIG(K), $1\leq K\leq NZ$, then RNORIG(K)=i and CMORIG(K)=j. The code for calculating ANORM1=$\|A\|_1$ by the use of this scheme and a working array W of length at least equal to N is given in **Fig. 9.1**.

The dynamic storage sheme, described in **Section 2.4**, is to be used in the further calculations (the solution of four systems with triangular matrices). The following information about the dynamic storage scheme is needed to understand the codes used in the calculation of an estimate for the condition number of matrix A. Both the lower triangular matrix L (without its diagonal elements) and the upper triangular matrix U (also without its diagonal elements) are stored in array ALU. The diagonal elements elements of L are never stored, while the diagonal elements of U are stored in array PIVOT (the pivot used at stage s, s=1,2,...,n-1, of the GE being stored in the s'th location of PIVOT). Pointers about the positions of the rows of L and U are stored in the first three columns of array HA. The non-zero elements of row i in matrix L (without the diagonal element) are stored in array ALU from position HA(i,1) to position HA(i,2)-1. The non-zero elements of row i in matrix U (without the diagonal element) are stored from position HA(i,2) to position HA(i,3). There are no free locations in array ALU between HA(i,1) and HA(i,3), but there may be free locations both before HA(i,1) and after HA(i,3). If a non-zero element, either from L or from U, is stored in ALU(K), then its column number is stored in CNLU(K). The codes for solving the four systems with triangular coefficient matrices using the storage scheme described above are given in **Fig.9.2** - **Fig. 9.5**.

```
        ANORM1=0.0
        DO 10 I=1,N
           W(I)=0.0
    10  CONTINUE
        DO 20 I=1,NZ
           W(CNORIG(I))=W(CNORIG(I))+ABS(AORIG(I))
    20  CONTINUE
        DO 30 I=1,N
           ANORM1=AMAX1(W(I),ANORM1)
    30  CONTINUE
```

Figure 9.1
Calculation of ANORM1 by the use of the input storage scheme.

```
        W(1)=W(1)/PIVOT(1)
        DO 50 I=2,N
           DO 40 J=HA(I-1,2),HA(I-1,3)
              W(CNLU(J))=W(CNLU(J))-W(I-1)*ALU(J)
    40     CONTINUE
           W(I)=W(I)/PIVOT(I)
    50  CONTINUE
```

Figure 9.2
Solution of the system $U^Tw=e$ by the use of the dynamic storage scheme.
It is assumed that vector e is stored in array W on entry.
The solution vector w is stored in array W on exit.

```
        DO 70 I=N,2,-1
           DO 60 J=HA(I,1),HA(I,2)-1
              W(CNLU(J))=W(CNLU(J))-W(I)*ALU(J)
    60     CONTINUE
    70  CONTINUE
```

Figure 9.3
Solution of the system $L^Ty=w$ by the use of the dynamic storage scheme.
It is assumed that vector w is stored in array W on entry.
The solution vector y is stored in array W on exit.

```
        DO 100 I=2,N
           DO 90 J=HA(I,1),HA(I,2)-1
              W(I)=W(I)-ALU(J)*W(CNLU(J))
 90        CONTINUE
100 CONTINUE
```

Figure 9.4
Solution of the system Lv=y by the use of the dynamic storage scheme.
It is assumed that vector y is stored in array W on entry.
The solution vector v is stored in array W on exit.

```
        W(N)=W(N)/PIVOT(N)
        DO 120 I=N-1,1,-1
           DO 110 J=HA(I,2),HA(I,3)
              W(I)=W(I)-ALU(J)*W(CNLU(J))
110        CONTINUE
           W(I)=W(I)/PIVOT(I)
120 CONTINUE
```

Figure 9.5
Solution of the system Lz=v by the use of the dynamic storage scheme.
It is assumed that vector v is stored in array W on entry.
The solution vector z is stored in array W on exit.

Let YNORM1 and ZNORM1 be the one-norms of vectors y and z. The calculation of these norms is the same as in the dense case. Codes for calculating YNORM1 and ZNORM1 are given in **Fig. 9.6** and **Fig. 9.7**. The code from **Fig. 9.6** should be used after the solution of $L^T y=w$, while that from **Fig. 9.7** should be used after the solution of Uz=v. Subroutines from **LINPACK** ([55]) may be attached to the condition number estimators for sparse matrices and use instead of the codes given in **Fig. 9.6** and **Fig. 9.7**.

```
        YNORM1=0.0
        DO 80 I=1,N
           YNORM1=YNORM1+ABS(W(I))
 80 CONTINUE
```

Figure 9.6
Calculating the one-norm of vector y.

```
        ZNORM1=0.0
        DO 130 I=1,N
           ZNORM1=ZNORM1+ABS(W(I))
130 CONTINUE
```

Figure 9.7
Calculating the one-norm of vector z.

The pieces of codes given in the above figures can be connected in an obvious way and the number κ^*=ACOND1=ANORM1*ZNORM1/YNORM1 will often be a good estimate of the magitude of the condition number of matrix A even if an arbitrary vector   e   is chosen in the beginning of the process (see, for example, [43], p. 371, or [173], pp. 7-8). The probability of achieving a good estimate κ^* is enhanced when   e   is chosen by the algorithms described in **Section 9.2**. The implementation of these algorithms in a sparse condition number estimator is discussed in the next section.

9.4. Algorithms for choosing a starting vector: implementation and comparison

The implementation, in **Y12M**, of the algorithms for choosing a starting vector   e   from   [43,173] is discussed below. This is not an easy task. The difficulties arise because   the components of   e   must be determined dynamically during the solution of system $U^T w = e$ (and, therefore, the code given in **Fig. 9.2** has to be modified).

Let   **X**   be any subroutine for calculating an estimate of the condition number of a **dense** matrix. Then the following strategy seems to be very straightforward in the efforts to obtain a version of subroutine   **X**   for sparse matrices.

Strategy 9.1. Use the following two rules in the transition from the dense matrix subroutine   **X**   to a sparse matrix subroutine for calculating an estimate of the condition number of a matrix:

(i) Replace the loops involving the two-dimensional array in which matrix A is stored by loops involving one dimensional arrays. The loops corresponding to the dynamic storage scheme used in package   **Y12M**   are given   in   **Fig. 9.1 - Fig. 9.5**.

(ii) Leave the other parts of subroutine X unchanged. For example, loops like these given in **Fig. 9.6** and **Fig. 9.7** will not be changed according to this rule. The same is true for the loops carried out to scale vectors in [55].   ∎

Strategy 9.1   has successfully been applied to obtain a version for sparse matrices from the algorithm used in **DECOMP** ([99]). This version is called **Algorithm 1** in this chapter.

Strategy 9.1   has also been used to obtain a version for sparse matrices from the algorithm used in subroutine   **SGECO**   ([55]). The sparse matrix version found from   **SGECO**   by using   **Strategy 9.1**   is referred to as **Algorithm 2***. This version is not efficient because the sums   (9.13)   require too many, $2(n-s+1)$,   arithmetic operations at stage   s, $s=2,3,\ldots,n$. For **dense** matrices the relations $t_i^+ \neq t_i$ and $t_i^- \neq t_i$ are normally satisfied for $\forall i \in \{s+1,s+2,\ldots,n\}$ at stage   s, $s=2,3,\ldots,n$. Therefore the sums   (9.13) **must**   be calculated at each stage when matrix   A   is dense. If   A   is sparse, however, then $t_i^+ = t_i$ and $t_i^- = t_i$ for many   i, $i \in \{s+1,s+2,\ldots,n\}$,   at any stage   s, $s=2,3,\ldots n$,   because many elements u_{ki} are equal to zero. Therefore the calculation of t_i^+ and t_i^- by (9.12)   is not justified if A   is sparse. A variable   SUM   is introduced.   SUM=0.0   is set before the beginning of the solution of $U^T w = e$.   At any stage   s, where $s=2,3,\ldots,n$, if $t_i^+ \neq t_i$,   then   SUM is updated by   SUM:=SUM-ABS(t_i)+ABS(t_i^+).   The question whether $t_i^+ \neq t_i$ or $t_i^+ = t_i$ is never asked, because by the use of the pointers stored in the first three columns of array   HA   (see **Chapter 2** or the previous section) only indices for which $t_i^+ \neq t_i$ are considered. The operations involving t_i^- are considered in a similar way. The version so found is called **Algorithm 2**.

The modifications advocated in [135] can easily be inserted in **Algorithm 2** by following the instructions given in [135] on p. 385. This has been done and the algorithm so found is called **Algorithm 3**. An attempt to reduce the number of scalings (performed in **LINPACK** and, therefore, also in **Algorithm 2** and **Algorithm 2***), is carried out in **Algorithm 3**.

The subrourines in which the above algorithms are applied were run on a wide range of test examples. The number of test matrices used was several thousands. The order of the matrices tested varies from 32 to 10000. Both matrices that arise in practical problems and matrices that are produced by subroutines that generate sparse test matrices were used in the experiments. Matrices arising in fluid dynamics ([59]), in nuclear spectroscopic theory ([211-214]), in thermodynamics ([325]), in chemistry ([32]) and the Harwell set ([76]) were used in the runs, some of which are described below.

All experiments in this chapter were run on a **UNIVAC 1100/82** computer at **RECKU** (Regional Computing Centre at the University of Copenhagen). All computing times are given in seconds. The subroutine **Y12MFE** from package **Y12M** was used. This subroutine carries out the factorization of matrix A and solves the system of linear algebraic equations by iterative refinement (see **Chapter 5** or [331]). In all experiments the right-hand side of the system was generated so that $x_i=1$ for $i=1,2,...,n$.

The results in **Table 9.1** show that **Algorithm 2*** is not suitable for sparse matrices. Therefore this algorithm is not used in the other comparisons. **Algorithm 2*** performs worse than the other algorithms both when the matrix is large (see the results for the first two matrices in **Table 9.1**) and when the sparsity is perfectly preserved (see the results for last two matrices in **Table 9.1**).

Source	Order	NZ	COUNT	Y12MFE	Alg 1	Alg 3	Alg 2	Alg 2*
[59]	1000	6400	16769	10.45(21)	0.23(0)	0.34(5)	0.34(6)	10.16(6)
[76]	1176	9864	18552	17.81(3)	0.23(0)	0.23(4)	0.34(4)	11.16(4)
[76]	822	4028	4865	2.19(3)	0.09(0)	0.13(8)	0.18(35)	5.55(35)
[330]	300	710	710	0.28(4)	0.02(0)	0.03(7)	0.04(11)	0.77(11)

Table 9.1
Results obtained by four algorithms for computing an estimate of the condition number of a matrix. Algorithm 1, Algorithm 3, Algorithm 2 and Algorithm 2* are abbreviated to Alg 1, Alg 3, Alg 2 and Alg 2* respectively. The numbers in brackets in the last four columns are the the numbers of scalings. The computing times needed to solve the system by using the IR option of Y12M are given under Y12MFE (the numbers of iterations are given in brackets). The matrix from [330] is A=F2(300,300,100,2,100.0).

The computing times for **Algorithm 2** are normally larger than the computing times both for **Algorithm 3** and for **Algorithm 1**, but the difference, as a rule, is not very large. Moreover, the computing times for

all three algorithms are normally much smaller than the computing time needed
to solve the system Ax=b. Nevertheless, it must be emphasized that there
exist matrices for which **Algorithm 2** is rather inefficient. Such matrices
are given in [135]. An experiment with matrices of class E(n,c) was carried
out with n=100(100)5000 and c=2(14) for each n. In each of these runs
Algorithm 2 performed about n scalings and, therefore, is very expensive.
It is even more expensive than the solution of system Ax=b with **IR**.
However, it must be mentioned that the matrices of class E(n,c) are positive
definite matrices with narrow bandwidths for small values of c and this fact
has been exploited in these runs by using a special option in package **Y12M**
in which the pivotal search is suppressed. Some of the results obtained in
this experiment are given in **Table 9.2**.

Order	c	NZ	COUNT	Y12MFE	Algorithm 1	Algorithm 3	Algorithm 2
	2	4994	4994	0.95(5)	0.10(0)	0.14(5)	2.18(998)
1000	3	4992	5989	1.32(5)	0.11(0)	0.16(5)	2.20(995)
	4	4990	7978	1.71(5)	0.13(0)	0.19(5)	2.22(992)
	2	9994	9994	2.13(5)	0.19(0)	0.27(5)	8.38(1998)
2000	3	9992	11989	2.64(5)	0.22(0)	0.33(5)	8.43(1995)
	4	9990	15978	3.41(5)	0.26(0)	0.38(5)	8.47(1992)
	2	14994	14994	3.20(5)	0.29(0)	0.41(5)	18.61(2998)
3000	3	14992	17989	4.29(5)	0.34(0)	0.49(5)	18.67(2995)
	4	14990	23978	5.49(5)	0.40(0)	0.57(5)	18.74(2992)

Table 9.2
Comparison of computing times found in runs with matrices of class E(n,c).
Under Y12MFE are given the computing times needed to solve the systems
by the IR option of Y12M (the iterations are given in brackets).
In the last three columns the numbers of scalings are given in brackets.

Many runs were carried out to compare the accuracy of the estimates
calculated by the three algorithms for matrices:

   (i) **with different orders,**
  (ii) **with different sparsity patterns,**
 (iii) **with different numbers of non-zeros,**
  (iv) **with non-zeros of different magnitude.**

Matrices of class $F2(m,n,c,r,\alpha)$ were applied in one of the experiments
with m=n, n=200(50)300, c=20(20)n-20, r=2(7)30, and $\alpha=10^k$ (k=0(1)9).
The **LINPACK** subroutine **SGECO** was also applied in this experiment. "Exact"
values of the condition numbers were calculated by using the **LINPACK**
subroutine **DGEDI** (where double precision calculations are used). Some
results from this experiment are given in **Table 9.3 - Table 9.5**. The results
of this experiment, in which **1750 matrices were treated with several
different algorithms, are summarized in the following remarks.**

Remark 9.1. The estimates obtained by **Algorithm 2** and **Algorithm 3**
were the same for all 1750 matrices used in this experiment. Therefore only
results calculated by **Algorithm 2** are given in the tables (together with the
results obtained by **Algorithm 1**, by the **LINPACK** subroutine **SGECO** and

with the "exact" condition numbers). The number of scalings in **Algorithm 2** was much smaller than n for these matrices and, therefore the computing times for **Algorithm 2** and **Algorithm 3** were of the same order (being much smaller than the computing times needed to obtain the factorization). ∎

α	Algorithm 1	Algorithm 2	LINPACK	$\kappa(A)$
1.0E0	7.50E01 (0.24)	7.50E01 (0.24)	7.50E01 (0.24)	3.04E02
1.0E1	3.72E02 (0.44)	3.72E02 (0.44)	3.72E02 (0.44)	8.46E02
1.0E2	3.88E04 (0.61)	3.84E04 (0.61)	3.84E04 (0.61)	6.40E04
1.0E3	4.12E06 (0.66)	4.12E02 (0.66)	4.12E06 (0.66)	6.19E06
1.0E4	4.15E08 (0.67)	4.15E08 (0.67)	4.15E08 (0.67)	6.17E08
1.0E5	4.16E10 (0.67)	4.16E10 (0.67)	4.16E10 (0.67)	6.17E10
1.0E6	4.16E12 (0.67)	4.16E12 (0.67)	4.16E12 (0.67)	6.17E12
1.0E7	4.16E14 (0.67)	4.16E14 (0.67)	4.16E14 (0.67)	6.17E14
1.0E8	4.16E16 (0.67)	4.16E16 (0.67)	4.16E16 (0.67)	6.17E16
1.0E9	4.16E18 (0.67)	4.16E18 (0.67)	4.16E18 (0.67)	6.17E18

Table 9.3
Estimates of the condition number of matrices F2(300,300,100,2,α).
The subroutine SGECO is used to get the LINPACK estimates.
"Exact" condition numbers $\kappa(A)$ are obtained
by using the subroutine DGEDI from LINPACK.
The ratios (estimated condition number)/("exact" condition number)
are given in brackets.

α	Algorithm 1	Algorithm 2	LINPACK	$\kappa(A)$
1.0E0	2.05E03 (0.30)	2.11E03 (0.31)	2.38E03 (0.35)	6.87E03
1.0E1	2.02E03 (0.29)	2.12E03 (0.31)	2.26E03 (0.33)	6.87E03
1.0E2	2.09E03 (0.30)	2.15E03 (0.31)	3.75E03 (0.54)	6.91E03
1.0E3	4.41E04 (0.45)	5.80E04 (0.60)	5.91E04 (0.61)	9.70E04
1.0E4	3.62E06 (0.64)	3.40E06 (0.60)	3.65E06 (0.64)	5.68E06
1.0E5	3.37E08 (0.63)	3.44E08 (0.65)	3.44E08 (0.65)	5.32E08
1.0E6	3.71E10 (0.64)	3.67E10 (0.63)	3.74E10 (0.63)	5.81E10
1.0E7	8.74E12 (0.61)	8.74E12 (0.61)	8.78E12 (0.61)	1.44E13
1.0E8	2.19E14 (0.64)	2.21E14 (0.65)	2.14E14 (0.63)	3.40E14
1.0E9	3.98E16 (0.70)	3.97E16 (0.70)	3.98E16 (0.70)	5.66E16

Table 9.4
Estimates of the condition number of matrices F2(300,300,100,23,α).
The subroutine SGECO is used to get the LINPACK estimates.
"Exact" condition numbers $\kappa(A)$ are obtained
by using the subroutine DGEDI from LINPACK.
The ratios (estimated condition number)/("exact" condition number)
are given in brackets.

Remark 9.2. No relationship between the order n of the matrix and the accuracy of its condition number estimate was observed. Neither was there a relationship between the sparsity pattern (determined by parameter c) of the matrix and the accuracy of its condition number estimate. Results obtained with n=300 and c=100 are given in **Table 9.3 - Table 9.5.** ∎

Remark 9.3. If $r=2$, then the estimates of the condition numbers found by the different algorithms are practically the same; the results for $n=300$ are given in **Table 9.3**. This is not surprising and can be explained as follows. The number of non-zeros in nearly all rows and columns of a matrix of class $F2(m,n,c,r,\alpha)$ is equal to 2 when $m=n$ and $r=2$. Moreover, the sparsity is perfectly preserved for such a matrix when the GE is performed; COUNT=NZ=710 for $n=300$. Therefore **Algorithm** 1 and **Algorithm** 2 should perform in quite similar ways; see (9.12) and (9.13) and use the fact that for nearly all values of k only one t_i is different from t_i^+ and t_i^-. An element $a_{ij}^{(k)}$ can be chosen as a pivot in **Y12M** if

$$(9.15) \quad |a_{ij}^{(k)}| \geq \max_{k \leq j \leq n} (|a_{ij}^{(k)}|)/u,$$

where u is the stability factor (see **Chapter** 4). The runs discussed in this section were carried out with $u=4$. Since many rows contain only 2 elements and since the ratio of the absolute values of these two elements is normally larger than 4, the sparse pivotal strategy with $u=4$ (used in **Y12M**) and the partial pivoting (used in **LINPACK**) perform in a quite similar way for $r=2$. The above arguments are not true for $r>2$ and the results obtained by the three algorithms were different for all $r>2$ used in this experiment. For $r=23$ this is illustrated in **Table 9.4**. It should be mentioned that the differences are usually not very large; see **Table 9.4** and **Table 9.5**. ■

Remark 9.4. Let R be the ratio of the estimated condition number to the "exact" condition number. For all three algorithms this ratio increases when α is increased from 10^0 to 10^4. For $\alpha > 10^4$ R varies normally in the interval $[0.6,0.7]$ This means that the behaviour shown on **Table 9.4** is typical for the matrices used in this experiment. ■

r	Algorithm 1	Algorithm 2	LINPACK
2	[0.24,0.67]	[0.24,0.67]	[0.24,0.67]
9	[0.29,0.68]	[0.14,0.66]	[0.26,0.63]
16	[0.15,0.67]	[0.15,0.67]	[0.30,0.63]
23	[0.29,0.70]	[0.31,0.70]	[0.33,0.70]
30	[0.29,0.71]	[0.29,0.71]	[0.33,0.71]

Table 9.5
The intervals for the ratios R for the matrices
$A=F2(300,300,100,r,\alpha)$ with $\alpha=10^k$ for each r.
R=(estimated condition number)/("exact" condition number).
The "exact" condition numbers are obtained by using
the subroutine DGEDI from LINPACK.

Remark 9.5. In all runs $R \geq 0.1$ was observed for all algorithms. For the case $n=300$ and $c=100$ this is demonstrated in **Table 9.5**. This shows that the estimates of the condition number are acceptable according to the criterion introduced in [43], p.375. ■

Remark 9.6. The parameter γ used in **Algorithm** 3 ([135]) to determine the proper scaling factors was set equal to 0.001 in this experiment. This is the value recommended in [135]. ■

9.5. Sparse pivotal strategies versus accuracy of the condition number estimates

The pivotal element that is to be used at step s (s=1,2,...,n-1) has to be chosen so that (see also **Chapter 4**):

(i) the sparsity of matrix A is preserved as well as possible,

(ii) the stability is preserved as well as possible.

These two requirements work in opposite directions and cannot be satisfied simultaneously. Therefore a compromise is necessary. Any compromise leads to a relaxation of the stability requirements during the factorization (compared with the stability requirements for dense matrices; see **Section 4.1**). The influence of the relaxed stability requirements in the sparse pivotal strategies on the accuracy of the estimates of the condition number will be discussed in this section. It is assumed that pivotal strategies of Markowitz type are used, but many of the conclusions are also valid when other sparse pivotal strategies (as, for example, those discussed in **Section 4.3** and **Section 4.5**) are applied.

Assume that an **IGMS** (improved generalized Markowitz strategy) is used. Let the stability requirements be imposed by (9.15). Then among all elements in several "best" rows $p(s)$ that have the optimal Markowitz cost (4.20), the largest (in absolute value) will be chosen as a pivot. Denote by M_s and M_s^* the optimal and the minimal Markowitz cost. It is clear that if $M_s^* \leq M_s$. If $M_s > M_s^*$, then either there is no non-zero element with a minimal Markowitz cost in the $p(s)$ "best" rows (i.e. $p(s)$ is too small) or there are non-zero elements that have minimal Markowitz cost, but they do not satisfy the stability requirement (9.15) (i.e. the stability factor is too small). This shows that one should expect to preserve sparsity better either by increasing the number of "best" rows $p(s)$ or by increasing the stability factor u. However, it is not very efficient to carry out the GE with a large $p(s)$, because this leads to a considerable increase of the calculations in the pivotal search. Moreover, the elements with minimal Markowitz cost are normally in rows with a small number of non-zeros in their active parts, but these cannot be found because of the stability restriction (9.15). These arguments as well as many numerical tests indicate that as a rule it is not justified to use $p(s) > 4$. On the other hand, sparsity can sometimes be preserved better by choosing a larger stability factor u (when the classical manner of exploiting the sparsity is used; see **Section 5.6**). However, the use of a large u may cause large errors in the factors L and U. Indeed, it is easy to obtained the following bounds (see, for example, [197,295]):

$$(9.16) \quad |a_{ij}^{(s+1)}| \leq (1+u) \max_{\substack{s \leq i \leq n \\ s \leq j \leq n}} (|a_{ij}^{(s)}|), \qquad s=1,2,\ldots,n-1,$$

$$(9.17) \quad b_n \leq (1+u)^{n-1} b_1, \qquad (\; b_m = \max_{\substack{1 \leq s \leq m \\ s \leq i \leq n \\ s \leq j \leq n}} (|a_{ij}^{(s)}|), \qquad m=1,2,\ldots,n \;),$$

(9.18) $|e_{ij}| \leq 3.01 b_n \epsilon n$ ($e_{ij} \in E = LU-PAQ$),

where ϵ is the computer precision. The above bounds could be improved by replacing n with the number of non-zeros that participate in the calculations; see [107]. However, even the bounds from [107] show clearly that the norm of the perturbation matrix $E=LU-PAQ$ may be very large (the permutation matrices must be used in this section). The bounds (9.16)-(9.18), the last one with 2.01 instead of 3.01 ([271]) are also valid when partial pivoting is used for dense matrices. This shows that $\|E\|$ can be large also when A is dense. Moreover, examples, where $b_n=2^{n-1}b_1$, can be constructed ([271], see also **Section 4.1**). However, it is well known that in practice $\|E\|$ is much smaller than $\|A\|$ and the computation of an estimate κ^* of the condition number of matrix A (PAQ+E being actually used in a part of the computations instead of PAQ) is based on the assumption that this is true; see §5 in [43]. The conclusion is: **the above bounds are too pessimistic even if A is dense and partial pivoting is used. The experiments carried out indicate that this is also true for sparse matrices when the stability factor u varies in a quite large range.**

Stability factor u	Algorithm 1	Algorithm 2
4.0	[0.31,0.67]	[0.12,0.67]
16.0	[0.28,0.66]	[0.30,0.66]
64.0	[0.30,0.68]	[0.15,0.68]
256.0	[0.31,0.66]	[0.31,0.67]
1024.0	[0.30,0.67]	[0.30,0.68]

Table 9.6
Intervals for R=(estimated condition number)/("exact"condition number)
for matrices A=F2(200,200,100,8,α) with $\alpha=10^k$, k=0(1)9.

The experiment from **Section 9.3**, where u=4 was used, was repeated with u=1024. For all matrices (1750 matrices are used in this experiment) $0.1 < R < 1$ was found, where R=(estimated condition number)/("exact"-condition number) and the "exact" condition number is obtained by computing A^{-1} in double precision by using subroutine **DGEDI** from **LINPACK**.

Another experiment was performed by using several different values of u. The matrices were A=F2(200,200,100,8,α) with $\alpha=10^k$, k=0(1)9 (i.e. 10 systems are solved for each value of u). The values of the stability factor u were $u=2^k$, k=2(2)10. The results from this experiment are shown in **Table 9.6**. The **LINPACK** subroutine **SGECO** was also used in this experiment. The values of R were in the interval [0.27,0.67] when the subroutine **SGECO** was applied to solve the systems with the same matrices as those used in **Table 9.6**.

An explanation of the fact that **the condition number estimators give good results even if the stability factor is rather large** can be given as follows. Let u^* be such that $u^*|a_{ij}^{(s+1)}| = max(|a_{ij}^{(s+1)}|)$, s=1,2,...,n-1, holds for the element that will be chosen as pivotal at stage s. The relationship $u^* \ll u$ occurs often in the GE when u is large. Thus, the fact that a large stability factor u is specified may have no influence on the choice of pivots at many stages of the GE.

The possibility of having $u^* \ll u$ is enhanced by the fact that an **IGMS** is used. The largest (in absolute value) element is chosen as a pivot if there are several elements with an optimal Markowitz cost when an **IGMS** is used, while any of these elements may be chosen as a pivot when a **GMS** is applied, see **Section 4.4**. In the latter case the accuracy of L and U could be very poor for some matrices even for small values of the stability factor u. If this is so, then the solution of Ax=b may be quite wrong (this is illustrated in **Chapter 4** and **Chapter 5**). It is surprising that in the experiments carried out with matrices of class $E(n,c)$, $n\in[1000,10000]$ and $c\in[4,n-100]$, the calculated condition numbers are not very bad even when the solution is quite wrong. It is not clear why this is so, and further investigation of this phenomenon is needed. At the same time, it is important to emphasize here that the statement from **Section 9.1** ("one should expect a solution computed in t-digit arithmetic to be accurate to about t-p digits") is no longer true in this situation. Of course, this is not a surprise, because the first condition needed for this statement is not true (to observe this one can monitor the growth factors). The problem is that many people often forget the conditions under which the statement at the end of **Section 9.1** is made and believe that the condition number alone can give all answers about the sensitivity of the solution to rounding errors. The experiments made by several thousands of matrices show that one should, in some situations at least, be very careful and checking also other factors (the growth factor b_n/b_1 and the size of the minimal pivot) may be very useful. Some examples are given in **Table 9.7**.

c	GMS used	IGMS used	LINPACK (SPBCO)	GROW
200	5.25E+2 (8.34E+0)	2.89E+2 (8.77E-5)	2.89E+2 (6.21E-4)	7.1E+ 7
300	1.89E+3 (6.31E+2)	1.36E+2 (3.15E-5)	1.36E+2 (2.69E-4)	2.2E+10
400	8.01E+1 (9.18E-1)	8.07E+1 (2.19E-5)	8.06E+1 (8.69E-4)	6.0E+ 7

Table 9.7
Estimates of the condition numbers of matrices A=E(3600,c) obtained by two sparse pivotal strategies (a GMS and an IGMS) and LINPACK. The largest errors in the solution vectors found are given in brackets (all components of the exact solution vectors are equal to one). GROW is the growth factor b_n/b_1 when the GMS is used (the value of GROW is always equal to one for this class of matrices when an IGMS is in use).

The results in **Table 9.7** show, once again, the advantages of using an **IGMS**. At the same time, it must also be emphasized that in general if the stability factor is very large, then both the accuracy of the solution of Ax=b and the accuracy of the estimate of the condition number of matrix A may be very poor even if an **IGMS** is used. If this is so, then the growth factor b_n/b_1 and the size of the minimal pivot will normally give a warning signal. An experiment with very large u, $u=2^{14}$, has been carried out; the results are given in **Table 9.8**. The values of the growth factor are given in **Table 9.8**. The experiment was carried out on UNIVAC 1100/82 with $\epsilon \approx$ 1.49E-8. Therefore $\epsilon(b_n/b_1) \approx 10^3$ for the matrices in **Table 9.8**. In all experiments that were carried out the results were satisfactory when $\epsilon(b_n/b_1) < c$, where c is of order $O(1)$. More experiments are needed in order to decide if such a criterion can be applied in the judgement of the accuracy of the results obtained by the sparse condition number estimators.

The experiments in this section show that good estimates of the condition numbers of sparse matrices can be obtained when the stability factor   u varies in a quite large interval. If u is very large, then it is useful to check the growth factor and/or the size of the minimal pivot. If this is not done, then the estimate of the condition number calculated may sometimes give a wrong impression of the sensitivity of the system to rounding errors (this is illustrated in **Table 9.7**). The stability factor is normally varied in the interval [4,16] and for such values of u   the main pivotal strategy in **Y12M** (which is an **IGMS**) gave good results in all experiments. Therefore the conclusion is that if   u ∈ [4,16],   then the sparse pivotal strategy in **Y12M** ensures as good results as the partial pivoting in the   **LINPACK** subroutines (to demonstrate this the **LINPACK** subroutines were used in all experiments).

α	Algorithm 1	Algorithm 2	LINPACK	$\kappa(A)$	GROW
1.0E0	3.65E5	3.55E6	2.69E3	8.74E3	2.35E11
1.0E1	9.75E4	8.33E4	2.55E3	8.74E3	1.73E10
1.0E2	3.12E6	6.84E5	3.49E3	8.74E3	5.11E11
1.0E3	8.01E5	2.63E5	4.68E4	7.65E4	1.55E10
1.0E4	8.01E7	2.63E7	4.68E6	7.65E6	3.75E10

Table 9.8

Estimates of the condition numbers of matrices A=F2(300,300,100,30,α) obtained by two algorithms for sparse matrices with u=2^{14} and with the LINPACK subroutine SGECO.
The "exact" condition numbers $\kappa(A)$ are obtained by using the LINPACK subroutine DGEDI (double precision computations).
GROW is the growth factor b_n/b_1 (an IGMS is used in these runs).

9.6. Drop-tolerance versus accuracy of the condition number estimates

In all experiments in the previous sections a very small value of the drop-tolerance T, T=10^{-25}, was used. In many cases this is equivalent to the use of T=0.0 (and, thus, to the use of the classical manner of exploiting sparsity). The computer-oriented manner of exploiting sparsity will be studied in this section in connection with the accuracy of the computed condition number estimates. It will be assumed that the computer-oriented manner is applied together with iterative refinement (**IR**), but all conclusions are also valid when other iterative methods are used (as, for example, the conjugate gradient-type methods, which will be studied in **Chapter 11**).

The matrices run in the previous sections, matrices A=F2(300,300,100,r,α) with $\alpha=10^k$, k=0(1)9, and with r=2(7)30, used in **Section 9.4** with T=10^{-25} were also run with some large values of T, T=10^{-k}, k=3(-1)1, and with u=4.

If $\alpha \le 10^4$, then the estimates of the condition number were worse than those calculated with T=10^{-25} but nevertheless not very bad. The intervals in which the ratios R, R=(estimated condition number)/("exact" condition number), vary are given in **Table 9.9** for **Algorithm 1** and **Algorith 2**; the estimates calculated with **Algorithm 3** were again the same as those calculated with **Algorithm 2**.

If $\alpha > 10^4$, then the estimates are often very bad. Some results obtained with r=9 and T=10^{-1} are given in **Table 9.10**. Normally, R > 1 was found when $\alpha > 10^4$. However, R < 0.1 has also been observed sometimes; see the second example in **Table 9.10**.

T	r	Algorithm 1	Algorithm 2
1.0E-3	2	[0.24,0.67]	[0.24,0.67]
	9	[0.13,0.62]	[0.06,0.65]
	16	[0.12,0.61]	[0.11,0.62]
	23	[0.06,0.59]	[0.04,0.59]
	30	[0.39,0.60]	[0.05,0.59]
1.0E-2	2	[0.24,0.67]	[0.24,0.67]
	9	[0.13,0.60]	[0.13,0.59]
	16	[0.17,0.60]	[0.17,0.60]
	23	[0.34,0.64]	[0.16,0.64]
	30	[0.33,0.64]	[0.39,0.65]
1.0E-1	2	[0.24,0.67]	[0.24,0.67]
	9	[0.09,0.72]	[0.09,0.68]
	16	[0.34,0.60]	[0.33,0.69]
	23	[0.07,0.59]	[0.07,0.67]
	30	[0.04,0.64]	[0.08,0.64]

Table 9.9
Interavals for R=(estimated condition number)/("exact" condition number)
when matrices A=F2(300,300,100,r,α) are run with α=10^k,
where k=0(1)4 for every value of r.

α	Algorithm 1	Algorithm 2	LINPACK	κ(A)
1.0E5	6.45E09	6.45E09	2.20E09	3.48E09
1.0E6	3.24E10	3.20E10	2.20E11	3.48E11
1.0E7	5.23E19	5.23E19	2.20E13	3.28E13
1.0E8	4.93E20	4.93E20	2.20E15	3.48E15
1.0E9	1.15E22	1.15E22	2.20E17	3.48E17

Table 9.10
Estimates of the condition numbers of matrices A=F2(300,300,100,9,α)
obtained with two algorithms for sparse matrices with T=0.1
and with the LINPACK subroutine SGECO.
The "exact" condition numbers κ(A) are found by using DGEDI.

The conclusion from this experiment and many others (all matrices run in the previous sections were also run with large drop-tolerances) is that the sparse condition number estimators are unreliable when the drop-tolerance T is large. More precisely, the accuracy of the estimates found may be poor for not very ill-conditioned matrices and the estimates are usually very bad when the matrices are very ill-conditioned. However, the use of large values of T has to be combined with the use of some iterative method (the simple IR is used here). If this is done, then reliable information about the result of the solution process is as a rule available from the iterative method ([99],

p. 49). If no iterative method is used, then T must be very small (normally T=0 is to be used) and the sparse condition number estimators work very satisfactorily in this situation (assuming also that the stability factor u is not very large); see **Section 9.4.**

9.7. Concluding remarks and references concerning the condition number estimators for sparse matrices

Three sparse condition number estimators have been developed and attached to package **Y12M.** This makes **Y12M** the only package for the treatment of large and sparse general matrices with which the user has a possibility to obtain an estimate of the condition number of the matrix handled. The condition number estimators have been tested with a large set of matrices (several thousand matrices). The work done in this direction was first reported in [338]. The results of the tests described in this chapter (and in [338]) can be summarized as follows.

Remark 9.7. **Algorithm 1** is the most efficient algorithm with regard to computing time. **Algorithm 2** may be very inefficient for some matrices, because too many scalings are carried out. However, this happens very seldom. It was very difficult to find matrices different from those given in [135] for which the number of scalings was large. **Algorithm 3** is more expensive than **Algorithm 1** but in general cheaper than **Algorithm 2.** Moreover, if the number of scalings was large for **Algorithm 2,** then it was always reduced considerably by **Algorithm 3.** ∎

Remark 9.8. **Algorithm 2** is the safest. A very careful attempt to avoid overflows and divisions by zero is carried out by scaling appropriate vectors when this algorithm is applied. **Algorithm 1** is not so robust. The computations in this algorithm are carried out without any scaling. **Algorithm 3** is in the middle: it is not so robust as **Algorithm 2,** but it is more robust than **Algorithm 1.** ∎

Remark 9.9. The fact that the accuracy requirements are relaxed in the sparse pivotal strategies (compared with the accuracy requirement in the commonly used partial pivoting for dense matrices) seems to be not very important when both the stability factor u and the drop-tolerance T are sufficiently small. The results obtained in runs of several thousand matrices with u=4 and T=10^{-25} were acceptable in the sense that $0.1 < R < 1$ was found for all matrices tested and for all three sparse condition number estimators; R = (estimated condition number)/("exact" condition number). ∎

Remark 9.10. It is not easy to answer the question: what algorithm is the best? Examples where **Algorithm 1** fails can be constructed ([43]). However, examples in which **Algorithm 2** fails to calculate a good estimate can also be constructed ([42]). This means that no algorithm, the statement being also true for **Algorithm 3,** will guarantee a good estimate of the condition number for every matrix. On the other hand, experience, including here the results described in the previous sections and in [338], indicates that failures to obtain a good condition number estimate occur very seldom (this being true for all three algorithms; especially when the matrix is large). The situation is probably similar to that for partial pivoting: one knows that examples where the partial pivoting fails can be constructed, but experience indicates that this situation will happen very seldom in practice.

Therefore partial pivoting is the only pivotal strategy used in subroutines in scientific libraries. More experiments are needed to confirm or reject such a similarity between condition number estimators and partial pivoting; i.e. to answer the following question (in the case where large and sparse matrices are treated): do failures of the condition number estimators occur as seldom as failures of partial pivoting?   ■

   Remark 9.11.   It is also difficult to answer the question: which algorithm should be used in a package for sparse matrices?   Algorithm 1 should be chosen if one wants to keep the cheapest algorithm.   Algorithm 2 should be chosen if the reliability is the most desired property.   Algorithm 3 is a rather good compromise between the two. Thus each algorithm may be the best one in special circumstances. Therefore all three algorithms are implemented in package Y12M. ■

   Remark 9.12.   Some interesting results concerning condition number estimates have recently been reported by Higham [147-148] and, what is even more important, Higham also published a code, [148], that can be applied in connection with condition number estimators for real and sparse matrices. ■

   Remark 9.13.   The concept of an incremental condition number was introduced and studied by Bischoff [18] in connection with orthogonal decompositions based on the Householder reflections; see also [19,20,220]. Similar ideas could also be applied in connection with the LU factors of a general matrix. Moreover, when a positive drop tolerance is used, it is probably possible to apply the incremental condition number estimates adaptively in order to decide whether the decomposition, QR or LU, should be aborted or continued (and, if the latter case takes place, how precisely the decompostion process should be continued). Roughly speaking, when the incremental condition number estimation becomes greater than a prescribed level, then the computations could be stopped or some other action could be taken (for example, the drop-tolerance might be reduced). Of course, such an adaptive application of the incremental condition number would be of interest only if the computation of the incremental condition number estimation is not expensive. ■

   Remark 9.14.   The concept of "expected conditioning" of a system of linear algebraic equations is introduced and discussed in Fletcher [98]. The expected conditioning can be estimated and used in the evaluation of the sensitivity of the solution of the system of linear algebraic equations to rounding errors (or, more generally, to arbitrary perturbations). It would be worthwhile to try to implement the ideas described in [98] for large and sparse systems of linear algebraic equations. ■

CHAPTER 10

PARALLEL DIRECT SOLVERS

There are three main difficulties that have to be overcome when a parallel code for the treatment of **general** sparse matrices is developed:

(i) the matrix structure is irregular,
(ii) the loops involve short vectors,
(iii) the nested loops are not well-balanced (for example, assume that an outer loop scans the target rows and consider the inner loop; the work needed to modify a target row will in general vary from one row to another because the target rows contain different numbers of non-zero elements).

None of these difficulties is present in the case of dense matrix technique. Therefore, it is not surprising that parallel codes for dense matrices have been developed and extensively studied during the last decade, while the problem of developing a robust, efficient and well-documented package of subroutines for general sparse matrices is still open.

The main ideas applied in the development of three sparse codes for parallel computers with shared memory and with a few tightly connected processors will be described. The first of them, **Y12M1**, is based on the traditional use of Gaussian elimination (this code is closely related to the original **Y12M** that has been studied in **Chapter 4 - Chapter 6**). The second one, **Y12M2**, is based on the simultaneous use of several pivots. The third code, **Y12M3**, is based on an a priori reordering of the matrix to an upper block-triangular form with rectangular diagonal blocks. In all three versions a switch to dense matrix technique can be made. Many numerical results, obtained by using matrices from the well-known **Harwell-Boeing set** [71,72], will be presented **(1)** to demonstrate the efficiency of the new codes in comparison with three sequential codes: **MA28**, **SPARSPAK-C** and **Y12M**, [63,122,123,331,341], and **(2)** to illustrate the fact that none of the new versions is the best one in all cases (and, thus, if many systems with similar matrices are to be solved, it is worthwhile to test all three codes in the beginning and then to choose the best one for the particular matrices that are to be handled).

The portability of parallel codes is an important, but still unsolved problem. In all three codes a separate subroutine is written for every part of the computation that can be performed in parallel. Thus, although the codes have been written for and tested on an **ALLIANT** computer, the adjustment for any parallel machine that allows concurrent calls of subroutines is easy. Moreover, since the parts of the computation that can be run in parallel are well separated, it will be easy to optimize the codes even on computers that do not allow concurrent calls of subroutines.

10.1. Difficulties in the development of parallel codes

Parallel and vector computers have been commonly used during the last decade and such computers will probably be even more used in the future. Many codes for the treatment of dense matrices on parallel and/or vector computers have been proposed (see, for example, [57,58,102,103]). Codes for sparse matrices with special features (such as bandedness) have also been reported in the literature (see, [3,4,60,263,309,310,317,328]). For **general** sparse matrices (i.e. sparse matrices that have neither some special property, such as symmetry or positive definiteness, nor a special structure, such as bandedness) only a few codes have been developed until now ([50,51,280], see also the references given in **Davis and Yew,** [51]). This should not be surprising. It is very difficult to develop efficient codes for general sparse matrices because of the lack of regularity of the structure and because short arrays only are involved in the computations. Some ideas, which can be used in sparse matrix codes for parallel and/or vector machines and which have been implemented in three experimental versions of package **Y12M,** will be discussed in this chapter. It will be assumed that a parallel computer with a shared memory and not too many processors is in use. It is assumed that each processor has a vector capability. The optimization is carried out on two levels: **(i)** try to distribute the computations to all processors available, and **(ii)** try to vectorize the work that has to be done by every processor. All experiments reported in this chapter were carried out on an **ALLIANT** computer with 8 processors.

It should be stressed that the main attention was concentrated on identifying the parallelism in the problem solved (a system of linear algebraic equations with a general sparse matrix). For each part of the computation that could be performed concurrently in a non-trivial way, a separate subroutine has been written and this subroutine is called in parallel. Of course, this procedure is not the most efficient one. However, if the code is written in this way, then it is clear how to optimize it on many other parallel computers. If the compiler of the computer used allows one to call subroutines concurrently, then the work is nearly trivial. Thus, **the main goal has been not only to obtain a concurrent code, but also to find and separate the parts of the work that could be done in parallel.**

Three main ideas will be discussed. First, an attempt to exploit parallelism in the case where Gaussian elimination is performed in the traditional manner will be described; the code based on this attempt is called **Y12M1.** Then the use of parallel pivots and the corresponding code, **Y12M2,** will be described. Finally, a code, **Y12M3,** based on an a priori partitioning of the matrix, will be presented. **In all three cases switching to dense matrix technique is worthwhile.** Therefore the discussion will be started with a short description of the idea of switching. Numerical results, obtained by using these three codes, will be given at the end of the chapter.

10.2. Switching to dense matrix technique

The idea of switching to dense matrix technique is not a new one. It has been used, with some mixed success, even on scalar computers. For such computers it has first been proposed probably by **Gustavson** in 1976 (see [65]), but it has not been used in many of the well-known codes for general sparse matrices (such as **MA28** [63], **SPARSPAK-C** [122,123], **YALEPACK** [82-83],

Y12M [331,341] and the code described in [124,125]). The main problem with this approach is that it is not very clear when to switch from sparse matrix technique to dense matrix technique.

The situation changes when vector machines are in use. In this case subroutines which achieve a very high speed of computation have been developed (see, for example, [57,58]). Let **MFLOPS** be an abbreviation for million floating point operations (additions or multiplications) per second. Then a good code for matrix decomposition will achieve about **240 MFLOPS** on an **AMDAHL VP1100** computer when it is run in a vector mode and when the matrix is sufficiently large (say $n > 512$). If the same code is run in a scalar mode, then the speed will be about **8 MFLOPS**. Thus, the speed up is about **30**. This is possible because one works with long vectors. If the matrix is sparse, however, the sparse vectors involved in the computations are normally not very long; even at the end of the factorization. Therefore, it is easy to construct a strategy by which one will nearly always win by switching to dense matrix technique on a vector computer: **the code will perform more operations but with a much greater speed.**

Switching to dense matrix technique is used in the three experimental parallel versions of package **Y12M** that will be discussed in the following sections. These versions are, as mentioned above, developed for parallel machines with shared memory (first and foremost for the **ALLIANT**), but switching to dense matrix technique is also efficient when a single vector processor, such as the **AMDAHL VP1100**, is applied. Several criteria are used to decide when to switch, but in general the switch is made when the active submatrix A_s contains more than 10% non-zero elements.

10.3. Parallel codes based on the traditional use of Gaussian elimination

Assume that Gaussian elimination (GE) is carried out in the traditional way; i.e. at each stage s, $s=1,2,\ldots,n-1$, one determines the pivot, makes interchanges, and then performes a rank-1 modification of the active part A_s of matrix A. If a sequential computer is in use, then it is natural to go sequentially through the appropriate elements of A_s:

(i) one modifies the element $a_{ij}^{(s+1)}$ when both $a_{is}^{(s)} \neq 0$ and $a_{sj}^{(s)} \neq 0$,

(ii) one inserts the fill-in if $a_{ij}^{(s)} = 0$,

(iii) one proceeds with the next element for which the elements in the pivotal row and column are non-zeros.

While this is a good strategy for a sequential machine, it performs rather badly on a vector and/or parallel computer, because the whole procedure has to be carried out sequentially.

A simple way by which the drawback of the above strategy can be avoided can be described as follows. Instead of performing the rank-1 modification by carrying out the computations as sketched above, one can first perform "symbolic" modifications to determine the pattern of the structure by reserving locations for all fill-ins that will appear during the actual computations, and then perform the actual computations in an efficient manner.

Symbolic modifications are also used [115,117], but there they are performed **globally** (for the whole GE process before its start). Here they are carried out **locally** at each stage of GE, and the main parts of the computations **at stage s** (s=1,2,...,n-1) of the GE process are:

(i)   pivotal search,
(ii) interchanges,
(iii) "symbolic" modifications in the column ordered list,
(iv) "symbolic" modifications in the row ordered list,
(v)   actual modifications in the active part of the matrix.

The **pivotal search** can easily be organized in a vector-concurrent mode on the **ALLIANT** if at each stage one searches for the "best" row (or rows). The rows with minimal number of non-zero elements in their active parts are called "best" rows. Assume that only one "best" row is to be found (many experiments indicate that in general there is no great gain when several rows are used, although particular matrices for which the gain is considerable can be found). A simple vectorizable code (on the **ALLIANT**) that finds the "best" row has been described in **Chapter 6**. It is necessary here to emphasize that the cost of such a pivotal search is $O(n^2)$. Another possibility is to order the "best" rows in the beginning and to update this structure after every stage of ths GE (see again **Chapter 6**). The cost of such a search (implemented in the original **Y12M**) is, roughly speaking, $O(npq)$, where p is the average number of target rows per stage, while q is defined as follows. Let r_{new} and r_{old} be the numbers of the non-zeros in the active part of row i after stage s and before stage s respectively. Set $q_{si}=|r_{new}-r_{old}|$. Then q is taken to be the average number over all q_{si}. The bound $O(npq)$ is correct only when the number of target rows is equal to p at all stages and if all q_{si} are equal to q. Thus, the conditions under which this bound is obtained are very restrictive. Nevertheless, it indicates (because both p and q will normally be much smaller than n) that on sequential machines updating the ordering of the "best" rows should be preferred (and in fact this is the choice made both in **MA28** and in **Y12M**). However, on vector and/or parallel machines the situation is not very clear. In the experiments the direct search at each stage was often more successfull. There are at least two reasons for this: **(1)** the code for the direct search is very simple, which indicates that the constant c_1 in the term $O(n^2)$ is less than the corresponding constant c_2 in the term $O(npq)$, and **(2)** the direct search can be performed in a vector-concurrent mode, while the updating procedure is sequential. Nevertheless, for very big matrices the updating procedure should probably be preferred. Therefore, in the parallel code, **Y12M1**, that is based on the ideas discussed in this section, both pivotal searches are implemented (in two different versions). ■

The **interchanges** can be carried out in parallel. Assume that the pivot chosen at stage s has indices i and j. The operations in the row ordered list can be organized as follows. Consider a target row k > s (a row that has a non-zero element in the pivotal column j is called a target row at stage s). This row should be scanned in order to: **(i)** find the element with a column number j, **(ii)** replace j with s, and **(iii)** put this element (and its column number) in the front of the active part of row k. This scan of a target row does not depend on the other target rows and, therefore, can be performed concurrently on a parallel computer. One also has to scan all rows that have a non-zero element in the active part of column s in order to: **(i)** find the element with column number s, and **(ii)** replace s with j. This operation can also be performed concurrently.

Similar interchanges are carried out in the column ordered list. Consider a target column m > s (a column that has a non-zero element in the pivotal row i is called a target column at stage s). This column should be scanned in order to: **(i)** find the element with a row number i, **(ii)** replace i with s, and **(iii)** put this row number in the front of the active part of column m. This operation does not depend on the other target columns and, therefore, can be performed concurrently. One also has to scan all columns that have a non-zero element in the active part of row s in order to: **(i)** find the element with row number s, and **(ii)** replace s with i. This operation can also be performed concurrently.

In this way, the actual interchanges (sometimes the term physical interchanges is used) **are** carried out in code **Y12M1**. This is in fact not necessary and there are codes where actual interchanges are not carried out. The gain, however, is not very great, because a large part of the work described above (to find the elements in the target rows and columns and to move them in the front of their active parts) has to be done anyway. ■

It is difficult to optimize the performance of local **"symbolic"** **modifications** by which one determines the sparsity pattern of the **column ordered list** after stage s (but before the actual performance of the rank-1 modifications in the active part A_s of the matrix). The main problem is connected with copying columns at the end of the column ordered list when fill-ins appear and when there are no free locations for them: **several processors may want to make a copy at the same time.** There are several ways to avoid this problem. The simplest is to use a buffer. If a copy is to be made by processor i (i=1,2,...,NPROC, where NPROC is the number of processors), then processor i makes the copy in its own part of the buffer. The disadvantage of this method is that at the end of stage s the modified columns have to be moved back to the column ordered list (i.e. some extra work has to be done, but it could be done in vector-concurrent mode). When a switch to dense matrix technique is performed, then the storage for the dense matrix, which is not needed during the sparse computations, can be used as a buffer. Thus, normally no extra storage for the buffer will be needed when a switch to dense matrix technique is made, but the method requires some extra storage for the buffer when no switch to dense matrix technique is to be made. The advantage of the method is the fact that it is portable: no special property of the machine in use is exploited (this piece of code is written as a separate subroutine and the subroutine is called concurrently through the target columns; thus, the only requirement is that the compiler used should allow the user to call a subroutine concurrently). Another possibility is to **(i)** lock the arrays of the column ordered list when a copy of a column is to be made, **(ii)** calculate the pointer where the first free location after the copy will be, and **(iii)** unlock the arrays of the column ordered list. Note that the structure is locked only until the new pointer is calculated (not in the whole time when the copy is made). The disadvantages of the first method (using a buffer) are avoided when the second method is used. However, the code is no longer portable (or at least not easily portable). Exeperiments with both methods have been carried out. There is no big difference in the performance. At present the first method is used in **Y12M1**. In a **CEDAR** version of **Y12M1** the second method is applied (see, for example [161] concerning the **CEDAR** computer). ■

The problems with the concurrent performance of **the local "symbolic" modifications in the row ordered list** are solved in a very similar way. If a buffer is to be used, then it is necessary to have a bigger buffer than that used when "symbolic" modifications in the column ordered list are carried out, because both the non-zero elements end their column number are to be stored in the buffer when a copy of a row is made (only the row numbers of the non-zero elements in a column are to be stored in the buffer when "symbolic" modifications in the column ordered list are made, and when a column is to be copied; see above). Also here the lock-unlock device sketched above can be used and it has been applied in a **CEDAR** version of **Y12M1**. ■

Once the "symbolic" modifications in the row ordered list are made, it is easy to perform **the actual modifications of the appropriate elements in the active part A_s of the matrix**. These are carried out in a nested loop (the outer loop being concurrent, the inner loop being vectorized). Thus, there is no problem with the actual rank-1 modifications; the difficult part of the job is to find out how the local "symbolic" modifications can be carried out. On machines like the **ALLIANT** it may be even more efficient to combine the last two parts of the computational work per stage. ■

The performance of **Y12M1** during each of the five steps could eaisily be evaluated, because separate subroutines are written for each step (the operations in the last two steps are carried out together on the **ALLIANT**). Profiling can be used to perform such an evaluation. This has been done for many matrices. Results for two matrices from the Harwell-Boeing set of sparse test-matrices, [71,72], are given in **Table 10.1**. The first matrix, **west2021**, is very sparse and remains so. It is of order 2021 and the code switches to dense matrix technique when A_s is of order 120. The second matrix, **hwatt_2**, is not very sparse and, moreover, many fill-ins are produced during the factorization. Its order is 1856 and the code switches to dense matrix technique when A_s is of order 596 (which shows that the number of non-zeros in the active part A_s of the matrix increases rather quickly to more than 10%). It is seen from **Table 10.1** that while the speed-up for the pivotal search is about 7 for both matrices, the other main parts of the sparse factorization are performed more efficiently (the speed-ups are greater) when the matrix, **hwatt_2**, that produces many fill-ins is decomposed.

Experiments with **Y12M1** (some of them will be described in **Section 10.6**) indicate that the results shown in **Table 10.1** are typical: the algorithms used work quite satisfactorily if the matrix is not very sparse or if the matrix is very sparse, but many fill-ins are created during the calculations. If the matrix is very sparse and stays very sparse during the whole factorization, then the algorithms do not work very well. The main task which is distributed to the processors is **the modification of the non-zero elements of a row (column)**. If the matrix is very sparse and stays very sparse, then there will not be sufficiently many tasks for all processors and some processors will be idle most of the time. This is illustrated in **Fig. 10.1**. If a_{11} has been chosen as a pivot, then there are only three target rows and, thus, only three processors will be busy during the first stage of GE.

Step	west2021		hwatt_2	
	1 processor	8 processors	1 processor	8 processors
Piv. search	1.02	0.15	0.86	0.14
Permut.	0.46	0.12	1.67	0.29
Symb. col.	0.89	0.42	2.63	0.71
Symb. row	1.05	0.45	2.87	0.82

Table 10.1
The computing times for the most time-consuming parts
of the computations during the sparse factorization.
The last two steps (symbolic modifications
in the row ordered list and actual computations)
are combined.

```
      1 2 3 4 5 6 7 8 9 0 1 2 3 4 5 6
  1 | ■ 0 ■ 0 0 ■ 0 0 0 ■ 0 0 0 ■ 0 0
  2 | 0 ■ 0 0 ■ ■ ■ 0 0 0 0 0 0 0 0 ■
  3 | 0 0 ■ 0 0 0 0 ■ 0 0 0 ■ ■ 0 0 0
  4 | 0 0 0 ■ 0 0 0 0 ■ 0 ■ 0 0 0 ■ 0
  5 | 0 0 0 0 ■ 0 0 ■ ■ ■ 0 0 0 0 0 ■
  6 | ■ ■ 0 0 ■ ■ ■ 0 0 0 0 0 0 ■ 0 ■
  7 | 0 ■ ■ 0 0 0 0 ■ 0 0 0 ■ ■ 0 0 0
  8 | 0 0 ■ ■ 0 0 0 0 ■ 0 ■ 0 0 ■ ■ 0
  9 | 0 0 0 ■ ■ 0 0 ■ ■ ■ 0 0 0 0 0 ■
 10 | ■ 0 ■ 0 0 ■ 0 0 0 ■ 0 0 0 ■ 0 0
 11 | 0 ■ 0 0 ■ ■ ■ 0 0 0 0 0 0 0 0 ■
 12 | 0 0 ■ 0 0 0 0 ■ 0 0 0 ■ ■ 0 0 0
 13 | 0 0 0 ■ 0 0 0 0 ■ 0 ■ 0 0 0 ■ 0
 14 | 0 0 0 0 ■ 0 0 ■ ■ ■ 0 0 0 0 0 ■
 15 | ■ ■ 0 0 ■ ■ ■ 0 0 0 0 0 0 ■ 0 ■
 16 | 0 ■ ■ 0 0 0 0 ■ 0 0 0 ■ ■ 0 0 0
```

Figure 10.1
Sparsity pattern of a matrix which has only three target rows
when the element a_{11} is chosen as a pivot at stage 1 of the GE.

If the elements a_{11}, a_{22}, $a_{33,}$, a_{44} and a_{55} are used simultaneously
as pivots (which is possible, because these pivots form an upper triangular
matrix) the number of target rows increases from 3 to 11 and, thus, much
more parallelism is present. The use of several pivots simultaneously will be
studied in the next section.

10.4. Parallel codes based on a simultaneous use of several pivots

Instead of using a single pivot per stage and a rank-1 modification of
A_s, one can try to find several pivots and to perform in parallel all
computations corresponding to these pivots. The simplest case is to find
pivots that form a diagonal submatrix: in this case a rank-m modification has

to be performed (m being the number of pivots found). One can also use pivots that form an upper triangular matrix (the 5x5 submatrix in the upper left corner of the matrix given in **Fig. 10.1** is an upper triangular matrix). In such a case the computations become slightly more complicated. On the other hand, one should expect to obtain some advantages (comparing with the case where a requirement for pivots that form diagonal matrix is imposed):

(i)   more pivots per generalized stage can be determined,
(ii)   the search for simultaneous pivots is simpler,
(iii)   the sparsity can be preserved better.

The reason for the first two advantages is obvious: if the simultaneous pivots form an upper triangular matrix, then one has to ensure that the elements under the diagonal of the submatrix formed by the pivots are zeros, while both the elements under and over the diagonal must be zeros if a requirement for pivots that form a diagonal matrix is imposed. More stringent requirements for preservation of sparsity can be used (comparing with the case where the pivots form a diagonal matrix), because the pivotal search is much easier; there are many more sets of potential pivots that form an upper triangular matrix than the sets of pivots that form a diagonal matrix. This is the reason for the third advantage. At the same time it must be emphasized that both the sparsity and the accuracy requirements are relaxed in order to facilitate the pivotal search (in comparison with the case where single pivots are searched), but not so much as in the case where the pivots form a diagonal matrix.

The main parts of the work at each generalized stage (a stage in which several pivots are used simultaneously) are:

(i) find a set of pivots that can be used simultaneously,
(ii) move the pivots from the row ordered list to array PIVOT,
(iii) perform the modifications in the active part of the matrix,
(iv) reorder the fill-ins by columns,
(v) perform "symbolic" modifications in the column ordered list.
(vi) update the structure, in which the rows are ordered in an increasing number of non-zero elements.

The first task is to determine **a set of pivots that can be used simultaneously**. The rows are ordered, before the start of the GE, in an increasing number of non-zero elements. At each generalized stage one starts the search with the "best" row (the row with minimal number of non-zeros in its active part). The search is carried out until 24 pivots and 80 target rows are found, but there is no guarantee that this condition will be satisfied at the end of the search. The code may perform a search of all active rows without finding so many pivots and/or so many target rows; this will normally happen at the end of the factorization process. It should be stressed, however, that normally the code will not search all active rows; the search is terminated if no pivot can be chosen during the search of five successive rows. Finally, it should be noted that the above figures, 24 and 80, were found experimentaly. While this seems to be a good choice for the **ALLIANT**, the figures should probably be changed if another computer is used. The figures can be related to the number of processors. Let this number be NPROC. Then 3*NPROC and 10*NPROC will probably be a good choice.

Both the accuracy requirements and the sparsity requirements are relaxed in order to facilitate the search of pivots after the first one. The accuracy requirements are relaxed by using a stability factor which is 4 times greater than that used to find the first pivot during a generalized stage. The sparsity requirements are relaxed by accepting pivots that have Markowitz cost less than 2*R+30, where R is the Markowitz cost of the first pivot found during the generalized stage in consideration.

When a pivot is found, the code marks all rows in its column: these rows should not be used in the subsequent searching. The code also negates the row numbers, stored in array RNLU, of the non-zero elements in the pivotal row (these have to be removed at the end of the modifications in the column ordered list). Finally, the code stores information about the interchanges, but the actual interchanges are not carried out. This is one of the major differences between the code **Y12M2** with simulatneous pivots and the code **Y12M1** where single pivots are used. If one pivot only is used, then it is natural to perform the pivotal interchanges: the elements in the pivotal column and row are needed anyway and these are to found; therefore it is not very expensive to perform the actual interchanges. Here the situation is different, because at each generalized stage many pivots are in use and it is not worthwhile to perform the pivotal interchanges.

The first step is performed by subroutine **PARPIV**, which calls some other subroutines. ∎

The second step is to **move the pivots found during the generalized stage to array PIVOT**, and to perform the appropriate modifications in the row ordered list. No modifications in the column ordered list are necessary, because all pivotal columns are removed at the end of the generalized stage (these are not needed in the further computations). This is a relatively easy step, because only modifications in the rows are performed and therefore the removal of each pivot is an independent task.

Step (ii) is performed by subroutine **REMPIV**. ∎

When the pivots are found and moved to array PIVOT, the actual computations, **the modifications in the active part of the matrix**, may start. The main task here is **the modification of a target row**. The code goes through the target rows (the order can be arbitrary) and modifies each of them by using the appropriate pivots. The target row is scattered (another major difference: in **Y12M1** the pivotal row is scattered) in an array of length at least equal to n and the actual computations are performed in this array. During the modification of a target row the code prepares: **(i)** information about the separator (the pointer of the start of the active part of this row for the next generalized stage), and **(ii)** a list of the row numbers and column numbers of all fill-ins. A buffer is used to make it possible to carry out all these computations in parallel (in a similar way as in **Y12M1**). At the end of this part of the computations some of the rows are in the buffer; these are moved back to the row ordered list (in the arrays ALU and CNLU).

The third step is carried out by subroutine **PSYMRW**. ■

The next step is **to reorder the fill-ins by columns**. The fill-ins have been collected in the buffer during the previous step, but they are ordered by rows; all computations in the previous step are performed by rows.

The fourth step is done in subroutine **REORDF**. ■

The fifth step is to perform **the local "symbolic" modifications of the column ordered list**. The code (i) removes the row numbers in the pivotal columns (these have been negated during the first step when the simultaneous pivots have been determined), and (ii) adds the fill-ins (which have been ordered by columns in the fourth step). Thus, only the active parts of the columns are kept after each generalized stage.

Subroutine **PSYMCL** performs the "symbolic" modifications in the column ordered list; the fifth step. ■

The final task during the generalized stage is to **reorder the structure where the rows are kept in an increasing number of non-zero elements in their active parts**; such a structure will be needed during the determination of the set of simultaneous pivots in the next generalized stage. This part of the work is carried out as in the original **Y12M** (the description of the arrays by which the stucture is reordered is given in **Chapter 6**; see also **Østerby and Zlatev [341]**).

The subroutine that performs the last step is called **REORDT**. ■

Subroutine	west2021		hwatt_2	
	1 processor	8 processors	1 processor	8 processors
PARPIV	0.40	0.29	0.81	0.52
REMPIV	0.04	0.01	0.03	0.01
PSYMRW	1.16	0.23	10.61	2.12
REORDF	0.04	0.04	0.74	0.76
PSYMCL	0.19	0.05	1.81	0.44
REORDT	0.08	0.09	1.30	1.29

Table 10.2
The computing times for the most time-consuming parts
of the computations during the sparse factorization.

The performance of Y12M2 during each of the six major steps could eaisily be evaluated, because a separate subroutine is written for each step (the subroutine **PARPIV** that searches for simultaneous pivots calls some other subroutines also). This has been done for many matrices. Results for two matrices from the Harwell-Boeing set of sparse test-matrices, [71,72], that have been used in the previous section are given in **Table 10.2**. The first matrix, **west2021**, is very sparse and stays very sparse; it is of order 2021 and the code switches to dense matrix technique when A_s is of order 99. The second matrix, **hwatt_2**, is not very sparse and, moreover, many fill-ins are produced during the factorization; its order is 1856 and the code switches to dense matrix technique when A_s is of order 705 (which shows that the number of non-zeros in the active part A_s of the matrix increases rather quickly to more than 10%).

It is seen that the best speed-up is achieved for **PSYMRW**. This is also most desirable, because this subroutine performs the most expensive part of the computations during the sparse factorization. The speed-up for **PSYMCL** and **REMPIV** is also quite good (for the last subroutine this is not very important, because it is very cheap).

REORDF and **REORDT** are carried out in a sequential mode (no speed-up). It is probably not possible to get some improvement for these subroutines; the processes performed by them are sequential.

The improvement for subroutine **PARPIV** is modest. This subroutine searches for pivots sequentially; subroutine **FINDPV** is called to find a pivot (some loops in **FINDPV** are optimized). Then information about the interchanges is stored and the target rows are marked; this work is also carried out sequentially. Finally, the row numbers of the non-zero elements of the pivotal rows are negated, and, even more important, the lengths of the active parts of the target columns are updated (reducing them by one). This part is performed in parallel (the task being a modification of a column). This is not the only way to organize the search for parallel pivots. One could also search for pivots in parallel (calling concurrently a subroutine similar to **FINDPV** in a loop). If a pivot is found, then the structure should be locked in order to **(i)** store information about the interchanges, **(ii)** mark the target rows, **(iii)** negate the row numbers of the non-zero elements in the pivotal row, and **(iv)** update the lengths of the target columns. When these four tasks are performed, the structure is unlocked and the pivotal search is continued. However, there are several objections against this approach:

(1) the concurrent call of the modified **FINDPV** is degraded by locking and unlocking operations,

(2) it is necessary to check if the pivot found is consistent (if its row number has not been marked during the search by another processor that succeeded in finding first a pivot),

(3) in the search of pivots the Markowitz cost cannot be correctly applied (one processor may find a pivot before the column lengths have been updated by another processor, which succeeded in finding first a pivot),

(4) there will be some redundant calls of **FINDPV** (when the pivot found is not consistent with the other pivots; its row number has been marked in connection with another pivot). It should be mentioned here that **FINDPV** is a relatively cheap subroutine (when the algorithms described in **Chapter 4** and **Zlatev [291]** are used),

(5) the process of negating the row numbers in the pivotal row has to be carried out sequentially by the processor that has found the pivot (this process is performed concurrently by the algorithm used in **Y12M2**).

For the matrix **west2021** the computing times for **FINDPV** are 0.12 when one processor only is in use and **0.08** when eight processors are used (about **30%** and **28%** of the computing time for **PARPIV**, while the percentages are about **6%** and **11%** from the total time needed for the sparse factorization). Thus, **(1)-(5)** indicate that there is not much to win by a concurrent call of **FINDPV**. Note that if the matrix is not very sparse and/or if many fill-ins are produced, then the relative part of the work done in the first step is much smaller; see the second example in **Table 10.2**. Nevertheless, the approach where concurrent calls of **FINDPV** are used will be efficient when it is properly implemented. Such an approach is used in the code developed by **Davis and Yew** [51], which is a very efficient code for parallel machines with shared memory.

10.5. Parallel codes based on an a priori reordering the matrix

If the matrix is sparse, then it is natural to try to reorder it to an upper block-triangular form. In order to get more well-balanced blocks, it is also natural to drop the requirement for square diagonal blocks, which is normally imposed (see, for example, [7,8,90,91,144,145]). An algorithm based on these ideas consists of the following steps:

(1) initial reordering,
(2) partitioning,
(3) first phase of the factorization,
(4) second phase of the factorization,
(5) second reordering,
(6) third phase of the factorization.

The initial reordering is performed in two parts. The columns are ordered during the first part (by a subroutine called **ORDCOL**). Let c_j be the number of non-zero elements in column j. Assume that the first part of the initial reordering is finished, and that the column interchanges are performed (a subroutine called **PERMUT** is used for this purpose). Then the columns of the matrix are ordered so that:

(10.1) $j < k$ $\Rightarrow$ $c_j \leq c_k$.

The rows are ordered during the second part (by a subroutine called **ORDROW**). Let r_i be the column number of the first non-zero element in row i. Assume that the second part of the initial reordering is finished and that the row interchanges are performed (subroutine **PERMUT** is used also when the row permutations are to be carried out). Then the rows of the matrix are ordered so that:

(10.2) $i < m$ $\Rightarrow$ $r_i \leq r_m$.

It is clear that, by reordering the matrix in this way, most of the zero elements are moved to the lower left-hand corner of the matrix. This can be used to partition the matrix into an upper block-triangular form. ■

The **partitioning** is carried out by subroutine **BLOCKS**. The idea is to partition the matrix into **p parts**, each of them containing a certain number of rows. In all experiments that will be discussed in the next section p is set to be equal to 8 (the number of processors on the **ALLIANT**). However, such a restriction (that the number of parts be equal to the number of processors) is not necessary; p may also be either larger or less than the number of processors.

Assume that p is equal to the number of processors. Then each processor will carry out the work on one **part** of the matrix. Therefore it is important to partition the matrix so that the work done (by the different processors) on each **part** is approximately the same. It is difficult to find an easy criterion by which such a partitioning can be achieved. A simple idea is to try to get **parts** that have approximately the same number of rows. This is the criterion used in **Y12M3**. It is easy to make such a partitioning because there is no requirement for square diagonal blocks (the block is called "diagonal" if all elements under it are zeros).

One could try to partition the matrix so that the different **parts** contain approximately the same number of non-zero elements (instead of approximately the same number of rows). This partitioning is slightly more expensive, but for some matrices it performs better. ∎

Once the matrix is partitionned, **the first phase of the factorization** can be started. Each processor produces zeros under the main diagonal of the diagonal block in its **part** of the matrix. This is a straightforward operation and may be carried out by calling a slightly modified version of one of the subroutines described in the previous sections (or of the factorization subroutine in the original package **Y12M**). If each **part** of the work is performed by a cluster of processors, then the computational work during the first phase of the factorization can be optimized on three levels: (a) vectorization for each processor, (b) concurrency between the processors within a cluster, and (c) parallel computations on the different clusters.

Assume that the partitioning is performed with p=4. The blocks obtained by such a partitioning can be depictured as in **Fig. 10.2**.

$$
\begin{vmatrix}
A_{11} & A_{12} & A_{13} & A_{14} \\
0 & A_{22} & A_{23} & A_{24} \\
0 & 0 & A_{33} & A_{34} \\
0 & 0 & 0 & A_{44}
\end{vmatrix}
$$

Figure 10.2
Partitioning with p=4.

Assume that A_{11} is a rectangular qxr block with q>r. Then an upper triangular matrix $U_{11} \in \mathbf{R}^{rxr}$ will be obtained after the factorization of this block. The last q-r rows of A_{11} will contain zeros only. Thus, the block A_{11} can, after the factorization, be partioned into matrix U_{11} (containing the first r rows) and a zero matrix (containing the last q-r rows). The same partitioning can be performed for the other three blocks in the first **part** of the matrix; A_{12}, A_{13} and A_{14} (however, the lower blocks in these two matrices will not be zero matrices like the lower block of A_{11}). The first phase of the factorization in the second and the third **parts** is carried out in a quite similar way. The last block, A_{44}, has another structure. In general it is also rectangular, but the number of its columns, r, is greater than the number of its rows, q. An upper triangular matrix $U_{44} \in R^{q \times q}$ and a matrix $X_{44} \in R^{q \times (r-q)}$ (formed by the last r-q columns) will be obtained after the first phase of the factorization. It is convenient to perform a similar partitioning in the whole last block-column of the matrix in **Fig. 10.2**. After the first phase of the factorization the matrix can be partioned as shown in **Fig. 10.3**.

$$
\begin{vmatrix}
U_{11} & U_{12} & U_{13} & U_{14} & X_{14} \\
0 & W_{12} & W_{13} & W_{14} & V_{14} \\
\\
0 & U_{22} & U_{23} & U_{24} & X_{24} \\
0 & 0 & W_{23} & W_{24} & V_{24} \\
\\
0 & 0 & U_{33} & U_{34} & X_{34} \\
0 & 0 & 0 & W_{34} & V_{34} \\
\\
0 & 0 & 0 & U_{44} & X_{44}
\end{vmatrix}
$$

Figure 10.3
**Partioning of the matrix at the end of the first phase
of the factorization. U_{ii} (i=1,2,3,4) are upper
triangular matrices. The elements in the fourth part
as well as the elements in the first block-rows of the
other three parts will not be modified in the further
calculations.**

The calculations in the different **parts** of the matrix can be performed concurrently. The transition from the matrix in **Fig. 10.2** to the matrix in **Fig. 10.3** can be carried out on four processors (or four clusters of processors). ∎

During the **second phase of the factorization** zeros are produced in the blocks W_{ij} (i=1,2,3, j=2,3,4). The process is carried out **"by diagonals"**. First, zeros are produced in the blocks W_{12}, W_{23} and W_{34} by using the pivots in the blocks U_{22}, U_{33} and U_{44}, respectively. These computations can be carried out in parallel on three processors (or on three clusters of processors). When the computations with the first diagonal are completed, the computations with the second diagonal can be started. Zeros are produced in the blocks W_{13} and W_{24} by using the pivots in the blocks U_{33} and U_{44} respectively. These calculations can be carried out in parallel on two processors (or on two clusters of processors). Finally, zeros are produced in block W_{14} by using the pivots in block U_{44}. These computations are to be carried out on one processor (or on one cluster of processors).

Only the computations in the different blocks within a diagonal can be performed concurrently. Indeed, when the computations to produce zeros in the blocks in the first diagonal are carried out, all other blocks (to the right of the blocks transformed) are modified. Therefore, it is not possible to avoid the loss of concurrency during the second phase of the factorization. On computers like the **ALLIANT** it is better to perform the computaions during the second phase of the factorization not by blocks, but "by target rows". Consider the block W_{12}. Each row of this block can be modified (using appropiate pivots in block U_{22}) independently of the other rows. The same statement holds also for the other blocks. This technique is used on the **ALLIANT**.

The result after performing the second phase of the factorization is given in **Fig. 10.4.** ∎

After the fourth step only the elements in the blocks V_{14}, V_{24} and V_{34} need to be modified. These blocks, if they are gathered together, form a square matrix. Moreover, it is reasonable to expect them to be rather dense. Therefore it is worthwhile to reorder the matrix again by pushing these blocks to the lower right-hand corner of the matrix and then switching to dense matrix technique. This is the fifth step, **the second reordering**. This is a straightforward step; the result is given in **Fig. 10.5.**

$$
\begin{vmatrix}
U_{11} & U_{12} & U_{13} & U_{14} & X_{14} \\
0 & 0 & 0 & 0 & V_{14} \\
\\
0 & U_{22} & U_{23} & U_{24} & X_{24} \\
0 & 0 & 0 & 0 & V_{24} \\
\\
0 & 0 & U_{33} & U_{34} & X_{34} \\
0 & 0 & 0 & 0 & V_{34} \\
\\
0 & 0 & 0 & U_{44} & X_{44}
\end{vmatrix}
$$

Figure 10.4
Partioning of the matrix at the end of the second phase
of the factorization. U_{ii} (i=1,2,3,4) are upper
triangular matrices. Only the elements in the blocks
V_{i4}, i=1,2,3, will be modified in the further computations.

The amount of work during the fifth step is very small in comparison with the work during the other steps. ∎

The matrix D_{22} has to be factorized during the last step, **the third phase of the factorization**. As mentioned above, this is done by a dense matrix subroutine. The subroutine used in this step is the same as the subroutine used in the two other versions. On the **ALLIANT** it is worthwhile to use the subroutines from package **PARALIN** (and this package is used in the experiments in this section and in the following section). ∎

$$
\begin{vmatrix}
U_{11}^* & U_{12}^* \\
\\
0 & D_{22}
\end{vmatrix}
$$

Figure 10.5
Partioning the matrix after the second reordering.
U_{11}^* is upper triangular (formed by the blocks U_{ij}, i=1,2,3,4, j=1,2,3,4).
U_{12}^* is rectangular (formed by the blocks X_{i4}, i=1,2,3).
D_{22} is square (formed by the blocks V_{i4}, i=1,2,3).

A simple example has been used here to explain how the factorization process is performed. However, this is made only in order to facilitate the exposition of the results. It is clear that the same kind of computations are to be carried out in the case where the number of **parts** is not four.

Stability problems may arise when this method is used. Indeed, the initial reordering ensures that at least one non-zero element appears at each row of a diagonal block. However, some columns in the diagonal blocks may contain zeros only. Therefore such columns (if they exist) are moved to the end by subroutine **BLOCKS**. Thus, subroutine **BLOCKS** not only performs partitioning, but also guarantees that no structurally singular diagonal block occurs. Moreover, the stability check during the first phase of the factorization is carried out by using the longer dimension of the diagonal blocks (by columns in all **parts** excepting the last one, where the stability check is carried out by rows). The fact that the blocks are (in general) rectangular decreases the possibility of the occurence of structural singularity, but, unfortunately, does not remove it. Therefore special care is taken to avoid blocks that are structurally singular. Assume that a diagonal block has dimensions q and r (with $q \geq r$). The code tries to find a diagonal whose elements are non-zeros. If such a diagonal exists, then the diagonal block is structurally non-singular. In fact, a stronger requirement is imposed: the code tries to find a diagonal, whose elements (in absolute value) are greater than the product of 0.01 and the largest element (in absolute value) in their rows. This could be considered as a stability requirement. There is no guarantee that such a diagonal exists. However a diagonal that has s elements that satisfy the stability requrement (with $s \leq r$) can always be found. If $s < r$, then the code will move the $r-s$ columns for which the requirement is not satisfied to the end. This is a heuristic device that works pretty well in practice.

No stability check is performed during the second phase of the factorization (it is assumed that the pivots found during the first phase are good enough also for the second phase). In this way some computations are saved and, what is even more important in this case, the computations can easily be performed concurrently. However, this may cause problems. Therefore, it is strongly recommended to use this method together with some iterative improvement even if no "small" elements are dropped during the computations.

	west2021		hwatt_2	
Step	1 processor	8 processors	1 processor	8 processors
1	0.37	0.36	0.60	0.54
2	0.11	0.12	0.13	0.11
3	1.01	0.26	6.78	3.99
4	0.58	0.13	22.25	5.11
5	0.08	0.02	0.12	0.04
6	0.22	0.11	1.01	0.38

Table 10.3
The computational work during the six steps
(see the beginning of this section)
of the code based on partitioning.

The method may lead to poor preservation of sparsity. During the first phase of the factorization the pivots are restricted to the diagonal blocks. No sparsity check is carried out during the second phase of the factorization: it is assumed again that the pivots found during the first phase are good enough (not only in regard to numerical stability, but also in regard to preserving sparsity) also for the second phase. The experiments show that the method is competitive for many problems with the other methods. It should be mentioned here that if "small" elements are removed, then the method leads to a good preservation of sparsity also in the cases where this is not so when all elements are kept. Dropping small elements will be considered separately in the next chapter.

The two typical examples (a very sparse matrix that stays sparse and a matrix that creates many fill-ins), which have been used in the previous sections, have been tested also for this method. The results are given in **Table 10.3**.

It is seen that the last four steps are relatively well optimized, while the first two are not. It is probably possible to get some better optimization of these two steps, but in many cases the work performed in these steps is much less than the work performed in the other four steps (see the results for **hwatt_2**).

10.6. Numerical results

If a code for performing operations with dense matrices is optimized for parallel and/or vector computers, then there is a simple way to measure the efficiency of the optimization: one measures the speed of computations in **MFLOPS** (millions of floating point operations, additions or multiplications, per seconds). This is a good measure for dense matrices, because nearly all operations that are to be carried out by the code **are** floating point operations. For sparse matrices the situation is different. Now only a very little part of the computations are floating point operations. There are extreme cases where the whole factorization is completed without performing any floating point operations, for example the matrices SHLL from the Harwell-Boeing set of test matrices, **[71,72]**. This is so, because these matrices are permuted triangular matrices. This is, of course, an extreme example. However, it explains in an excellent way why counting **MFLOPS** can be irrelevant for sparse matrices, even though it is a simple and elegant tool for evaluating the performance of codes for dense matrices.

Another simple way to measure the performance of a concurrent code is to compare the performance of the code run on one processor only with the performance of the code run on many processors. However, this **may** be insufficient. The danger is that a rather good speed-up could be obtained with a code that performs very badly on a sequential computer.

The above discussion shows clearly that there is no simple way to check the performance of a sparse code. Therefore the check must be done in two steps: **(1)** compare the sequential version of the code with other good sequential codes and **(2)** if the sequential version of the code compares well with other good sparse code, then compare the sequential version of the code with the parallel versions. This procedure will be followed in this section. First the test-matrices will be presented. Then the sequential version of **Y12M1** will be compared with results obtained with two well-known codes, **MA28**

[63] and **SPARSPAK-C** [122,123]. Finally, results for the three parallel codes will be given. A preliminary version of the code **Y12M1** has already been used in the literature (as, for example, in [51,104,105]). The newest version of this code performs much better. Some results obtained with this preliminary version, which will be called here **Y12M1P**, will also be presented in order to illustrate the much better performance of the newest version and to avoid any confusion that may appear if results from this chapter are compared with results in, for example, [51,104,105].

10.6.1. Test-matrices used

Test-matrices from the well-known Harwell-Boeing set for sparse-test matrices, [71,72], will be used in this section. We selected 27 matrices. The requirement used in the selection was that the matrices should have order greater than 900. There are 25 such matrices in the subset of general matrices. Two matrices of order less than 900 were added; these matrices have more than 9 non-zeros per row. The matrices chosen together with some of their characteristics (order, number of non-zeros and conditioning) are listed in **Table 10.4**.

Matrix	Order	Non-zeros	COND
sherman3	5005	20033	1.8E+05
gemat11	4929	33108	1.5E+06
gemat12	4929	33044	1.8E+06
lns_3937	3937	25407	1.2E+06
lns3937a	3937	25407	1.2E+06
saylr4	3564	22316	1.2E+07
sherman5	3312	20793	6.0E+03
or6781hs	2529	90158	2.1E+06
orsreg_1	2205	14133	8.1E+03
west2021	2021	7310	4.5E+07
hwatt_2	1856	11550	3.1E+05
hwatt_1	1856	11360	4.7E+04
west1505	1505	5414	5.8E+07
nnc1374	1374	8588	2.6E+13
mahistlh	1258	7682	1.7E+04
pores_2	1224	9613	3.6E+05
gre_1107	1107	5664	1.8E+07
sherman4	1104	3786	1.6E+03
gaff1104	1104	16056	3.4E+10
sherman2	1080	23094	2.8E+03
orsirr_1	1030	6858	1.2E+04
sherman1	1000	3750	5.5E+03
jpwh_991	991	6057	2.0E+02
west0989	989	3518	4.2E+07
pde_9511	961	4681	1.4E+07
mc_fe	765	24382	7.7E+01
steam2	600	5660	3.2E+00

Table 10.4
The set of the test-matrices used.

The estimates of the condition numbers of the matrices, given under **COND** in **Table 10.4**, are obtained by using **Y12M**; see **Chapter 9**. **Y12M** is the only package for general sparse matrices that allows the user to calculate an estimate of the condition number of the matrix. It should be mentioned here that the matrices are scaled; this leads to a considerable reduction of the condition numbers for some of the matrices; see [104,105].

10.6.2. Results obtained with sequential codes

The matrices listed in **Table 10.4** were run by using **MA28**, **SPARSPAK-C** and **Y12M1**. Since the first two codes are sequential, the routines in **Y12M1** were compiled with the   -Og   option of the **ALLIANT FORTRAN** compiler (this means that both vectorization and concurrency were suppressed). Moreover, to be fair to the first two codes where there is no switching to dense matrix technique, the routines of **Y12M1** were also run without switching to dense matrix technique.

Matrix	MA28	SPARSPAK-C	Y12M1 (-Og)	Y12M1P
sherman3	354.90	225.70	258.79	79.02
gemat11	25.15	70.93	39.23	8.62
gemat12	32.39	70.77	41.02	9.22
lns_3937	1217.59	169.84	923.48	268.93
lns3937a	1359.81	167.34	905.83	264.04
saylr4	1054.51	292.51	385.25	123.04
sherman5	253.54	143.10	195.97	57.79
or6781hs	86.67	321.68	225.52	52.49
orsreg_1	117.71	205.28	163.82	54.21
west2021	31.95	54.70	6.48	2.17
hwatt_2	104.58	128.43	106.78	35.32
hwatt_1	130.26	135.25	108.75	35.92
west1505	18.16	52.68	4.03	1.59
nnc1374	237.95	63.40	27.99	9.77
mahistlh	5.56	58.36	7.38	2.51
pores_2	71.53	62.83	43.69	14.58
gre_1107	40.37	85.55	37.12	12.49
sherman4	5.98	53.27	6.23	2.40
gaff1104	49.57	77.13	56.81	18.85
sherman2	610.53	947.34	292.89	96.83
orsirr_1	28.34	78.99	32.45	11.35
sherman1	13.60	56.87	8.97	3.44
jpwh_991	39.51	102.03	42.63	15.79
west0989	8.13	51.14	2.04	1.00
pde_9511	6.73	54.32	7.53	3.16
mc_fe	54.68	344.48	43.62	13.58
steam2	17.66	60.76	10.13	3.69

Table 10.5
Computing times for three sequential sparse codes
and the partially optimized code Y12M1P.

As mentioned above, a preliminary version of **Y12M1**, called here **Y12M1P**, is also used in the experiments. This code is partially optimized. The main difference between **Y12M1** and **Y12M1P** is the optimization of the "symbolic" factorizations (see **Section 10.3**): these steps are performed in parallel in the former code, but sequentially in **Y12M1P**. The results in **Table 10.5** show that even by this partial optimization much better results than those obtained with the sequential codes, including **Y12M1** when this code is run in a sequential mode, can be obtained. However, if **Y12M1P** is compared with **Y12M1** (run in parallel and with switching to dense matrix technique), then the much better performance of **Y12M1** is clearly seen.

Two remarks are needed in connection with the first two codes. **MA28** is run with the default values of the parameters. This means, among other things, that the old pivotal strategy is used. This strategy is very inefficient for some matrices. There is another strategy in **MA28**, quite similar to the strategy described in [291]. The computer time for some matrices will be reduced considerably if this strategy is used. In **SPARSPAK-C** a parameter is provided by which the user can tell the code when a row should be considered as dense. In this experiment this parameter was set to 25.

Matrix	MA28	SPARSPAK-C	Y12M1 (-Og)	Y12M1P
sherman3	250985	358046	210394	210394
gemat11	51516	62740	45987	45987
gemat12	51780	70653	49118	49118
lns_3937	376754	331197	467697	467697
lns3937a	389749	331428	454424	454424
saylr4	482387	489252	308455	308455
sherman5	164852	229912	158691	158691
or6781hs	109378	321536	133715	133715
orsreg_1	151129	312650	158085	158085
west2021	10707	10574	9355	9355
hwatt_2	124922	200967	114771	114771
hwatt_1	127880	111250	117103	117103
west1505	8006	8098	6978	6978
nnc1374	92225	66599	47360	47360
mahistlh	11488	20667	12066	12066
pores_2	48862	49913	60812	60812
gre_1107	46845	92635	51364	51364
sherman4	16632	21320	14225	14225
gaff1104	71106	82758	74032	74032
sherman2	168539	842195	178506	178506
orsirr_1	52097	90087	50456	50456
sherman1	23158	21941	18259	18259
jpwh_991	52497	115398	51419	51491
west0989	5049	5076	4371	4371
pde_9511	21633	29981	20139	20139
mc_fe	67674	294911	61419	61419
steam2	26093	40809	21207	21207

Table 10.6
Numbers of nonzeros in LU for three sequential codes and the partially optimized code Y12M1P.

The computing times are given in **Table 10.5**. It is seen that **Y12M1** is at least comparable with the other two codes (and for some matrices better). This indicates that it is reasonable to assume that a parallel version of this code, which gives much better results than its sequential version, is indeed a good code Such a conclusion could not be made if, for example, the results obtained by **SPARSAK-C** were 10 times better.

The numbers of non-zeros in L and U are given in **Table 10.6**. These numbers provide information about the storage that was actually used. However, this information is of no great use. The number of fill-ins is in general not known in advance. Therefore both in **MA28** and in **Y12M1** it is customary to reserve much more space than the space actually needed. However, if the problem has to be solved many times (as often happens in simulation processes), then after the first run one can reserve space in a nearly optimal way. In **SPARSPAK-C** the code tries to find an upper bound on the storage needed. One can do some preprocessing to determine this bound and, thus, if the bound found is close to what is actually needed, one has the possibility to reserve an optimal amount of space. However, even if the bound found is close to the space actually needed (which is not always the case), the space needed for this code will often be considerably larger than the space needed for the other two codes. This is due to the fact that partial pivoting is used in this code, while Markowitz-type pivotal strategies, [63,122,123,331,341], are applied in the other two codes).

Matrix	MA28	SPARSPAK-C	Y12M	Y12M1P
sherman3	1.46E-12	3.97E-13	8.76E-14	8.76E-14
gematl1	1.51E-11	1.44E-10	3.00E-12	3.00E-12
gematl2	6.67E-12	9.53E-11	1.96E-12	1.96E-12
lns_3937	1.82E-11	1.74E-04	2.54E-12	2.54E-12
lns3937a	2.38E-11	1.81E-04	2.37E-12	2.37E-12
saylr4	1.25E-10	8.01E-12	5.55E-11	5.55E-11
sherman5	4.64E-13	9.39E-14	4.95E-14	4.95E-14
or6781hs	1.33E-13	4.06E-14	4.75E-14	4.75E-14
orsreg_1	1.10E-13	3.05E-13	2.17E-13	2.17E-13
west2021	2.21E-10	7.10E-10	7.94E-11	7.94E-11
hwatt_2	1.68E-13	7.56E-14	1.24E-14	1.24E-14
hwatt_1	1.53E-14	2.41E-14	8.44E-15	8.44E-15
west1505	4.51E-10	2.18E-10	3.34E-11	3.34E-11
nnc1374	1.19E-02	2.63E-03	5.33E-05	5.33E-05
mahistlh	8.34E-10	1.49E-10	9.55E-15	9.55E-15
pores_2	2.71E-13	8.53E-12	2.99E-13	2.99E-13
gre_1107	1.39E-07	1.68E-09	1.13E-08	1.13E-08
sherman4	2.89E-14	1.35E-14	5.33E-15	5.33E-15
gaff1104	9.83E-08	9.90E-08	9.83E-08	9.83E-08
sherman2	3.32E-11	3.04E-07	1.53E-13	1.53E-13
orsirr_1	2.39E-13	6.02E-13	4.63E-13	4.36E-13
sherman1	6.68E-13	2.63E-13	1.50E-13	1.50E-13
jpwh_991	4.39E-14	1.08E-14	1.89E-15	1.89E-15
west0989	3.13E-10	2.51E-10	1.36E-10	1.36E-10
pde_9511	6.44E-15	2.22E-15	2.44E-15	2.44E-15
mc_fe	1.93E-14	1.84E-14	2.11E-14	2.11E-14
steam2	7.03E-13	1.14E-12	1.22E-15	1.22E-15

Table 10.7
The accuracy of the solution for three sequential codes
and the partially optimized code Y12M1P.

The results given in **Table 10.6** indicate that **Y12M1** will in general need no more storage than the other two codes when it is used without switching to dense matrix technique (the storage being aproximately equal to that used by the original **Y12M**). However, switching to dense matrix technique does lead to a very substantial reduction of computing time. Therefore it is always used in **Y12M3** and recommended in the other two codes. This implies that the parallel codes with switching to dense matrix technique will require more storage than the sequential codes.

Matrix	SEARCHING ALL ROWS		UPDATING THE BEST ROWS	
	WITHOUT DMT	WITH DMT	WITHOUT DMT	WITH DMT
sherman3	37.57	14.59	46.14	13.71
gemat11	8.76	7.88	7.72	7.01
gemat12	7.26	8.00	8.10	7.26
lns_3937	111.87	21.87	112.43	25.12
lns3937a	120.86	20.62	113.31	25.75
saylr4	55.73	20.83	70.56	22.30
sherman5	24.27	8.18	36.50	8.52
or6781hs	8.63	22.61	33.82	23.33
orsreg_1	24.18	8.61	34.04	9.80
west2021	2.53	2.34	2.34	2.26
hwatt_2	16.38	6.02	22.56	7.06
hwatt_1	16.59	5.91	20.91	6.77
west1505	1.85	1.74	1.71	1.69
nnc1374	4.65	2.86	5.14	2.87
mahistlh	2.27	1.94	2.32	1.78
pores_2	6.01	3.31	9.62	3.85
gre_1107	5.22	2.36	7.06	2.49
sherman4	1.71	1.24	1.82	1.10
gaff1104	9.86	4.55	13.30	4.90
sherman2	33.41	5.42	45.66	6.03
orsirr_1	5.59	2.56	8.24	2.74
sherman1	2.47	1.27	2.36	1.16
jpwh_991	6.79	1.78	8.61	1.71
west0989	1.15	1.09	1.11	1.07
pde_9511	2.13	1.59	2.48	1.55
mc_fe	7.26	2.99	8.53	3.25
steam2	2.15	1.09	2.87	1.26

Table 10.8
Computing times obtained by Y12M1 run on 8 processors
in four different modes. "SEARCHING ALL ROWS" means that
all appropriate rows are investigated to find a row with
minimal number of non-zeros in its active part; the other
option, "UPDATING THE BEST ROWS", updates at each stage
the array where the rows are ordered in increasing number
of the non-zero elements in their active parts (three
"best" rows are used in the pivotal search). "WITH DMT"
and "WITHOUT DMT" refer to the options where the code
switches to dense matrix technique and where no such a
switch is made.

The accuracy obtained in the runs is given in **Table 10.7**. The right-hand side vectors of all problems were created so that the exact solution was $x_i = 1$, $i = 1, 2, \ldots n$. The accuracy obtained depends, of course, on the condition number of the problem solved. However, the accuracy achieved by the three codes is as a rule approximately of the same order. It should be noted that there are a few cases, where the accuracy achieved by **SPARSPAK-C** is considerably poorer than that achieved by the two other codes; see, for example, the results for matrix **lns_3937** in **Table 10.7**.

10.6.3. Results obtained with the three parallel codes

Results obtained with four options of **Y12M1** are given in **Table 10.8**. This code is run with and without a switch to dense matrix technique (**WITH DMT** and **WITHOUT DMT** being used in the table). It is seen that the switch to dense matrix technique is very profitable for many problems.

Matrix	Y12M1	Y12M2	Y12M3
sherman3	14.59 (4.6)	19.74 (4.7)	30.59 (3.8)
gematl1	7.88 (3.4)	5.17 (3.0)	160.31 (6.5)
gematl2	8.00 (3.4)	5.24 (3.1)	157.02 (6.5)
lns_3937	21.87 (5.1)	25.27 (5.0)	44.76 (5.1)
lns3937a	20.62 (5.1)	25.17 (4.9)	39.70 (5.5)
saylr4	20.83 (5.4)	35.32 (4.8)	89.65 (6.1)
sherman5	8.18 (4.5)	8.34 (4.2)	11.72 (4.3)
or678lhs	22.61 (4.7)	9.54 (3.0)	9.30 (2.7)
orsreg_1	8.61 (4.6)	11.88 (4.2)	13.33 (5.6)
west2021	2.34 (2.7)	1.19 (2.3)	2.04 (3.7)
hwatt_2	6.02 (4.3)	9.99 (4.1)	13.28 (4.3)
hwatt_1	5.91 (4.4)	9.81 (4.2)	13.15 (4.2)
west1505	1.74 (2.6)	0.86 (2.3)	1.31 (3.3)
nnc1374	2.86 (3.4)	3.21 (3.5)	5.80 (4.9)
mahistlh	1.94 (2.8)	0.98 (2.3)	1.27 (2.6)
pores_2	3.31 (3.8)	3.42 (3.7)	3.66 (3.8)
gre_1107	2.36 (3.5)	2.17 (3.5)	2.67 (4.5)
sherman4	1.24 (3.0)	1.03 (3.1)	1.27 (3.3)
gaff1104	4.55 (4.4)	5.38 (4.0)	5.23 (3.7)
sherman2	5.42 (4.4)	6.22 (3.9)	8.37 (4.4)
orsirr_1	2.56 (3.8)	3.12 (3.6)	2.55 (4.3)
sherman1	1.27 (3.2)	1.04 (3.0)	1.44 (2.9)
jpwh_991	1.78 (3.4)	1.65 (3.5)	2.42 (4.2)
west0989	1.09 (2.4)	0.54 (2.3)	0.73 (2.7)
pde_9511	1.59 (3.1)	2.00 (3.5)	2.36 (4.3)
mc_fe	2.99 (3.9)	3.28 (3.3)	3.22 (3.6)
steam2	1.09 (3.2)	1.15 (3.2)	1.70 (3.5)

Table 10.9
Computing times obtained with three codes.
The speed-up factors are given in brackets
(the computing time obtained when 1 processor is in use
is divided by the computing time obtained with the same
code when 8 processors are used to get the speed-up factor).

Both the option in which the code searches through all appropriate rows
to find a row with minimal number of non-zeros in its active part, and the
option, in which the code updates the contents of the array where the rows
are ordered in increasing number of non-zero elements in their active parts,
are used ("SEARCHING ALL ROWS" and "UPDATING THE BEST ROWS" being used in
the table). The results do not differ too much, but, nevertheless, "SEARCHING
ALL ROWS" is often the better choice. This effect will be even more profound
when more processors are available.

Matrix	Y12M1	Y12M2	Y12M3
sherman3	8.90E-14	1.27E-11	1.62E-07
gemat11	2.30E-12	1.37E-12	8.34E-12
gemat12	7.19E-12	5.54E-11	2.20E-11
lns_3937	1.15E-11	3.23E-11	2.25E-09
lns3937a	1.21E-11	3.41E-11	9.34E-10
saylr4	1.90E-10	6.83E-11	8.24E-10
sherman5	2.07E-14	7.02E-13	1.20E-11
or6781hs	2.16E-14	5.52E-14	2.82E-14
orsreg_1	1.52E-13	3.96E-13	5.85E-13
west2021	5.09E-11	2.30E-11	3.71E-10
hwatt_2	8.88E-15	1.72E-12	5.10E-10
hwatt_1	4.22E-15	2.92E-11	2.79E-11
west1505	6.74E-11	9.76E-11	5.52E-09
nnc1374	5.71E-05	1.69E-04	1.65E-04
mahistlh	6.13E-14	1.33E-13	7.99E-15
pores_2	2.22E-12	5.54E-12	3.74E-12
gre_1107	8.43E-09	2.13E-10	4.43E-09
sherman4	4.66E-15	2.38E-13	6.84E-08
gaff1104	9.83E-08	9.89E-08	9.85E-08
sherman2	4.21E-13	4.13E-13	4.71E-07
orsirr_1	1.66E-13	2.34E-13	2.78E-13
sherman1	1.60E-13	2.31E-12	1.69E-13
jpwh_991	1.78E-15	1.48E-12	1.34E-12
west0989	7.09E-11	1.60E-11	2.53E-11
pde_9511	2.22E-15	1.14E-12	3.21E-05
mc_fe	1.57E-14	5.33E-14	4.22E-15
steam2	8.88E-16	1.33E-15	2.33E-15

Table 10.10
Accuracy of the solutions obtained with three codes.
The right-hand side vectors for all systems of
linear algebraic equations are produced so that
all components of the solution vector are equal
to one.

The computing times obtained with the three parallel codes are given in
Table 10.9. Y12M1 is run with switching to dense matrix technique and with
"SEARCHING ALL ROWS", which means that the figures for this code in Table
10.9 are the same as the figures in the second column of Table 10.8. It is
possible to run Y12M2 without switching to dense matrix technique, but it
is clear that this will not be efficient (see also Section 10.4). Therefore
this code is run with switching to dense matrix technique only. In Y12M3 the
code automatically switches to dense matrix technique in the third phase of

the factorization (i.e. in this code there is no option for performing the computations without switching to dense matrix technique and, moreover, the point where switching is to be made is fixed during the partitioning).

The speed-ups (the time obtained when the code under consideration is run on one processor over the time obtained by the same code on 8 processors) are given in brackets in **Table 10.9**. The speed-ups for the third code are in general greater than the speed-ups for the other two codes. However, this does not lead always to the best results: the third code obtains a speed up greater than 6 for the two matrices **gemat**, while the computing times for these matrices are worse than those obtained by the other two codes. This illustrates why one should not look only at this parameter (the speed-up) when sparse codes for general matrices are compared.

Accuracy results are given in **Table 10.10**. There is no surprise here. The results obtained by **Y12M1** are the most accurate. The results with the other two codes are not so accurate. This is so because the accuracy requirements were relaxed in order to facilate the pivotal search; see **Section 10.4** and **Section 10.5**. However, the results obtained with **Y12M2** are pretty good and it should be accepted that the pivotal strategy used in this code is quite reliable. **Y12M3** also gives very accurate results for many matrices. Nevertheless, the pivotal strategy used in this code is not so reliable as the pivotal strategy in the other two codes. Therefore, this code should probably be run with some kind of iterative improvement. The experiment with this code was repeated by using **ORTHOMIN**, a method discussed in the next chapter. The accuracy for all systems becomes comparable with that obtained when **Y12M1** is used. The computing times were slightly increased (for many of the problems with no more than 1%).

10.7. Concluding remarks and references concerning parallel codes for general sparse matrices

The results presented in this section as well as many other results obtained in runs with several hundreds of test-matrices (including here test matrices used in [59,152,211-214] as well as matrices produced by the matrix generators introduced in **Chapter 3**) allow us to draw the following conclusions.

Remark 10.1 Switching to dense matrix technique seems to be a powerful tool for improving the performance of the code (with regard to computing time). Therefore switching may optionally be used in the first two codes discussed here. The main device for switching in these two codes, **Y12M1** and **Y12M2**, is very simple: the code switches to dense matrix technique (when switching is requested) if the density of the active part of the matrix becomes more than 10%. In the third code, **Y12M3**, the code switches always to dense matrix technique during the third phase of the factorization. ∎

Remark 10.2 The three codes that are optimized for the **ALLIANT** computer perform much better than the sequential codes with regard to the computing time used. For some problems the computing time is reduced by a factor more than 60 (see the results for **lns_3937** obtained by **MA28** and **Y12M1** in **Table 10.5** and **Table 10.9**). A price has to be paid for this greater efficiency: the parallel codes use more storage than the sequential codes. The accuracy of the first two codes, **Y12M1** and **Y12M2**, is normally comparable with that of the best sequential codes, while the third code, **Y12M3**, sometimes produces solutions with rather poor accuracy. ■

Remark 10.3 If the matrix decomposed is not very sparse, or if it is very sparse but many fill-ins are produced during the factorization, then **Y12M1** is usually better than the other two codes with regard to computing time. Moreover, **Y12M1** will normally give best accuracy among the three parallel codes (see **Table 10.10**) and it is also less storage consuming. ■

Remark 10.4 If the matrix is very sparse and if it stays sparse during the computations, then the second code, **Y12M2**, performs much better than **Y12M1** with regard to computing time (see, for example, the results for matrix **west2021** in **Table 10.9**). The accuracy of this code is normally quite acceptable. There is a potential danger that **Y12M2** will not preserve sparsity very well (the requirements for the preservation of sparsity are relaxed; see **Section 10.4**). However, all experiments indicate that this does not happen in practice. ■

Remark 10.5 The third code, **Y12M3**, produces best results for some matrices (as, for example, for the matrix **or6781hs** in **Table 10.9**). However, this code may lead to a very poor preservation of sparsity (see the results for the matrices **gemat11** and **gemat12** in **Table 10.9**) or to poor accuracy in comparison with the two other codes (see the results for **pde9511** in **Table 10.10**). There are several reasons for this: **(i)** the pivoting is restricted to the diagonal blocks, **(ii)** it is difficult (and, perhaps impossible) to achieve a good load balance since this depends on the fill-ins that appear in the different parts and these are not known in advance, **(iii)** the pivots found in the first phase of the factorization are accepted as good pivots also during the second factorization (see **Section 10.5**). ■

Remark 10.6 Since no code performs best in all situations, it is worthwhile to have all three codes (or codes based on the three ideas used in this chapter) in a good **parallel** package for general sparse matrices. If this is done, and if a long sequence of systems of linear algebraic equations with the same matrix (or similar matrices) is to be handled, then one will be able to choose the code which is best for the particular matrix (matrices) involved in the computational process. ■

Remark 10.7 In all three codes the most time-consuming parts of the computations were optimized. However, when this was done, then some parts of the computational work which were relatively cheap in the original **Y12M** became rather significant in the parallel versions. As a typical example the initial reordering of the structure (from an arbitrary order given by triples, a non-zero together with its row and column numbers, to structure ordered by rows) could be mentioned. The next step should be to try to optimize these parts of the computations (this is an important task for the cheap problems; such as, for example, the matrix **west2021**) ■

CHAPTER 11

PARALLEL ORTHOMIN FOR GENERAL SPARSE MATRICES

Consider again the system of linear algebraic equations Ax=b and assume, as in **Chapter 4** - **Chapter 10**, that A is a **general sparse matrix**; i.e. **(i)** A has no special property, such as symmetry or positive definiteness, **(ii)** A has no special structure, such as bandedness, **(iii)** A is large and contains many zeros. It has been shown that the simple iterative refinement with some kind of dropping of "small" non-zero elements during the factorization (**IR+D**) can successfully be used to improve the performance of a code for general sparse matrices on sequential machines (see **Chapter 5** - **Chapter 8**). On parallel machines, however, the situation changes. Assume that a parallel computer with shared memory and a few tightly connected processors is to be used. Assume also that the sparse code is highly parallelized in order to exploit the capabilities of such a computer. Then **IR+D** is not very efficient in comparison with the direct solution (without any dropping of "small" elements) of the system by Gaussian elimination, **DS**. This is so because the factorization process can be optimized quite well, while it is difficult to improve very much the performance of the back solver. The factorization time is by far the most expensive part when the **DS** is used, while very often the solution time dominates when the **IR+D** is used. In order to improve the performance when parallel computers are available, it seems to be necessary to find other iterative methods, in which the LU factorization is used as a preconditioner and which converge faster than simple **IR** (of course, faster convergence is also an advantage with sequential computation). The implementation of an **orthomin** algorithm in parallel sparse software will be discussed in this chapter. An important property of this method is that the number of Krylov vectors used is **not fixed in advance** (if some i, $1 \leq i \leq$ n, is given, then the Krylov vectors form the orthonormal basis of the Krylov subspace $\{x, Ax, A^2x, \ldots, A^{i-1}x\}$). The code starts with a prescribed number of Krylov vectors and increases their number by one (until some upper limit of the number of Krylov vectors is reached), whenever the convergence is judged, by the code, to be too slow with regard to number of iterations. In this way the code tries adaptively to form an optimal Krylov subspace that is most suitable for coefficient matrix of the particular problem solved. It is shown that such a non-traditional implementation of the classical orthomin algorithm performs nearly always better than **IR+D** and very often better than the **DS**.

In this chapter the analysis is restricted to a particular conjugate gradient-type method, a modified orthomin algorithm, in order to facilitate the exposition of the main ideas (an adaptive way to improve the accuracy of the preconditioner, an attempt to find the optimal number of Krylov vectors, and a careful choice of stopping criteria). However, these ideas are fairly general and can also be used in conjunction with other conjugate gradient-type methods. In fact, some experiments with two other such methods, GMRES and CGS, have already been carried out; see [105].

11.1. Iterative improvement for parallel sparse software

Many scientific and engineering problems lead, perhaps after some kind of space and/or time discretization, to the solution of systems of linear algebraic equations

(11.1)   Ax=b   ($A \in R^{n \times n}$, $b \in R^{n \times 1}$, $rank(A) = n$).

     Assume that A is **a general sparse matrix**, i.e. **(i)** A has no special
property, such as   symmetry and positive definiteness,   **(ii)** A has no
special structure, such as bandedness and **(iii)** A is large and contains
many zeros. It is appropriate then to apply some sparse matrix technique in
solving (11.1). The packages described in [77,82-84,122-125,331,341]   are
well-known. In this chapter the parallel versions of package **Y12M** that are
described in **Chapter 10** will be used to calculate in a cheap way a
preconditioner for a conjugate gradient-type method. Many of the ideas could
immediately be applied to the packages described in [77,82-84], while it is
not very clear how dropping of "small" non-zero elements (see below) could be
implemented in the software described in [122-125].

     The most straightforward method is to solve (11.1) directly by using
Gaussian elimination (**GE**). Three different implementations of **GE** were
discussed in **Chapter 10**. The abbreviation **DS** will be used in connection
with **any** of these three implementations, but mainly the second one (where
parallel pivots are used) will be discussed in this chapter.

     Sometimes it is worthwhile to try to avoid the use of "small" non-zero
elements during the factorization process (or, in other words, to apply the
computer-oriented manner of exploiting sparsity; see **Chapter 1**). Such
elements are, as soon as these are created, removed from the structure and not
used in the further computations; i.e. the "small" elements are **dropped**.
Both storage and factorization time could be saved when dropping is used.
Sometimes the savings are very considerable (see the previous chapters).

     The factors L and U obtained by **GE** are normally inaccurate when
dropping is applied and therefore some kind of iterative improvement of the
first approximation x_1 is to be performed, where

(11.2) $x_1 = (LU)^{-1}b$

     In **Chapter 5 - Chapter 8** it was demonstrated that simple **IR**
(iterative refinement) performed quite efficiently on sequential machines: **the
reduction in the factorization time was normally greater, and sometimes much
greater, than the extra time needed to perform the iterations.**

     However, the situation is different when the code is optimized for a
parallel machine. In this case the factorization time is reduced very
considerably (especially when a switch to dense matrix technique is applied);
see **Chapter 10**. At the same time the solution time, the time needed to
perform the iterative process, is not changed too much (compared with the
solution time in the case where the code is run in a sequential mode). This
implies that two major tasks must be solved in the process of obtaining an
efficient parallel code for iterative improvement of x_1:

     (i) The iterative process must be optimized. The crucial part is the
optimization of the substitutions $(LU)^{-1}z$, where L and U are inaccurate
factors obtained by some kind of dropping and z is some vector obtained
during the iterative process. Normally many substitutions are to be carried
out until an acceptable approximation is obtained and this is the most
expensive part of the computational work per iteration.

(ii) Iterative algorithms which converge quickly, also when the factors L and U are rather inaccurate, are to be used. This is importatant, because simple iterative refinement is either not convergent when the drop-tolerance is large or converges very slowly.

The solution of these two tasks, and especially of the second task, is discussed in the following sections. A modified version of an algorithm, developed by **Anderson and Saad [4]** and implemented in a code by **Anderson [3]**, is described in **Section 11.2**. A preconditioned orthomin algorithm is presented in **Section 11.3**. The problem of choosing the stopping criteria is discussed in **Section 11.4**. The use of dropping to get the preconditioner is sketched in **Section 11.5**. Numerical results are given in **Section 11.6**. Several conclusions are drawn in **Section 11.7**.

11.2. A modified Anderson-Saad algorithm

The first task is solved by using the algorithm proposed by **Anderson and Saad [4]**. The **Anderson-Saad** algorithm is implemented in a code by **Anderson [3]** under the assumption that an incomplete LU factorization (ILU) is used. An **ILU** factorization is, in principle at least, a factorization obtained by removing some non-zero elements, and if this definition is accepted then the computer-oriented manner of exploiting sparsity might be considered as an **ILU** factorization. However, the truth is that in the literature this name is used for a factorization that satisfies at least one of the following three conditions (and it must be emphasized that the problem is simplified considerably when these conditions are imposed).

(i) No **pivoting** is used.

(ii) The sparse storage scheme used is **static** and the non-zero elements are ordered by rows; in every row first the non-zero elements of L are stored, then the non-zero elements of U.

(iii) The structure is **compact and ordered**; there are no free locations within a row and the non-zero elements of row i+1 are stored immediately after the last non-zero element of row i, i=1,2,...,n-1.

Of course, any iterative process for **general** matrices which is based on the **ILU may** fail because: (a) a diagonal element may be zero or at least become very small and (b) large non-zero elements may be removed only because they do not appear within the initial sparsity pattern of the matrix.

The basic idea in the **Anderson-Saad** algorithm is very simple. Assume that a forward substitution is to be performed. The equations are grouped by "levels". All equations in level i (i=1,2,...,q) depend only on the variables of the equations of the previous levels j=1,2,...,i-1. Therefore these equations can be processed in parallel when all equations of the previous levels have been processed. If matrix L is sparse, then normally q<<n and, therefore, many equations can be distributed to the processors available. It is necessary to sort the equations into different levels, to initialize pointers for the first equations in each level and to set up linked lists for the equations of each level. This can be done by a special routine that has to be called before the start of the iterative process. Similar ideas can be used for the back substitution.

Matrix	Traditional solver	New solver - 8 p.	New solver - 1 p.	Iters
sherman3	8.98	2.84	13.21	33
gemat11	1.39	0.33	1.48	3
gemat12	1.05	0.25	1.08	2
lns_3937	1.26	0.53	2.00	3
lns3937a	5.35	1.94	7.91	17
saylr4	9.87	1.91	10.74	36
sheramn5	3.70	1.39	6.02	23
or6781hs	1.24	0.72	2.59	5
orsreg_1	3.52	0.60	3.52	21
west2021	0.49	0.09	0.46	3
hwatt_2	3.48	1.00	4.30	25
hwatt_1	2.50	0.71	3.10	18
west1505	0.57	0.13	0.60	5
nnc1374	0.45	0.19	0.65	3
mahistlh	0.34	0.10	0.38	4
pores_2	3.09	0.61	3.32	34
gre_1107	0.56	0.29	0.96	5
sherman4	0.92	0.32	1.39	18
gaff1104	3.54	1.43	5.55	41
sherman2	0.66	0.22	0.87	7
orsirr_1	1.83	0.33	1.89	24
sherman1	0.59	0.17	0.72	10
jpwh_991	0.56	0.15	0.64	8
west0989	0.29	0.06	0.29	4
pde_9511	0.63	0.18	0.73	9
mc_fe	0.35	0.17	0.56	4
steam2	0.08	0.02	0.07	1

Table 11.1
Computing times for the solution part of the computations
with the orthomin algorithm, which are
obtained when Harwell-Boeing matrices are run with the
traditional forward-back solver (on 8 processors) and the modified
Anderson-Saad algorithm on 8 processors and on 1 processor.
The numbers of iterations used are given in the last column.
Some characteristics of the matrices are given in Table 10.4.

The method is in general not very efficient when the system is solved directly: the gain obtained by performing the forward and back substitutions in parallel is not very great because some extra work is needed to order the equations by levels. The process of setting up the levels is sequential. However, if an iterative process is used, then many iterations are normally needed and the gain from the parallelization of the algorithm more than compensates for the time spent in reordering the structure by levels.

In package **Y12M** pivoting is used and, moreover, the structure is **dynamically** changed (see **Chapter 2**). The use of dynamical changes (needed to store new non-zero elements, **fill-ins**, when these are large) and pivoting (needed because the matrix is general; see **Wilkinson [270]**) leads to destruction of both the compactness and the order of the structure. That is there may be free locations between any two rows, and the elements of row i+1 are not necessarily stored after the elements of row i, i=1,2,...,n-1).

Moreover, in the version optimized for the **ALLIANT** (**Chapter 10**) the situation is even more complicated, because a **switch to dense matrix technique** may be carried out. Therefore the forward-back substitution must be calculated in three steps: **(i)** perform a forward substitution by using the sparse part of matrix L (which is a unit lower trapezoidal matrix), **(ii)** perform both forward and back substitution by using the dense parts of L and U (which are a unit lower triangular matrix and an upper triangular matrix respectively) and **(iii)** perform a back sustitution by using the sparse part of matrix U (which is an upper trapezoidal matrix).

All necessary modifications in the **Anderson-Saad** algorithm were made so that the new algorithm works in the case where: **(i)** the rows are not ordered in the storage structure (i.e. the non-zeros of row $i+1$ are not necessarily stored immediately after the last non-zero of row i; in fact, these may be stored even before the first non-zero of row i, $i=1,2,...n-1$), **(ii)** the storage structure is no longer compact (there may be free locations between some rows), and **(iii)** trapezoidal sparse factors are to be treated (and some computations with dense triangular factors between the sparse forward sustitution and the sparse back substitution are to be performed).

The effect of the optimizing the solution part (the iteration process) is illustrated in Table 11.1. The matrices from the Harwell-Boeing set [71,72], which were used in **Chapter 10**, are also used in this experiment. The preconditioned orthomin algorithm, to be discussed in the following sections, is used in the runs. The preconditioners are obtained by using an initial drop-tolerance $T=2^{-6}$, and the code can sometimes reduce T. The use of such a variable drop-tolerance will be described in the following sections. In the second column the computing times obtained by the **Y12M2** with the old back solver (see **Chapter 10**) are given. The third column presents the computing times obtained when **Y12M2** is used together with the modified **Anderson-Saad** algorithm. The fourth column contains the results obtained by running the latter algorithm on one processor. Finally, the numbers of iterations (these are the same for the three runs) are given. It should be noted that it was required to achieve an accuracy of $ACCUR=10^{-4}$, and that this requirement was satisfied in all runs. In fact, because of the determinism of the algorithms used, the same accuracy was achieved for the three runs with every matrix (it is perhaps desirable to have a deteministic parallel algorithm, but not all parallel algorithms are deterministic; the parallel pivotal search in [50] is an example for a non-deterministic algorithm).

The comparison of the results in columns two and three in **Table 11.1** shows the superiority of the modified **Anderson-Saad** algorithm over the traditional performance of forward-back substitutions. The efficiency of the modified **Anderson-Saad** algorithm for parallel architectures is seen by comparing the third and the fourth columns.

The forward-back substitutions are the most time-consuming parts of the computational process, but there are also other parts that should be taken into account: the matrix-vector multiplications and the inner products. However, these operations are very simple and can easily be optimized. Therefore these parts will not be discussed here (some algorithms from [1,49,52,54,56-58,61,65,102-103,328] can be applied in these parts and also in the dense part of the forward and back substitutions).

11.3. Modification of the traditional orthomin algorithm

Assume that the factors L and U are to be used as preconditioners in some iterative method. Then the well-known iterative refinement could be considered as the simplest preconditioned iterative method. Indeed, the iterative refinement (see **Chapter 3** and **Chapter 5**) is defined by

$$(11.3) \quad r_i=b-Ax_i, \quad d_i=(LU)^{-1}r_i, \quad x_{i+1}=x_i+d_i, \quad i=1,2,\ldots,p,$$

where p is the iteration at which the process is stopped because the accuracy requirement imposed by the user is satisfied (or because the iterative process either is not convergent or the convergence speed is very slow). It is clear that (11.3) can be rewritten as

$$(11.4) \quad x_{i+1}=x_i+(LU)^{-1}b-(LU)^{-1}Ax_i,$$

which is an implementation of the functional iteration to the preconditioned system of linear algebraic equations $(LU)^{-1}Ax=(LU)^{-1}b$. While on sequential machines iterative refinement gives good results (compared with the direct solution; see **Chapter 5**), this is in general not true on parallel machines like the **ALLIANT**. On such machines one can improve the performance of the direct sparse solver up to 20-30 times (mainly by using parallel pivots and by switching to dense matrix technique; see **Chapter 10**). At the same time, the improvement in the major tasks during the iterative process (substitutions and matrix vector-multiplications) is not so great; even when the two major tasks are thoroughly optimized. Therefore it is crucial to attempt to find iterative methods that converge faster than iterative refinement. Preconditioned conjugate gradient-type methods can efficiently be used. The orthomin algorithm proposed originally in [258] (see also [81]) can successfully be used instead of the iterative refinement.

The orthomin algorithm is normally used with a prescribed number of Krylov vectors; the notation **orthomin(k)** being used to indicate that the last k Krylov vectors are kept and used in the calculations. Very often the **orthomin(k)** with a fixed in advance value of k is not the best choice. There are at least two reasons for this:

> **(i)** it is difficult to find the optimal value of k (a small value may lead to a divergent process, while a too large value leads to excessive computing time),

> **(ii)** if a time-dependent problem is solved, then the optimal value of k may be different in the different parts of the time-interval (and in many examples it was different; as a rule k=1 was optimal in the beginning, while it was necessary to increase gradually k after the starting phase).

Therefore **it is important to leave the task for a determination of a good value of k to the code itself.** A modification of the classical orthomin algorithm, in which the code tries in a self-adaptive way to find the value of k that is best for the problem being solved, will be discussed in this section.

Step 1 - Initialize: $r_0 = (LU)^{-1}(b-Ax)$, $p_0 = r_0$, $q_0 = (LU)^{-1}Ap_0$ and $i=0$.

Step 2 - Perform iteration step i by carrying out the following computations:

$$\alpha_i = (r_i, q_i)/(q_i, q_i),$$

$$x_{i+1} = x_i + \alpha_i p_i, \quad r_{i+1} = r_i - \alpha_i q_i, \quad s_{i+1} = (LU)^{-1}Ar_{i+1},$$

$$\beta_j = -(s_{i+1}, q_j)/(q_j, q_j), \quad j = \max(i-k+1,1), \max(i-k+2,1), \ldots, i,$$

$$p_{i+1} = r_{i+1} + \sum_{j=i-k+1}^{i} b_j p_j, \quad q_{i+1} = s_{i+1} + \sum_{j=i-k+1}^{i} b_j q_j.$$

Step 3 - Perform stopping criteria:

    (a) If there is no hope that the process will converge sufficiently fast and the accuracy required will be achieved at a reasonable price, then stop the iteration, recalculate the preconditioners L and U and repeat the whole process from Step 1.

    (b) If the iterative process is slowly convergent but there is some hope that the convergence rate could be accelerated by taking more Krylov vectors into account, then increase both **k** and **i** by one and go to Step 2.

    (c) If the convergence rate is judged to be fast, but the accuracy required is not achieved yet then increase **i** by one and go to Step 2.

    (d) If the convergence rate is judged to be fast, and the accuracy required is achieved, then go to Step 4.

Step 4 - Calculate an error estimate and stop the computations.

Figure 11.1
The modified orthomin preconditioned algorithm.

The orthomin algorithm can be introduced by the scheme given in **Fig. 11.1.** It is assumed that some incomplete LU factorization of matrix A is calculated and the factors L and U are used as preconditioners (the calculation of L and U will be discussed in the following section). It is also assumed that one starts with some prescribed initial number of Krylov vectors, $k = k_{start}$, and stops increasing the number of these vectors when some prescribed number, $k = k_{end}$ is reached.

Remark 11.1. In the algorithm given in **Fig. 11.1** the stopping criteria are separated from the operations performed during the i'th iteration. This is done only in order to facilitate the exposition of the algorithm. In fact, many of the stopping criteria are performed before the calculation of some quantities during the i'th iteration. For example, α_i is calculated only if (q_i, q_i) is a positive number that is not too small; when this is not true the iteration is stopped with an error message. ∎

Remark 11.2. The main difference between the algorithm given in **Fig. 11.1** and the traditionally used orthomin algorithm is the check made in **Step 3.b.** By this check an attempt is made to adjust the number of Krylov vectors for the particular matrix treated. ∎

11.4. Main ideas used in the stopping criteria

It is important to select correctly stopping criteria for the algorithm presented in the previous section. It is desirable to stop the iteration as quickly as possible, but also to obatin a reliable evaluation of the error in the approximation accepted as a solution. These two requirements work in opposite direction and a compromise is necessary.

It must be emphasized that the stopping criteria are as a rule neglected in the literature on conjugate gradient-type methods. In many papers the authors merely state: **"Until convergence do"** without defining what is the meaning of this expression. In many other papers the authors do not try to estimate the error in the solution at all, but just state that the calculations were carried out until some norm of the residual vector becomes smaller than certain quantity.

In order to find out why it is difficult to construct a good stopping criteria, one should try to find out where the difficulties come from. It is clear that the key operation in the algorithm is $x_{i+1} = x_i + \alpha_i p_i$. Assume now that the process is convergent. Then

$$(11.5) \quad x_i \rightarrow x \text{ as } i \rightarrow \infty \quad \Rightarrow \quad x = x_0 + \sum_{i=0}^{\infty} \alpha_i p_i \quad \Rightarrow \quad \alpha_i p_i \rightarrow 0 \text{ as } i \rightarrow \infty .$$

Unfortunately, the reverse relations do not hold. Nevertheless, it is reasonable to ask when will it be likely to expect that

$$(11.6) \quad \alpha_i p_i \rightarrow 0 \text{ as } i \rightarrow \infty \quad \Rightarrow \quad x_i \rightarrow x \text{ as } i \rightarrow \infty .$$

It is clear that there are good reasons to expect (11.6) to hold when $\alpha_i = 1$ for $\forall i$. The simple **IR** is an iterative algorithm for which this condition is satisfied. Thus, **IR** is an algorithm in which rather safe stopping criteria can be selected (but the convergence rate of this algorithm could be too slow).

If α_i varies from one iteration to another, then there are at least two situations, $\alpha_i \rightarrow 0$ and $\alpha_i \rightarrow \infty$, where it is apparent that (11.6) will normally not hold. An example is given in **Table 11.2** to show that the behaviour of parameter α_i should be studied carefully, especially when the drop-tolerance T is large.

The results in **Table 11.2** show that it does not matter too much whether
the relative norm of the difference between two iterates is used or the
relative residual norm is used: both may be unreliable when the parameter α_i
is too small (the code may report good results, while the actual solution is
quite wrong). Similar results were obtained with the code from [3] where the
relative residual norm is used; see [105].

Exponent	Correction norm				Residual norm			
	Est.	Exact	α	Iters	Est.	Exact	α	Iters
-5	8.0E-5	7.1E 0	2.1E-4	11	1.9E-5	8.1E 0	3.5E-4	22
-8	1.3E-5	8.8E-3	9.5E-4	27	3.8E-5	1.8E-2	3.3E-4	15
-13	5.4E-5	8.9E-4	8.3E-2	4	5.1E-5	3.7E-4	8.9E-2	5

Table 11.2
Results obtained when the values of α_i are not taken into account
during the stopping criteria and the evaluation of the error made.
The accuracy required is given by ACCUR = 10^{-4}.
$T=2^{-m}$ is used and under "Exponent" the exponent m is given.
Under "Est." the error estimated by the code is given. The exact
errors are given under "Exact". The minimal value of α_i in the run
is given under "α". Under "Iters" the numbers of iterations are given.
"Correction norm" means that in the stopping criterion the code tries
to estimate the error by using $\|x_{i+1}-x_i\|/\|x_i\|$. "Residual norm" means
that in the stopping criterion the code tries to estimate the error
by using $\|r_i\|/\|r_0\|$.

The above discussion indicates that one should check the behaviour of
α_i. This parameter:

(1)	should not decrease too quickly,
(2)	should not increase too quickly,
(3)	should not oscillate too much.

Assume that these three conditions are satisfied and consider some i
at which we should like to stop the iteration. Assume also that the process
is convergent. Then the following estimate of the error at iteration i can
be given:

$$(11.7) \quad \|x-x_i\| \le |\alpha_i| \, \|p_i\| \, (1 + \sum_{j=i+1}^{\infty} |\alpha_j| \, \|p_j\| \, / \, |\alpha_i| \, \|p_i\|) \, .$$

Assume that

$$(11.8) \quad |\alpha_{j+1}| \, \|p_{j+1}\| \, / \, |\alpha_j| \, \|p_j\| = \text{RATE} < 1 \quad \text{for} \quad \forall \, j > i \, .$$

Then (11.7) reduces to

(11.9) $\|x-x_i\| \leq |\alpha_i| \ \|p_i\| \ [\ \sum_{j=0}^{\infty} (RATE)^j \ ]$

and, since RATE < 1,   (11.9) could be replaced by

(11.10) $\|x-x_i\| \leq |\alpha_i| \ \|p_i\| \ / \ (1-RATE)$.

In general, RATE from (11.8) will vary from one iteration to another. Denote by $RATE_i$ the value of RATE at iteration i. If $RATE_i$ do not vary too much during several sucessive iterations, this means that one can try to evaluate the error made by (11.10) with RATE replaced by some appropriate parameter (as, for example the mean value of $RATE_i$ during a few last iterations; see [105], p. 186)

It is not necessary to go into further details. It is sufficient to emphasize that an attempt to evaluate the contribution of the sum in (11.7) to the error estimate is made by using the ideas described above (in general this sum is completely neglected in the error estimates commonly used in the codes). It should also be emphasized that the particular implementation of this idea that is described in [105], pp. 185-186, is used in the code; numerical results will be given in the next section.

Remark 11.3. The same ideas could be applied to formulate stopping criteria in which the residual vector is involved. However, in this case one attempts to control the size of the residual instead of the error in solution vector. If the error in the solution is needed (which is natural), then one should be aware that in some situations it is neither true that small residuals imply that error in the solution is also small nor that large resuduals mean that the error in the solution is large. This is discussed in more detail in [105] p. 185.

11.5. How to calculate the preconditioners?

It is very popular (and very easy) to calculate the preconditioners by allowing fill-ins in prescribed positions only. For example, very often in the well-known ILU factorization (see Section 11.2) no fill-ins are allowed. Moreover, the factorization is normally carried out without pivoting when this algorithm is used. The **ILU** can be generalized by allowing some prescribed in advance number of fill-ins (one may allow at most, say, k fill-ins per row). Finally, one could also introduce numerical criteria: one may still keep at most k fill-ins per row, but try to select for keeping the k largest in absolute value fill-ins.

There are numerous variations of devices used, but the main ideas remain the same: **(i) one prescribes in advance some sparsity pattern and throws away elements that do not fit into this, and (ii) one omits pivoting for numerical stability.** Methods based on these ideas are not only much simpler, but also very efficient for some special matrices (even for some non-symmetric matrices). Moreover, matrices suitable for such methods often appear in the treatment of scientific and engineering problems. However, while there is no doubt about the usefulness of these methods for many applications, one should emphasize that they may fail when general matrices are treated. The authors of **ITPACK** and **NSPCG** [159-160,183] recognize this fact and try to collect a huge set of methods, which can be applied with different preconditioners, hoping that a potential user may find a method among the set of methods that is suitable for his/her particular problem.

However, it may be rather difficult to find the right method among a very big set of methods. Moreover, if a time-dependent problem is solved, which is often the case, then different methods may be suitable for the matrices arising in the different parts of the time interval. This shows that **it may be more profitable to try to adapt (automatically, in the code) the preconditioned matrix $C=(LU)^{-1}A$ to the method selected than to try to select the best method for the matrix treated.** This can be done by calculating the preconditioner LU using dropping based on the size of the elements that are neglected without putting any condition on the pattern that the preconditioner must have; **thus, recognizing the fundamental fact that it is impossible to predict the pattern of a good preconditioner for a general sparse matrix.** If this strategy is followed, then one has certain control on the accuracy of the preconditioner: the accuracy can in general be improved by reducing the drop-tolerance. Moreover, this improvement could be achieved automatically; by the code itself.

The particular algorithm for dropping is described in [104,105]. It could be outlined as follows. The matrix is scaled before the start of the computations. Elements that are small relative to the largest element in the active part of its row are dropped: this means that if the computations of stage s (s=1,2,...,n-1) of the Gaussian elimination are carried out and if

$$(11.11) \qquad \left| a_{ij}^{(s+1)} \right| \leq T \max_{s+1 \leq j \leq n} \left(\left| a_{ij}^{(s)} \right| \right), \qquad i = s+1, s+2, \ldots, n, \qquad 0 \leq T < 1,$$

then $a_{ij}^{(s+1)}$ is considered as a small element and dropped. If the LU preconditioner computed by using (11.11) is not sufficiently accurate (the iterative method used does not converge or converges very slowly), then T is reduced and the calculations are repeated (a new LU factorization is computed). This could happen several times. Thus, more and more accurate preconditioners are calculated and the matrix $C=(LU)^{-1}A$ of the preconditioned system will eventually become suitable to the iterative method used. This is carried out automatically by the code and the user is completely freed for thinking about the properties of the matrix treated and its suitability to one or another iterative method.

The main advantage of the method is its **robustness**. Pivoting for stability (which is essential for general matrices) is used during the factorization. Every fill-in which is judged to be large enough is kept. The method will store as many fill-ins per row as needed. In the methods where one tries to prescribe some pattern of LU in advance, the number of fill-ins per row is restricted to, say, k, even if an attempt to drop only small elements is made (if k+1 fill-ins are large, such a method will drop k+1'st just because there is no place for it). Of course, the iterative process may fail to converge **for general matrices,** even if the "small" fill-ins are removed as in the method discussed in this chapter (because parameter T, the drop-tolerance, is large). However, if this happens, then a well-determined action is prescribed. The method will correct itself; it will automatically improve the quality of the preconditioners by reducing the drop-tolerance T.

11.6. Numerical results

The effect of performing the operations with the preconditioner in parallel, together with using a modified orthomin algorithm in which the code adaptively tries to determine an optimal number of vectors for the Krylov subspace, will be illustrated in this section by several numerical experiments. The same test-matrices as in Chapter 10 will be used. These matrices together with some of their characteristics are listed in Table 10.4.

Only the version where parallel pivots are selected, Y12M2, will be used here (parallel pivots have also been applied by Davis and Yew [51]; see also [50]). It has been demonstrated in Chapter 10 that Y12M2 is especially efficient when sparsity is preserved well during the factorization. Precisely this is likely to happen when dropping is used.

The right-hand side vectors were computed so that all components of the solution vectors are equal to one. The code tries to estimate the error made and to stop the iteration if $\|x-x_i\|/\|x\| \le$ ACCUR. The accuracy required was given by ACCUR=10^{-4}. Both the estimated errors and the exact errors will be given.

In the basic run, the code starts the computations with $T=2^{-6}$ and reduces T by a factor of 2^5 when the iterative process is not convergent or slowly convergent, and by a factor of 2^{10} when the factorization is not completed (because a row or a column without non-zero elements in its active part is found at some stage of the factorization). Thus, the code may try several times to get a sufficiently accurate preconditioner. For each system the sums of the computing times for all trials are given in Table 11.3.

Six other runs were performed; with starting tolerances: $T=2^{-k}$ (k=1,2,3,4,5,7). The drop-tolerance was varied as for $T=2^{-6}$ (see above). The best computing times for every matrix (among the computing times obtained in the seven runs) are also given in Table 11.3 together with the values of the starting drop-tolerances T for which the best results are achieved.

It is seen from Table 11.3 that the results, which are obtained with the preconditioned orthomin algorithm, are often better than those obtained by the direct method. The runs with different starting values of T indicate that for many of the systems solved the use of just some large positive value of T gives good results. However, there are also matrices which are rather sensitive to the starting value of the drop-tolerance. This is an important fact, which can efficiently be exploited when time-dependent problems are to be solved. In such a case one can try to adjust the value of T at the beginning and then use this value during the treatment of the whole sequence of systems. This idea has been exploited in the solution of linear ordinary differential equations (ODE's) arising in nuclear magnetic resonance spectroscopy, Chapter 8, but in Chapter 8 IR is used instead of the preconditioned orthomin algorithm.

The storage required by the direct solver is compared with the storage required by the preconditioned orthomin algorithm (used with a starting drop-tolerance $T=2^{-6}$) in Table 11.4. COUNT is the largest number of non-zero elements kept in the array where the sparse part of the preconditioner is stored. DENSE is the order of the dense part of the preconditioner. Neither

COUNT nor DENSE is known in advance. Nevertheless, the results in **Table 11.4** show in a very clear way that **one should expect the preconditioned orthomin algorithm to be much less storage consuming.** Indeed, COUNT is reduced more than 10 times for some problems and the reduction of DENSE is normally very considerable. Thus, it is safe to reserve considerably less storage for the run when the preconditioned orthomin algorithm is in use.

The reliability of the accuracy test in the stopping criteria of the preconditioned orthomin algorithm is demonstrated on **Table 11.5**. It is seen that: **(i)** the accuracy requirement is satisfied for all 27 examples used in this experiment and **(ii)** the evaluated error estimate computed by the code is as a rule between the accuracy requirement and the accuracy actually achieved.

Matrix	Y12M2	ORTMIN: BASIC	ORTMIN: BEST	BEST T
sherman3	19.74	8.99 (1)	8.76 (1)	0.125
gemat11	5.17	10.99 (2)	10.91 (2)	0.0078125
gemat12	5.24	10.95 (2)	10.84 (2)	0.0078125
lns_3937	25.27	22.16 (2)	15.51 (2)	0.5
lns3937a	25.17	20.31 (2)	16.31 (2)	0.5
saylr4	35.32	5.76 (1)	5.67 (1)	0.0625
sherman5	8.34	5.61 (1)	4.81 (1)	0.125
or678lhs	9.54	14.92 (1)	14.64 (1)	0.0078125
orsreg_1	11.88	2.93 (1)	2.92 (1)	0.0078125
west2021	1.19	2.45 (2)	2.45 (1)	0.015625
hwatt_2	9.99	4.40 (1)	2.93 (1)	0.125
hwatt_1	9.81	4.21 (1)	2.79 (1)	0.25
west1505	0.86	1.90 (1)	1.63 (2)	0.0078125
nnc1374	3.21	5.63 (4)	5.63 (4)	0.015625
mahistlh	0.98	1.58 (1)	1.55 (1)	0.0625
pores_2	3.42	2.18 (1)	2.18 (1)	0.015625
gre_1107	2.17	8.71 (3)	6.85 (3)	0.5
sherman4	1.03	1.40 (1)	1.06 (1)	0.25
gaff1104	5.38	4.71 (1)	4.71 (1)	0.15625
sherman2	6.22	4.15 (1)	4.13 (1)	0.0078125
orsirr_1	3.12	1.45 (1)	1.37 (1)	0.0078125
sherman1	1.04	1.06 (1)	1.04 (1)	0.0625
jpwh_991	1.65	1.60 (1)	1.17 (1)	0.125
west0989	0.54	1.20 (1)	1.12 (2)	0.0625
pde_9511	2.00	1.77 (1)	1.20 (1)	0.25
mc_fe	3.28	4.12 (1)	3.96 (1)	0.5
steam2	1.15	0.95 (1)	0.91 (1)	0.0078125

Table 11.3
Computing times obtained by running Y12M2 in three modes.
Under "Y12M2" the results obtained by the direct solver are given.
Under "ORTMIN: BASIC" the results obtained by the preconditioned
orthomin method when a starting $T=2^{-6}$ is used to get the preconditioner.
Under "ORTMIN: BEST" the best results obtained by the preconditioned
orthomin method when seven starting drop-tolerances are used to get the
preconditioner. The numbers of trials are given in brackets.
Under "BEST T" the values of the starting drop-tolerances,
for which the best results are found, are given.

Matrix	DIRECT SOLVER: Y12M2		ORTHOMIN WITH T=0.015625	
	COUNT	DENSE	COUNT	DENSE
sherman3	163630	981	36123	250
gemat11	42789	320	43003	295
gemat12	43133	326	43288	319
lns_3937	199078	1078	110928	596
lns3937a	195183	1072	85682	450
saylr4	331418	1195	19715	7
sherman5	78418	665	22544	255
or678lhs	95464	521	89528	565
orsreg_1	135388	732	12310	15
west2021	8400	99	9249	61
hwatt_2	107320	705	23598	180
hwatt_1	109377	699	24128	177
west1505	6497	62	6651	125
nnc1374	34966	389	28835	325
mahistlh	9650	91	8268	71
pores_2	40013	404	8754	31
gre_1107	26390	364	26637	365
sherman4	11332	242	6971	102
gaff1104	64093	546	22971	204
sherman2	71465	552	22484	142
orsirr_1	42808	412	6067	12
sherman1	11320	241	4912	99
jpwh_991	19643	318	9900	98
west0989	4284	57	4400	49
pde_9511	25302	307	12490	115
mc_fe	43771	385	26073	240
steam2	15461	241	5110	50

Table 11.4
Storage characteristics obtained by running Y12M2 in two modes.
Under "COUNT" the largest numbers of non-zeros
kept during factorization are given.
Under "DENSE" the orders of the dense matrices factorized after
switch to dense matrix technique are given.

11.7. Concluding remarks and references concerning the parallel preconditioned conjugate gradient-type methods

The implementation of a preconditioned orthomin algorithm for solving systems of algebraic equations with general sparse matrices is discussed in this chapter. Since no assumption is made about the matrix, it is important to be able to increase gradually the accuracy of the LU preconditioner so that orthomin becomes convergent for the preconditioned matrix. It is easy to prove that such an algorithm will always succeed when it is implemented in a proper way: note only that if the exact LU is computed, then the preconditioned matrix is the identity matrix and the orthomin algorithm will converge in one iteration. The difficulty is the implementation itself, because of (1) the complexity of the sparse storage schemes used (no assumption about the sparsity pattern of the matrix is made) and (2) the stopping criteria (the orthomin performs rather erratically when it is applied to a general matrix and it is very important to decide correctly whether the iteration should be continued or stopped). However, if the implementation is properly done, then a very robust algorithm for treatment systems with general sparse matrices will be achieved.

Matrix	Est. error	Exact error	Iterations	Krylov vectors
sherman3	7.7E-5	7.3E-5	33	9
gemat11	7.3E-6	1.6E-7	3	1
gemat12	7.8E-5	3.8E-8	2	1
lns_3937	3.6E-6	1.8E-8	3	1
lns3937a	1.5E-5	3.1E-7	17	6
saylr4	3.6E-5	5.9E-7	36	13
sherman5	2.2E-5	1.2E-6	23	9
or6781hs	1.7E-5	4.1E-7	5	1
orsreg_1	8.0E-5	8.5E-6	21	4
west2021	1.7E-6	1.0E-8	3	1
hwatt_2	6.9E-5	1.9E-6	25	5
hwatt_1	2.7E-5	3.2E-6	18	4
west1505	2.5E-5	1.7E-6	5	1
nnc1374	2.1E-6	3.3E-7	3	1
mahistlh	6.5E-5	1.6E-6	4	1
pores_2	3.9E-5	9.7E-6	34	10
gre_1107	2.0E-6	2.4E-7	5	2
sherman4	4.6E-5	7.4E-6	18	4
gaff1104	1.9E-5	2.7E-5	41	18
sherman2	5.8E-5	9.5E-6	7	1
orsirr_1	7.4E-5	3.0E-6	24	2
sherman1	7.7E-5	1.4E-5	10	1
jpwh_991	1.0E-5	1.6E-7	8	1
west0989	2.7E-5	5.8E-7	4	1
pde_9511	4.8E-5	2.5E-6	9	1
mc_fe	8.8E-5	1.3E-6	4	1
steam2	6.7E-8	8.4E-16	1	1

Table 11.5
Accuracy obtained by running the orthomin algorithm.
Under "Est. error" the estimated by the code errors are given.
Under "Exact error" the exact errors are given.
The numbers of iterations and the numbers of Krilov vectors used
are given under "Iterations" and "Krylov vectors" respectively.
The accuracy required is 10^{-4} and relative errors are calculated.

Other algorithms, like the ILU factorization with all of its modifications, have been proposed in the literature (see, for example, [34,45,46,59,80,81,101,196,208-210,227-229,251-252,268-269]). While such algorithm are very efficient for some types of nonsymmetric matrices, they do not guarantee convergence for any general matrix. Most of these algorithms are implemented without pivoting. This means that any such algorithm will fail if a diagonal element of the matrix is zero (or at least the user is forced to check if this is so, and if it is, then some permutations, to get a zero-free main diagonal, are to be carried out). Moreover, even when the preconditioner is computed without problems, the iterative process may fail to converge, because large elements have been dropped only because these do not suit for the pattern that is chosen in advance (this is also true when numerical criteria for dropping are used in these schemes). It is not clear what the user should do when this happens. On the other hand, in the method studied in this chapter the user has at his disposal a parameter, the drop-tolerance, which could easily be used (by the code itself) to improve the accuracy of the preconditioner and, thus, to achieve convergence.

Many of the ideas applied to develop the method studied here can also be used with other iterative methods (as, for example, those that are discussed in [10-17,29,34,87-89,93-96,139,157,185-186,190,204-210,257,261,281-287]. In fact some tests with the **GMRES** algorithm [210] and with the **CGS** algorithm [158,229] have been carried out in [104]. However, the sparse code used in [104] was only partially optimized (parallel pivots are not used there, no switch to dense technique is made and the forward-back substitutions are not optimized). It would be worthwhile to try these two methods (and also some other iterative methods) together with the highly efficient parallel techniques (applied in **Y12M1** or in **Y12M2**) that are discussed in **Chapter 10** and in this chapter.

Similar ideas could also be applied to the method based on partitioning with rectangular diagonal blocks (a direct solver, **Y12M3**, based on this technique is discussed in **Chapter 10**).

CHAPTER 12

ORTHOGONALIZATION METHODS

Consider a matrix $A \in R^{m \times n}$ ($m \geq n$). Assume that $rank(A)=n$. There exists many methods by which matrix A is decomposed into

(12.1) $A = QDR$,

where $Q \in R^{m \times n}$ has orthonormal columns ($Q^T Q = I$, I being the identity matrix in $R^{n \times n}$), $D \in R^{n \times n}$ is a diagonal matrix and $R \in R^{n \times n}$ is an upper triangular matrix. Very often matrix D is the identity matrix and if this is so, then (12.1) is reduced to

(12.2) $A = QR$.

The methods in which either the decomposition (12.1) or the decomposition (12.2) is calculated are called **orthogonalization methods**. Several such methods will shortly be discussed in this chapter. The main purpose of this discussion is to show that **the method of plane rotations** (the Givens method, [127,128], or a modification proposed by Gentleman, [109,110,141]) is the best one among the orthogonalization methods when **general sparse matrices** are to be decomposed. An implementation for sparse matrices, in which the Gentleman version of the Givens method is applied, will be described in the following chapters. The problems concerning the choice of a pivotal strategy, the use of a drop-tolerance and the improvement of the starting approximation by some iterative method will also be discussed in the following chapters.

12.1. Householder orthogonalization

One of the most popular orthogonalization methods (at least when dense and band matrices are to be treated) is the Householder orthogonalization, which is based on the use of elementary reflectors. An elementary reflector is a matrix of the form

(12.3) $Q = I_m - 2qq^T$ ($I_m \in R^{m \times m}$, $Q \in R^{m \times m}$, $q \in R^{m \times 1}$, $q^T q = 1$),

where I_m is the identity matrix in $R^{m \times m}$.

It is easy to verify that if $Q \in R^{m \times m}$ is an elementary reflector, then the following conditions are satisfied:

(a) Q is symmetric ($Q^T = Q$),

(b) Q is orthogonal ($Q^T Q = I_m$),

(c) Q is involutary ($Q^2 = I_m$).

Assume that matrix $A \in R^{m \times n}$ ($m \geq n$) is already transformed (by carrying out multiplications with $s<n$ elementary reflectors) into a matrix $A_s \in R^{m \times n}$ which can be partitioned as

$$(12.4) \quad A_s = \begin{vmatrix} R_s & B_s \\ 0 & C_s \end{vmatrix} \qquad (\ R_s \in R^{sxs}, \qquad B_s \in R^{sx(n-s)}, \qquad C_s \in R^{(m-s)x(n-s)} \ ),$$

where R_s is an upper triangular matrix. Consider a vector q whose first s components are zeros. Partition this vector as

$$(12.5) \quad q = \begin{vmatrix} 0 \\ v \end{vmatrix}, \qquad v \in R^{(m-s)x1}.$$

Then the corresponding elementary reflector can be partioned as

$$(12.6) \quad Q_s = I_m - 2qq^T = \begin{vmatrix} I_s & 0 \\ 0 & (I_{m-s} - 2vv^T) \end{vmatrix}, \qquad v^Tv=1,$$

and the multiplication of matrix A_s by the elementary reflector Q_s (from the left) leads to a modification of matrix C_s only since

$$(12.7) \quad Q_sA_s = A_{s+1} = \begin{vmatrix} R_s & B_s \\ 0 & (I_{m-s} - 2vv^T)C_s \end{vmatrix}.$$

The components of vector v must be chosen so that all but the first element in the first column of matrix $(I_{m-s} - 2vv^T)C_s$ are equal to zero. Denote by

$$(12.8) \quad a^{(s)}_{s+1,s+1}, \ a^{(s)}_{s+2,s+1}, \ \ldots, \ a^{(s)}_{m,s+1},$$

the elements of the first column of matrix C_s and by

$$(12.9) \quad a^{(s+1)}_{s+1,s+1}, \ a^{(s+1)}_{s+2,s+1}, \ \ldots, \ a^{(s+1)}_{m,s+1},$$

the elements of the first column of matrix $(I_{m-s} - 2vv^T)C_s$.

Let

$$(12.10) \quad \sigma_{s+1} = sign[a^{(s)}_{s+1,s+1}]\{[a^{(s)}_{s+1,s+1}]^2 + [a^{(s)}_{s+2,s+1}]^2 + \ldots + [a^{(s)}_{m,s+1}]^2\}^{\frac{1}{2}},$$

$$(12.11) \quad w_1 = a^{(s)}_{s+1,s+1} + \sigma_{s+1}. \quad w_i = a^{(s)}_{s+i,s+1} \qquad (\ i=2,3,\ldots,m-s \ ),$$

$$(12.12) \quad w = (w_1, \ w_2, \ \ldots, \ w_{m-s})^T,$$

$$(12.13) \quad \rho_{s+1} = 0.5\|w\|^2_2 \ .$$

Set

$$(12.14) \quad v \overset{def}{=} w/\|w\|_2.$$

Then

$$(12.15) \quad I_{m-s} - 2vv^T = I_{m-s} - ww^T/\rho_{s+1}$$

is an elementary reflector and, moreover, the components in the first column of matrix $(I_{m-s} - 2vv^T)C_s$ are

$$(12.16) \quad a_{s+1,s+1}^{(s+1)} = \sigma_{s+1}, \quad a_{s+i,s+1}^{(s+1)} = 0 \qquad (\ i=2,3,\ldots,m-s\ ).$$

It is essential to choose the sign of σ_{s+1} the same as the sign of $a_{s+1,s+1}^{(s)}$; see (12.10). If these two quantities have opposite signs, then cancellation is possible in the first equality of (12.11). Since ρ_{s+1} can also be written as

$$(12.17) \quad \rho_{s+1} = \sigma_{s+1}(\sigma_{s+1} + a_{s+1,s+1}^{(s)})$$

(when the vector whose components are given by (12.8) is a non-zero vector) and since a division by ρ_{s+1} is performed in (12.15), it is clear that the cancellation in the first equality of (12.11) may lead to a disaster.

By the process described above the original matrix $A = A_0 \in R^{m \times n}$ $(m \geq n)$ can be transformed, after n multiplications with elementary matrices, into

$$(12.18) \quad A_n = Q_{n-1}Q_{n-2}\ldots Q_2 Q_1 Q_0 A_0 = \begin{vmatrix} R_n \\ 0 \end{vmatrix}, \qquad A = Q_0^T Q_1^T \ldots Q_{n-2}^T Q_{n-1}^T \begin{vmatrix} R_n \\ 0 \end{vmatrix},$$

where $R_n = R \in R^{n \times n}$ is an upper triangular matrix. If $m=n$, then Q_{n-1} is simply the identity matrix and the right-hand side of (12.18) is only the upper triangular matrix R_n (there is no zero matrix). If $rank(A)=n$ and if the computations are carried out without rounding errors, then $rank(R_n)=n$.

Let $\bar{Q} = Q_0^T Q_1^T \ldots Q_{n-2}^T Q_{n-1}^T$ and denote by Q the matrix that is formed by the first n columns of matrix $\bar{Q}$. Then (12.18) can be rewritten as

$$(12.19) \quad A = QR,$$

which is (12.2).

The method based on the above formulae is well-known as **the Housholder method**. It was proposed in **[150]** by Householder, but the elementary reflectors have been used even before this paper; see **[231]**, p. 244.

The matrix with orthonormal columns, Q, is normally not calculated explicitly. Information about the elementary reflectors in (12.18) is normally stored under the diagonal (where the elements $a_{s+2,s+1}^{(s)}, \ldots, a_{m,s+1}^{(s)}$, $s=0,1,\ldots,n-1$, are kept) and in two additional arrays of length n (where the numbers σ_s and ρ_s, $s=1,2,\ldots,n$, are kept).

When the method is used in practice it is important to try to avoid overflows and/or underflows during the calculation of σ_s $(s=1,2,\ldots n)$. It is also important to attempt to organize the calculations with the elements of the matrices C_s $(s=0,1,\ldots,n-1)$ in an efficient way.

The first of these two aims can be achieved by scaling the first column of matrix C_s before the calculation of σ_{s+1}. This is done by dividing all elements of the first column of C_s with the value, η, of the maximal (in absolute value) element in this column. If scaling is done, then $\eta\sigma_s$ should be stored instead of σ_s in one of the two additional arrays mentioned above.

Denote by $c_j^{(s)}$ and $c_j^{(s+1)}$ the j'th column of C_s before and after the multiplication with the elementary reflector Q_s. It is clear that

$$(12.20) \quad c_j^{(s+1)} = (I_{m-s} - 2vv^T)c_j^{(s)} = c_j^{(s)} - (\rho_{s+1})^{-1}[w^T c_j^{(s)}]w,$$

which shows that the elementary reflector need not be explicitly calculated. Formula (12.20) is very efficient. It requires a calculation of one inner product, a multiplication of a vector by a constant and a subtraction of two vectors. Thus, the second of the two aims stated above can also be achieved.

If the matrix treated is dense, then the Householder algorithm will require $O(mn^2 - n^3/3)$ multiplications. This is quite satisfactory for dense matrices. However, the situation changes when sparse matrices are to be decomposed. The case where a sparse matrix is decomposed will be discussed after the description of the other orthogonalization methods.

12.2. Givens orthogonalization

Consider a matrix Q_{ij} such that: **(i)** all its diagonal elements except the elements q_{ii} and q_{jj} are equal to one, **(ii)** $q_{ii} = q_{jj} = \gamma$, $q_{ij} = \delta$, $q_{ji} = -\delta$ with $\gamma^2 + \delta^2 = 1$ and **(iii)** all other elements of matrix Q_{ij} are equal to zero. In other words, the matrix Q_{ij} differs from the identity matrix only by the elements q_{ii}, q_{jj}, q_{ij} and q_{ji}. It is easy to verify that Q_{ij} is orthogonal. The matrices Q_{ij} defined by (i)-(iii) are called **plane rotations** or **elementary orthogonal matrices**. Givens has shown that plane rotations can be used to obtain an orthogonal decomposition of type (10.2); [127,128].

Assume that matrix $A \in \mathbb{R}^{m \times n}$ ($m \geq n$) has already been transformed into form (12.4). Let both $i > s$ and $j > s$. Assume also that

$$(12.21) \quad \gamma = a_{i,s+1}/(a_{i,s+1}^2 + a_{j,s+1}^2)^{\frac{1}{2}} \quad \wedge \quad \delta = a_{j,s+1}/(a_{i,s+1}^2 + a_{j,s+1}^2)^{\frac{1}{2}},$$

where $a_{i,s+1}$ and $a_{j,s+1}$ are elements in the first column of matrix C_s. Superscripts are needed here to show that the elements involved in the above two relations differ from the original elements in the non-transformed matrix, but are omitted for simplicity.

A multiplication of matrix A_s with the plane rotations matrix Q_{ij} (from the left) produces a zero at position $(j, s+1)$. Only the rows $i-s$ and $j-s$ of matrix C_s are modified by this multiplication. The last observation shows that the significant operations in the calculation of $Q_{ij}A_s$ can be presented by

$$(12.22) \quad \begin{vmatrix} \gamma & \delta \\ -\delta & \gamma \end{vmatrix} \begin{vmatrix} a_{i,s+1} & a_{i,s+2} & \cdots & a_{i,n} \\ a_{j,s+1} & a_{j,s+2} & \cdots & a_{j,n} \end{vmatrix} .$$

This means that the modified elements in these two rows can be found (for $k = 1, 2, \ldots, n-s$) by

$$(12.23) \quad a_{i,s+k}^* = \gamma a_{i,s+k} + \delta a_{j,s+k}, \quad a_{j,s+k}^* = -\delta a_{i,s+k} + \gamma a_{j,s+k}.$$

One can produce zeros under the element $a_{s+1,s+1}$ in the first column of matrix C_s by multiplying A_s with $m-s-1$ plane rotation matrices. In this way a matrix A_{s+1}, which is of the same type as A_s, is obtained. It is said that a **major step** s+1 is performed when these m-s-1 multiplications are performed, while each multiplication with a plane rotation matrix is called a **minor step**. The minor steps can be carried out in a **natural order** by multiplying successively (from the left) with the matrices

(12.24) $Q_{s+1,s+2}$, $Q_{s+1,s+3}$, $Q_{s+1,s+4}$, ..., $Q_{s+1,m}$.

However, the zeros can also be produced when the multiplications are performed by using the plane rotations in an arbitrary order. Therefore the method is very flexible. The flexibility can be exploited in a trivial way if some element of matrix C_s is zero; in such a case the multiplication with a plane rotation matrix is not carried out. This indicates that if the matrix decomposed is sparse, then the number of minor steps needed during the major step s+1 can be considerably smaller than m-s-1 and, thus, the method seems to be suitable for sparse matrices. This topic will be discussed further when the other orthogonalization methods are also shortly described.

While the preliminary analysis sketched above indicates that the Givens method is suitable for sparse matrices, the method is rather inefficient for dense matrices. It requires about twice as many multiplications as the Householder method; the approximate number of multiplications needed to complete the decomposition is $O(2mn^2-2n^3/3)$. Also the number of square roots is quite infavorable; mn for the Givens method and m for the Housholder method. Therefore an attempt to improve the performance of the Givens method (with regard to the number of arithmetic operations that are to be carried out) seems to be desirable. An improved version has been proposed by **Gentleman [109,110]**; see also **Hammarling [141]**. The improved version of the Givens method will be discussed in the following section.

12.3. The Gentleman version of the plane rotations

Consider again the matrix A_s from (12.4) and a plane rotation Q_{ij} defined as in the beginning of the previous section. Assume that matrix A has been decomposed, after s major steps (which will be described below), into a product of several plane rotations, a diagonal matrix D_s and matrix A_s partitioned as in (12.4). Then the multiplication of D_sA_s by the plane rotation matrix Q_{ij} (from the left) can be performed by calculating

$$(12.25) \quad \begin{vmatrix} \gamma & \delta \\ -\delta & \gamma \end{vmatrix} \begin{vmatrix} d_i & 0 \\ 0 & d_j \end{vmatrix} \begin{vmatrix} a_{i,s+1} & a_{i,s+2} & \cdots & a_{i,n} \\ a_{j,s+1} & a_{j,s+2} & \cdots & a_{j,n} \end{vmatrix},$$

where as in the previous section it is assumed that $\gamma^2+\delta^2=1$. It is obvious that (12.25) reduces to (12.22) if $D_s=I$. This means that the classical Givens method can be considered as a special case, obtained by $D_s=I$, of the method discussed in this section. If the requirement $D_s=I$ is not imposed, then the parameters d_i and d_j could be specified so that the arithmetic cost of the method becomes (theoretically at least) equal to the arithmetic cost of the Householder method from **Section 12.1**. This can be achieved by refactoring the first two matrices in (12.25) by the use of the relatioship:

$$(12.26) \quad \begin{vmatrix} \gamma & \delta \\ -\delta & \gamma \end{vmatrix} \begin{vmatrix} d_i & 0 \\ 0 & d_j \end{vmatrix} = \begin{vmatrix} d_i\gamma & 0 \\ 0 & d_j\gamma \end{vmatrix} \begin{vmatrix} 1 & \alpha \\ \beta & 1 \end{vmatrix},$$

Then the modification of the elements of the appropriate rows (row i-s and row j-s) of matrix C_s are carried out by

$$(12.27) \quad \begin{vmatrix} 1 & \alpha \\ \beta & 1 \end{vmatrix} \begin{vmatrix} a_{i,s+1} & a_{i,s+2} & \cdots & a_{i,n} \\ a_{j,s+1} & a_{j,s+2} & \cdots & a_{j,n} \end{vmatrix} ,$$

and, thus, the new elements of these two rows are to be calculated (for $k=1,2,\ldots,n-s$) by

$$(12.28) \quad a^*_{i,s+k} = a_{i,s+k} + \alpha a_{j,s+k}, \quad a^*_{j,s+k} = \beta a_{i,s+k} + a_{j,s+k} ,$$

while $d_i\gamma$ and $d_j\gamma$ are diagonal elements of the updated (at the minor step under consideration) matrix D_s.

Comparing (12.28) with (12.23) it is seen that only one multiplication per new element is needed when the new version of the Givens method is used, while two multiplications are needed when the classical Givens method is applied.

It is necessary to explain how to calculate α, β and γ. This can be done by using the following algorithm.

If

$$(12.29) \quad d_i^2 a_{i,s+1}^2 \geq d_j^2 a_{j,s+1}^2 ,$$

then the parameters α and β are determined by

$$(12.30) \quad \beta = -a_{j,s+1}/a_{i,s+1} \quad \wedge \quad \alpha = -\beta d_j^2/d_i^2 .$$

If

$$(12.31) \quad d_i^2 a_{i,s+1}^2 < d_j^2 a_{j,s+1}^2 ,$$

then the choice of the parameters α and β is made by

$$(12.32) \quad \alpha = -a_{i,s+1}/a_{j,s+1} \quad \wedge \quad \beta = -\alpha d_i^2/d_j^2 .$$

In both cases the third parameter, γ, is calculated by

$$(12.33) \quad \gamma^2 = 1/(1-\alpha\beta) .$$

Several remarks are needed here.

Remark 12.1. A zero in row j is produced when the parameters are determined by (12.30), while a zero in row i is produced when (12.32) is chosen. This means that $a^*_{j,s+k}=0$ when (12.30) is in use and $a^*_{i,s+k} = 0$ when (12.32) is applied. ∎

Remark 12.2. The diagonal elements of matrix D_s are never used in the algorithm; only the squares of the diagonal elements are needed. Therefore if the squares of the diagonal elements are stored (instead of the diagonal elements), then no square root will be needed. For this reason the method is sometimes called **a square root free Givens method.** ∎

Remark 12.3. Two different choices, (12.30) and (12.32), of the parameters α and β are introduced in an attempt to limit the decrease of the diagonal elements of the matrices D_s^2 $(s=1,2,\ldots,n)$ during the computations. With the proposed algorithm the diagonal elements involved in a minor step are decreased by a factor in the interval $[\frac{1}{2},1)$. However, this might be insufficient and scalings might be needed in order to prevent underflows and overflows. ∎

Remark 12.4. Weighted problems can also be treated when the modified Givens method is in use. If the problem solved is **not** weighted, then the calculations are to be started with $D_1=I$. If the problems is weighted, then the squares of the weights are to be used as diagonal elements of D_1 in the beginning. ∎

Remark 12.5. The refactorization (12.26) was first proposed by **Gentleman [109]**. Therefore the method is often called the Gentleman-Givens method or the Gentleman version of the Givens method. An error analysis of the method can be found in **Gentleman [110]** and **Hammarling [141]**. The flexibility of the method with regard to the order in which the minor steps are carried out was first observed by **Gentleman [111]**. **Gentleman [111]** also demonstrated how the flexibility can be exploited in some situations. Finally, the refactorization (12.26) is not the only possible refactorization; other examples are given in **Wilkinson [275]**. ∎

The arithmetic cost of the Genleman-Givens method is $O(mn-n^3/3)$ multiplications. Moreover, no square roots are needed. This is a very considerable reduction (compared with the classical Givens; see the previous section). On the other hand, one should be very careful when this method is implemented in a code, because the diagonal elements of matrix D_s are decreasing at each minor step. Therefore there is a danger for both underflows and overflows when this algorithm is used. It seems to be necessary to monitor the elements of the diagonal matrix during the whole computational process and to perform some scalings occasionally in order to prevent overflows and underflows. This is an overhead which should be taken into account when the method is compared with other methods. Some authors claim that the actual gain is very little when the Gentleman-Givens method is used instead of the classical Givens method. It is very difficult to confirm or to reject such a statement. More experiments are needed in this direction. This short discussion shows, however, that the name **fast Givens transformations**, which is also used in connection with the Gentleman-Givens method, must be justified not only by showing the arithmetic cost, but also by demonstrating that the method performs well in practice in spite of the overhead needed to prevent overflows and undeflows.

The arguments given above are connected with the implementation of the Gentleman-Givens method for **dense** matrices. For sparse matrices the situation is changed. Indeed, the flexibility of the classical Givens method with regard to the order of the minor steps is preserved when the Gentleman-Givens method is used. Moreover, pivotal interchanges for preserving the sparsity are to be carried out both with the classical Givens and with the Gentleman-Givens methods and this leads, in both cases, to a considerable overhead. When the Gentleman-Givens method is used, some operations needed in the pivotal search could be combined with operations needed to prevent overflows and underflows. Thus, the Gentleman-Givens method seems to be more efficient than the classical Givens method for sparse matrices. However, more experiments are needed in order to confirm or reject such a statement. The application of the Gentleman-Givens method for general sparse matrices will be further discussed in **Section 12.9** and in the following chapters.

12.4. The Gram-Schmidt method

Denote by $a_j \in R^{mx1}$ the j'th column of matrix $A \in R^{mxn}$ ($rank(A)=n$, m≥n). Consider (12.2). Denote by q_j the j'th column of matrix Q (this is a unit vector; i.e. $\|q_j\|_2 = 1$). Let r_{ij} $(i=1,2,\ldots,n, j=1,2,\ldots,n)$ be elements of matrix R. Since R is upper triangular, $r_{ij} = 0$ as $i>j$. In this notation (12.2) can be rewritten as

$$(12.34) \quad a_1 = r_{11}q_1, \quad a_2 = r_{12}q_1 + r_{22}q_2, \quad \ldots, \quad a_n = \sum_{k=1}^{n} r_{kn}q_k .$$

By exploiting the orthgonality of the columns of Q it is easy to determine both the orthonormal vectors q_j $(j=1,2,\ldots,n)$ and the elements of the upper triangular matrix R. One starts by setting

$$(12.35) \quad q_1 = a_1/\|a_1\|_2 \quad \wedge \quad r_{11} = \|a_1\|_2 .$$

Then the other calculations can successively be carried out by using the following formulae (under the assumption that all calculations for $i=1,2,\ldots,k-1$ have been performed before the start of the calculations for the quantities with index k):

$$(12.36) \quad s_k = a_k - \sum_{i=1}^{k-1} (q_i^T a_k)q_i, \quad r_{kk} = \|s_k\|_2, \quad q_k = s_k/r_{kk}, \quad r_{ik} = (q_i^T a_k)/r_{kk},$$

$$i=1,2,\ldots,k-1,$$

Both the orthonormal vector q_k (the k'th column of matrix Q) and the elements of the k'th column of matrix R can be computed by (12.36). After performing these operations one can proceed further (to calculate q_{k+1} and $r_{i,k+1}$, $i=1,2,\ldots,k+1$). Thus, after n stages both matrix Q and matrix R will be found. The method based on (12.35)-(12.36) is called **the Gram-Schmidt orthogonalization process** (see, for example, [99,133,164,272]).

The Gram-Schmidt method has at least one advantage (when it is compared to the orthogonalization methods studied in the previous sections): it explicitly calculates matrix Q and, therefore, may be useful when this matrix is needed. Nevertheless, the method is not very popular. This is not because the method is computationally expensive; it can be performed in $O(mn^2)$ multiplications, which is of the same order as the computational cost of the Householder method. However, the method may produce rather bad results due to the loss of the orthogonality of the vectors q_j $(j=1,2,\ldots,n)$. The performance can be improved by a simple rearrangement of the computations. The method so found will be discussed in the next section.

12.5. The modified Gram-Schmidt method

Consider (12.2) and denote again by q_j $(j=1,2,\ldots,n)$ the columns of Q and by r_i $(i=1,2,\ldots,n)$ the rows of R. It is clear that $A^{(k)}$ from the equality:

$$(12.37) \quad A^{(k)} = A - \sum_{i=1}^{k-1} q_i r_i^T = \sum_{i=k}^{n} q_i r_i^T \quad (k=2,3,\ldots,n), \qquad A^{(1)} \overset{\text{def}}{=} A,$$

is a matrix the first $k-1$ columns of which contain zeros only. Denote by $a_j^{(k)}$ ($j=1,2,\ldots,n$) the columns of $A^{(k)}$. Then the calculations with the Gram-Schmidt method, for an arbitrary k ($k=1,2,\ldots,n$), can be reorganized in the following way:

$$(12.38) \quad r_{kk} = \|a_k^{(k)}\|_2, \qquad q_k = a_k^{(k)}/r_{kk},$$

$$(12.39) \quad r_{kj} = q_k^T a_j^{(k)}, \qquad j=k+1,k+2,\ldots,n,$$

$$(12.40) \quad a_j^{(k+1)} = a_j^{(k)} - (q_j^T a_j^{(k)})q_k, \qquad j=k+1,k+2,\ldots,n.$$

The number of multiplications needed when the method is applied in this manner is again about $O(mn^2)$. The method may perform better (with regard to the accuracy) than the ordinary Gram-Schmidt method. This implementation is called **modified Gram-Schmidt orthogonalization.** The orthogonal basis is again obtained explicitly after the last step performed by (12.38)-(12.40). Therefore the modified Gram-Schmidt method should be preferred when the orthogonal basis is needed (and when the arithmetic cost is an important factor). The orthogonal basis can also be obtained by the methods studied in **Section 12.1 -Section 12.3,** but the price is some extra computational work, which is considerable (the number of arithmetic operations is nearly doubled when the orthogonal basis is obtained after the use of the Householder method). On the other hand, if the accuracy is an important factor, then the methods described in first three sections are to be used. This and other related problems are discussed in some more detail in **[26,133,272].**

12.6. Uniqueness of the orthogonal decomposition

Consider $A \in R^{m \times n}$ with $m \geq n$ and $rank(A)=n$. Consider the decomposition $A=QR$ ($Q^TQ=I \in R^{n \times n}$ and $R \in R^{n \times n}$ being upper triangular). Then **the orthogonal decomposition is unique if all diagonal elements of matrix R are positive.** It is not necessary to prove this statement here (a proof can be found in **[231]**). On the other hand it is important to emphasize that this statement tells us that all orthogonalization methods with $D=I$ lead to the same orthogonal decomposition if the computations are carried out in exact arithmetic and if the diagonal elements of R are positive. Furthermore, the results calculated by different orthogonal methods will not differ too much if the matrix decomposed is well-conditioned. However, if the matrix under consideration is ill-conditioned, then the rounding errors can have a great influence on the results for some of the methods studied. The influence of the rounding errors on the results will shortly be discussed in the following section.

12.7. Influence of the rounding errors
on the least squares solutions

The problem of finding the linear least square solution can be formulated as

$$(12.41) \quad Ax=b-r, \quad A^Tr=0 \quad (A \in R^{m \times n}, \ b \in R^{m \times 1}, \ r \in R^{m \times 1}, \ x \in R^{n \times 1}, \ m \geq n).$$

It is assumed here that the condition $rank(A)=n$ is satisfied. The case where this condition is not satisfied will shortly be discussed in next section.

Assume that the orthogonal decomposition $A=QR$ is calculated (either explicitly, for example by the Gram-Schmidt method, or implicitly, for example by the Householder method). Multiply the first equality in (12.41) by R^TQ^T. The result is

$$(12.42) \quad R^TQ^TAx = R^TQ^Tb-R^TQ^Tr \quad \Leftrightarrow \quad R^TQ^Tr = R^T(Q^Tb-Rx) \quad \Leftrightarrow \quad A^Tr = R^T(Q^Tb-Rx).$$

Since R is not singular (due to the requirement $rank(A)=n$), the third equality in (12.42) shows that $A^Tr=0$ is satisfied when $x=R^{-1}Q^Tb$ and, thus, that the QR decomposition can be used to solve linear least squares problems. Linear least squares problems can also be solved by the method of forming the normal equations (see **Example 3.2**) and the method of forming the augmented matrix (see **Example 3.3** and **Chapter 7**).

Assume that an orthogonal decomposition (calculated by any of the methods described in the previous sections of this chapter) is used to solve the linear least squares problem defined by (12.41). Let $x^* = x + \Delta x$ be the solution that is actually found (it differs from the exact solution x by an amount Δx because rounding errors are unavoidable when a computer is in use). The solution actually computed, x^*, is the exact solution of the perturbed problem

$$(12.43) \quad (A+\Delta A)x^* = b+\Delta b-r, \quad (A+\Delta A)^T(\Delta b-r) = 0,$$

where ΔA and Δb are a perturbation matrix and a perturbation vector respectively. Let $\kappa(A)$ be the spectral condition number defined by $\kappa(A)=\|A\|_2\|A^\dagger\|_2$. Assume that there exist two constants ϵ_1 and ϵ_2 such that

$$(12.44) \quad \|\Delta A\|_2 \le \epsilon_1\|A\|_2, \qquad \|\Delta b\|_2 \le \epsilon_2\|b\|_2, \qquad \kappa(A)\epsilon_1 < 1 .$$

Then the actually calculated solution of the linear least squares problem satisfies the following inequality:

$$(12.45) \quad \|\Delta x\|_2 \le \{\kappa(A)/[1-\kappa(A)\epsilon_1]\}[\epsilon_1\|x\|_2 + \kappa(A)\epsilon_1\|r\|_2/\|A\|_2 + \epsilon_2\|b\|_2/\|A\|_2] .$$

The bound (12.45) shows that, roughly speaking, the error in the actually calculated solution depends linearly on the condition number of matrix A when either the residual vector is equal to zero (the problem is **consistent**) or the residual vector is so small that the contribution due to the term containing $\|r\|_2$ is small in comparison with the other two terms. However, if the residual vector is not very small, then the rounding errors depend on the square of the condition number. If the normal equations are used to solve linear least squares problems, then the rounding errors depend on the square of the condition number of A even if the norm of the residual vector is small. The situation is better when the method of augmentation is used and the parameter α is carefully chosen (see **Section 9**).

The bound (12.45) is studied (in a general context) in [233]; see also [224-226,265-267]. Bounds for particular methods are given in [164].

If matrix A is not ill-conditioned ($\kappa(A)$ is not very large), then any method will give acceptable results when the accuracy requirements are not

very stringent. In this case even the method of the normal equations is often
successful and the truth is that this method is commonly used by scientists
and engineers. One should be careful when matrix A is ill-conditioned. In
this case the Householder and the Givens methods should be preferred, but the
modified Gram-Schmidt performs also quite well. An explanation of the good
performance of the modified Gram-Schmidt method even in the case where A is
rather ill-conditioned is given in [231], p. 245. Both the Householder
method and the Givens method are better than the modified Gram-Schmidt method
(with regard to accuracy) when an orthonormal basis is to be computed.
However, since matrix Q is stored in factored form when the former two
methods are in use, the calculation of the orthonormal basis requires extra
computations. Therefore the modified Gram-Schmidt method is more attractive
when an orthonormal basis is needed explicitly.

The general conclusion is that the orthogonal methods (excluding perhaps
the ordinary Gram-Schmidt orthogonalization) perform normally very well in the
solution of linear least squares problems in spite of the presence of the
square of the condition number in (12.45). Moreover, no pivoting for accuracy
is needed and in the subroutines from the standard libraries no pivotal
interchanges are carried out (pivotal interchanges may be necessary if the
matrix is very ill-conditioned).

The conclusions drawn in this section are only true when the condition
$rank(A)=n$ is satisfied. In fact the requirement is somewhat stronger: **the
matrix decomposed should not be close to a matrix A^* with $rank(A^*)<n$**. The
case $rank(A^*)<n$ will be discussed in the next section.

12.8. Rank deficiency

Consider again the linear least squares problem (12.41). Assume now that
$rank(A^*)=s<n$. Then there exists an orthonormal basis $\{q_1, q_2, \ldots, q_s\}$
determined by the columns of matrix A. Denote by $Q \in R^{mxs}$ the matrix whose
columns are vectors in the orthonormal basis and assume that this matrix has
been calculated in some way (or a sufficiently accurate approximation of this
matrix has been calculated). Then matrix A can be represented as $A=QR$
(where matrix R is not an upper triangular nxn matrix as in the previous
sections, but an upper trapezoidal matrix in R^{sxn}).

Assume that matrix R in the decomposition $A=QR$ ($Q \in R^{mxs}$, $R \in R^{sxn}$) is
partitioned as

(12.46) $R = (R_1, R_2)$, $R_1 \in R^{sxs}$, $R_2 \in R^{sx(n-s)}$,

while the solution vector x is partitioned as

(12.47) $x = \begin{vmatrix} y \\ z \end{vmatrix}$, $y \in R^{sx1}$, $z \in R^{(n-s)x1}$.

Then it can be shown (since R_1 is a non-singular matrix) that for any
vector z (for any set of n-s real numbers) a solution of the linear least
squares problem: **minimize the two norm of the vector** $r=b-Ax$ can be found
by solving the system of linear algebraic equations:

(12.48) $y = (R_1)^{-1}(Q^T b - R_2 z)$.

This means that the solution of the linear least squares problem is not unique in the case $rank(A^*)=s<n$. However, a unique solution can be defined by imposing the extra requirement: **find the linear least squares solution which has the minimal two-norm.**

In principle at least, some of the methods discussed in the previous sections can be modified to solve rank deficient linear least squares problems. Experience shows, however, that very often some difficulties arise. One of the difficulties is that column interchanges seem to be necessary in this situation and the introduction of column interchanges leads to extra computations. The second, and even more serious, difficulty is connected with the stopping criteria. It is difficult to find a reliable criterion that allows one to decide whether the orthonormal basis is already calculated or not. One can construct matrices A which are nearly rank deficient and such that the methods described in the previous sections will not detect this fact (even if column interchanges are in use); see, for example, [164].

Rank deficiency can be detected and handled in a reliable way only if the so called **singular value decomposition (SVD)** is in use. By the **SVD** matrix A is represented as

(12.49) $A = U \Sigma V^T$ $(U \in R^{m \times m}, \ \Sigma \in R^{n \times n}, \ V \in R^{n \times n})$,

where U and V are orthogonal matrices, while Σ is a diagonal matrix whose diagonal elements are non-negative real numbers $\sigma_i \geq 0$ $(i=1,2,\ldots,n)$. The diagonal elements of Σ are called **singular values of matrix A**, the columns of the orthogonal matrix U $(u_1, u_2, \ldots, u_m)$ are called **left singular vectors of matrix A** and, finally, the columns of the orthogonal matrix V $(v_1, v_2, \ldots, v_n)$ are called **right singular vectors of matrix A**. The singular values of matrix A are equal to the square root of the eigenvalues of $A^T A$.

Assume that the singular values of A are ordered as

(12.50) $\sigma_1 \geq \sigma_2 \geq \ldots \geq \sigma_n \geq 0$.

If the rank of the matrix is n, then all singular values are positive. If the rank is $s < n$, then the first s singular values are positive, while the other n-s are equal to zero. Of course, if the computations are carried out on a computer, then the last n-s singular values, which are actually calculated, will in general be non-zeros. Nevertheless, the fact that $rank(A^*)=s<n$ will normally be detected because the following estimate holds (see [133], pp. 174-175):

(12.51) $|\sigma_i - \sigma_i^*| \leq \epsilon^* \sigma_1$ $(i=1,2,\ldots,n)$,

where

(12.52) $\sigma_1^*, \ \sigma_2^*, \ \ldots, \ \sigma_n^*$,

are the singular values that are actually calculated on the computer used and ϵ^* is a small multiple of the machine accuracy (the minimal positive number ϵ such that $1+\epsilon \neq 1$ in the arithmetic of the computer used).

The bound (12.51) is very sharp and, therefore, the conclusion made in the book written by Golub and Van Loan ([133], p. 175): **"near rank deficiency cannot escape detection when the SVD of A is computed"** is at least not far from the reality. Thus, not only is the rank deficiency of matrix A detected when the SVD is used, but also the case where the matrix is close to a rank deficient matrix would be discovered.

Unfortunately, the SVD procedure (12.49) is clearly not suitable for sparse matrix codes. Indeed, even when the original matrix A is very sparse, the two orthogonal matrices U and V (the matrices containing the left and the right singular vectors of matrix A) will in general be rather dense. Therefore it seems to be necessary to apply some dense matrix technique in connection with the SVD. Even when some dense matrix technique is in use and when the SVD is compared with the other orthogonalization methods implemented with the same dense matrix technique, the comparison is rather infavorable for the SVD; both with regard to the computing time used and with regard to the storage needed. This means that the SVD should not be recommended as a general-purpose method. However the method can successfully be used when very ill-conditioned problems are to be treated. Moreover, the method is an excellent tool in many theoretical investigations.

12.9. Choice of an orthogonalization method for sparse matrices.

The matrix with orthonormal columns, matrix Q, which is used (directly or indirectly) in all orthgonalization methods, is in general rather dense even when the original matrix A is very sparse. Therefore it seems to be necessary to avoid the storage of Q when a sparse matrix technique is to be used. This could be done in the following way. Consider the orthogonal decomposition A=QR. Assume that $rank(A)=n$ and consider the linear least squares problem:

(12.53) $Ax=b-r$, $A^Tr=0$.

Multiply the first equality by A^T (or, in other words, consider the normal equations):

(12.54) $A^TAx=A^Tb$.

Replace A^T by R^TQ^T and A by QR in (12.54). Exploit the fact that $Q^TQ=I_n$. Then (12.54) is reduced to

(12.55) $R^TRx=A^Tb$.

The last equality shows that Q is in fact not needed in the process of determination of the linear least squares solution when a copy of matrix A (which is assumed to be sparse) is held.

The storage of matrix Q can also be avoided when the more complicated decomposition A=QDR (where D is a diagonal matrix; see the beginning of this chapter) is computed. In this case the equality

(12.56) $R^TD^2Rx=A^Tb$

can be obtained in the same way as (12.55). The equality (12.56) shows again that matrix Q is not needed, but that a copy of A is to be kept.

The real situation when sparse matrices are decomposed is slightly more complicated, because pivotal interchanges are normally to be used in order to preserve the sparsity during the computational process. Assume that both row and column interchanges are to be carried out during the orthogonalization and denote the permutation matrices by Y and Z. Then the orthogonal decomposition can be written as

(12.57) $YAZ=QDR$,

while the equality

(12.58) $ZR^TD^2RZ^Tx=A^Tb$

can be obtained from (12.56). An equality corresponding to (12.55) can be obtained from (12.58) by setting $D=I_n$. In both cases the solution does not depend on the row interchanges (on the permutation matrix Y). However, it must be stressed here that the amount of the computational work does depend on the choice of the row interchanges. This will be studied in more detail in one of the following chapters. Therefore one should be careful in the choice of the row interchanges.

If $rank(A)=n$, then $ZR^TD^2RZ^T$ is non-singular and its inverse is precisely the matrix H from (3.27). This means that the method described above is the same as the method in **Example 3.6** and, therefore it is a particular method in the general k-stage scheme for solving $x=A^Tb$. In fact, any of the orthogonalization methods could be represented in the form (12.58). The equality (12.58) represents a particular method only if one determines precisely how the upper triangular matrix R is computed. The equality (12.58) can be considered as a general description of the implementation of the orthogonalization methods in the particular case where the matrix under consideration is sparse. This equality shows that matrix Q which is normally rather dense is in fact not needed when the original matrix A, which is assumed to be sparse, is not destroyed during the computational process. Since matrix R is unique (see **Section 12.6**) this implies that if the column interchanges used are the same and if the computations are carried out without ronding errors, then all orthogonalization methods will lead to the same upper triangular matrix R. However, in the process of the computation of R one has to work with the non-zero under the diagonal. These non-zero elements gradually disappear and, at the end of the orthogonalization process, all non-zero elements are on or above the main diagonal. The non-zeros that are under the diagonal during the process of calculating the upper triangular matrix R will be called **intermediate fill-ins**. One cannot avoid the work with the intermediate fill-ins, but one can avoid the storage of the intermediate fill-ins when the classical Givens method or the Gentleman-Givens method is in use. This will be discussed in **Chapter 13**. The storage needed can be reduced significantly when one of these two methods is applied.

The use of the classical Givens orthogonalization or of the Gentleman version of this method leads not only to savings with regard to the storage needed, but also to savings with regard to the computing time used. This is due to the flexibility of these methods, which provides an obvious way of exploiting the sparsity. Indeed, if in the pivotal column under the main diagonal there are several zero elements then the corresponding minor steps (the corresponding multiplications with plane rotation matrices; see **Section 12.2** and **Section 12.3**) can simply be omitted. Thus, the procedure is extremely simple. This is not so when the other orthogonalization methods are in use. It is not clear (to the author at least) whether a reduction of the computational work like that achieved for the Givens methods can be obtained or not when any of the other orthogonalization methods is applied. However, it is quite clear that even if such a reduction can be achieved with any of the other methods, then this will not be so simple and obvious.

The use of the two versions of the Givens method has a third advantage over the other orthogonalization methods. This is the freedom in the order in which the minor steps (the multiplications with plane rotations) are performed. This freedom can be exploited in an efficient way to reduce the number of intermediate fill-ins by designing a suitable pivotal strategy.

The application of the classical Givens orthogonalization and of its Gentleman version will be discussed in the following chapters. Before this discussion, however, several remarks are needed in connection with (12.58), which is actually treated when sparse matrices are decomposed by orthogonalization methods.

Remark 12.6. The equality (12.58) is actually based on the use of the normal equation. However, the normal matrix, $A^T A$, is neither calculated nor used during the calculation of the upper triangular matrix R. Therefore the method is not so sensitive to perturbations due to rounding errors as the method of the normal equations. Experience confirms such a conclusion. Some theoretical results in this direction have been reported in [27]. Nevertheless, some care should be taken when (12.58) is used. ∎

Remark 12.7. The problems connected with the accuracy of the calculated solution arising by the fact that (12.58) is based on the normal equations can be avoided if the orthogonal transformations are carried out not only on the non-zero elements of the matrix decomposed, but also on the components of the vector $b \in R^{mx1}$. If this is done, then the vector $Q^T b \in R^{nx1}$ is available at the end of the decomposition and the linear least squares solution can be obtained by solving the system of linear algebraic equations:

(12.59) $Rx = Q^T b$.

Also in this case matrix Q need not be stored (vector $Q^T b$ is calculated during the decomposition). Moreover, no copy of matrix A is needed when this algorithm is applied (and when only one linear least squares problem only is to be handled). Finally, no loss of accuracy due to the implicit forming the normal equations takes place. However, the method cannot be applied when several problems with the same matrix but with different right-hand side vectors are to be treated (unless one decides to decompose the matrix for each right-hand side, which will be very expensive). ∎

Remark 12.8. One can attempt to regain the accuracy lost when (12.58) is applied by carrying out some iterative process (as, for example, iterative refinement). Assume the the iterative process applied converges. Then not only is the accuracy lost regained, but also a cheap error estimation is normally obtainable. This may be very important in the numerical treatment of many scientific and engineering problems where one is interested to stop the computations when certain accuracy is achieved (but not before this). It should be mentioned here that the iterative process may be used also in order to avoid the storage of some elements that are small in some sense. This has been discussed in **Chapter 3** in connection with the general k-stage scheme for solving linear algebraic problems. The use of iterative methods in connection with the solution of linear least squares problems will be discussed in detail in the following chapters. ■

  Remark 12.9. If an iterative method is used and if only one linear least squares problem is to be handled, then it is worthwhile to carry out the orthogonal transformations with the right-hand side during the decomposition. If this is done, then the starting approximation for the iterative process can be obtained by (12.59). In this way one has a chance to start the iterative process with a more accurate approximation and, thus, to reduce the number of iterations. ■

  Remark 12.10. One should be careful when the right-hand side vector of system (12.58) is calculated. The potential loss of accuracy connected with this operation can be avoided if some vectors are stored and accumulated in double precision. It should be emphasized here that double precision is used for a few vectors only and again a copy of the sparse matrix is to be held in single precision (instead of matrix Q). It should be emphasized also that the use of double precision to accumulate some inner products can have a great effect on the performance of the iterative process; especially when some residual vectors are calculated and stored in double precision. On the other side the use of a double precision arithmetic can be rather expensive on some computers. Therefore the use of double precision for some inner products should be avoided if the double precision arithmetic is inefficient on the computer available. ■

12.10. Concluding remarks and references concerning the theory of the orthogonalization methods

  The orthogonalization methods are described in many books and papers on matrix computations and numerical linear algebra; see, for example, [5,15,26,133,203,253-254,272]. Error analysis for many of the methods studied here is given in some of these references, and also in [24,110,141,153,164, 265-267,272]. The short discussion in this chapter is given in order to explain why the Givens method and/or the Gentleman-Givens method are to be preferred when general sparse matrices are to be handled.

  The Householder method was proposed in [150]. A short description of this method given in **Section 12.2** follows closely the exposition of the method in [231]. The Householder method is probably the best choice when dense matrices are to be decomposed by orthogonalization methods.

  The classical Givens method [127,128] (see **Section 10.3**) is perhaps the first orthogonalization method in which matrix Q is stored in factored form. It is not very popular for dense matrices because of its arithmetic cost (the number of multiplications is about twice greater than that for the Householder method).

  The Gentleman version of the Givens orthogonalization (see **Section 12.3**) was introduced in [109]. The arithmetic cost of this method is the same as the arithmetic cost of the Householder method. However, there are some difficulties with the implementation of the Gentleman version (as, for example the necessity to monitor the size of the diagonal elements of the diagonal matrix D in the course of the decomposition and, occasionally, to scale some vectors in order to avoid underflows and overflows). Nevertheless, when the implementation difficulties are overcome and the algorithm is correctly coded, then it performs rather satisfactorily. The Gentleman version of the Givens orthogonalization is a good choice for sparse matrices, while the other method with the same arithmetic cost, the Householder method, is not.

The bad accuracy properties of the Gram-Schmidt method (sketched in **Section 12.4**) are well-known (see, for example, [26,133,164,231,272]). However, if the matrix that is to be decomposed is not ill-conditioned and if an orthonormal basis is required, then the Gram-Schmidt method **may** be successfully applied. This is especially true if the real numbers can be written with many digits on the computer under consideration (this is the case for CDC Cyber computers). However, even in this situation it is probably better to switch from the Gram-Schmidt method to the modified Gram-Schmidt method.

The modified Gram-Schmidt method (described in **Section 12.5**) is the best choice when an orthonormal basis is required and when the arithmetic cost is an important factor. An orthonormal basis can also be calculated by the use of the Householder method or the Givens method. However, the arithmetic cost in the latter case is nearly doubled. Since the orthonormal basis is usually rather dense, some dense matrix technique is to be used when the modified Gram-Schmidt method is applied (or, more general, when an orthonormal basis is to be calculated). The modified Gram-Schmidt method is discussed in some more detail, for example, in [133].

The uniqueness of the orthogonal decomposition (see **Section 12.6**) is a very important fact. From the proof given in [231] it becomes quite clear that the upper triangular matrix R is the same as the Cholesky factor of the normal matrix A^TA. This shows that **if the storage of matrix Q and of the intermediate fill-ins can be avoided, then the orthogonalization method under consideration is no more storage consuming than the method based on the solution of the system of the normal equations.** It was commonly believed that the storage required is one of the greatest disadvantages of the orthogonalization methods; see, as an illustration only, [75]. The fact that the orthogonalization methods under the above conditions are comparable (with regard to the storage required) with the normal equations has been established by **George and Heath** [112]. The classical Givens method is implemented in [112], but the results could probably be extended for the Gentleman-Givens method.

The fact that the rounding errors are dependent on the **square** of the condition number of matrix A, see **Section 12.7**, is discussed in many papers and books (as, for example, [26,27,224-226,233,265-267]. The bound (12.45) shows that one should be careful, especially when the norm of the residual vector is large; some error estimate, obtained by applying iterative improvement of the direct solution, is perhaps necessary if no information about the condition number of matrix A is available.

Rank deficiency and methods for treating problems with nearly rank-deficient matrices, which were briefly discussed in **Section 12.8**, will not be discussed further in this book. The singular value decomposition method, which is most suitable for rank-deficient and nearly rank-deficient problems, is rather expensive and, what is more important, is beyond the scope of this book (because in general it leads to nearly dense orthogonal matrices and, therefore, some dense matrix technique has to be used when this method is applied). However, this is a very interesting topic, which is studied in many papers; see, for example, [130,132,253-254]. Other methods are also advocated for nearly rank-deficient problems; see, for example, [39-40,133,181-182].

The choice of a method for solving large and sparse linear least squares problems (discussed in **Section 12.9**) has been studied first in [75]. Many improvements in the sparse matrix algorithms have been proposed after the publication of [75] and, therefore, some of the recommendations made in [75] should probably be re-evaluated. Two of the newest algorithms for sparse least squares problems will be discussed in the **Chapter 13 - Chapter 16**. The main purpose in this introductory (for the orthogonalization methods) chapter was to justify the suitability of the Givens transformations for sparse matrix computations. Other orthogonal methods can also be applied (for further discussion see [111-121,129,134,143,164,195,231,259,272,277]. Pure iterative methods (see, for example, [29,164,185-186]) are also popular.

CHAPTER 13

TWO STORAGE SCHEMES FOR GIVENS PLANE ROTATIONS

The main conclusion drawn in the previous chapter was: **the Givens transformations (both in the classical form and in the Gentleman version) are best suitable, among the orthogonalization methods, for the case where linear least squares problems with large and sparse matrices are to be handled.** Two different implementations of sparse matrix techniques in connection with the Givens plane rotations will be described in this section. Both implementations are very attractive for certain classes of problems and/or for certain computer environments.

The first of the two implementations was described in [112]. The main advantage of this implementation is the economical storage scheme. The intermediate fill-ins are not stored when this technique is used and, therefore, it is very efficient in situations where the storage requirement is an important factor.

The second implementation was discussed in [296,320-322]. Intermediate fill-ins are stored when this implementation is in use. However, an attempt to reduce not only the number of intermediate fill-ins, but also the number of the elements in R is carried by using the algorithm that has been studied from a general point of view in **Chapter 3**. The idea is to drop some "small" elements during the decomposition and then to try to improve the direct solution iteratively. In this way not only is the storage considerably reduced for some classes of matrices, but also the computing time can be dramatically reduced in some cases.

It must be emphasized here that, while the storage needed is considerably reduced by the use of the first implementation (because no intermediate fill-in is stored), the computing time is not reduced (the computational work with the intermediate fill-ins is not avoided). It should also be emphasized that a very cheap error estimation can be obtained when the second implementation is in use. On the other side, the second implementation works very well only when many "small" elements are dropped during the computations (this has been discussed in **Chapter 3** in connection with the general scheme for solving linear algebraic problems $x=A^{\dagger}b$ and in **Chapter 5** in connection with the Gaussian elimination). The conclusion is that no implementation will be better than the other one in all situations.

13.1. The George-Heath implementation of the Givens method

Assume that $A \in R^{m \times n}$ $(m \geq n \wedge rank(A)=n)$ is a given matrix. Assume also that a matrix $Q \in R^{m \times n}$ with orthonormal columns $(Q^T Q=I_n \in R^{n \times n})$ such that $A=QR$, where $R \in R^{n \times n}$ is an upper triangular matrix, is to be calculated by the use of any of the orthogonalization methods discussed in the previous chapter. If the calculations are carried out in exact arithmetic, then the decomposition $A=QR$ is unique and the upper triangular matrix R is identical (neglecting some sign differences in the rows) with the upper triangular matrix L obtained by the Cholesky decomposition of the symmetric and positive definite matrix $B=A^T A$. Since B is symmetric and positive definite no pivotal intechanges for numerical stability are needed in the decomposition of this decomposition of this matrix. However, it is necessary to carry out some pivotal interchanges in order to better preserve the sparsity (to obtain a Cholesky factor L that is as sparse as possible). If pivotal interchanges are carried out for preserving the sparsity only, then the actual size of the elements transformed during the calculations is not important. Only the type of the elements (zero or non-zero) is essential for the computational process. This means that a symbolic Cholesky factorization can be carried out in order to determine the sparsity pattern of L. But the sparsity pattern of L is the same as the sparsity pattern of R. This observation shows that the storage needed for accumulating the non-zero elements of the upper triangular matrix R, obtained by any of the orthogonalization method, could be determined before the actual calculation of the non-zero elements of this matrix. This is performed by carrying out three preliminary steps (before starting the actual computation of the non-zero elements of R). The preliminary steps are:

(i)   Determine the sparsity pattern (not the numerical values of the non-zero elements) of matrix $B=A^TA$.

(ii) Find a permutation matrix P such that matrix P^TBP has a sparse Cholesky matrix L.

(iii) Determine the sparsity pattern (not the numerical values of the non-zero elements) of the Cholesky factor L of matrix P^TBP.

It is not very difficult to perform **the first of these three steps.** An element b_{ij} of matrix B is a non-zero element if there exists at least one index k (k=1,2,...,m) such that both $a_{ki} \neq 0$ and $a_{kj} \neq 0$; otherwise the element b_{ij} is a zero element (here it is assumed that a cancellation does not take place in the calculation of the inner product of the i'th column with the j'th column of matrix A). It is clear that the investigation can be stopped when the first index k, such that the k'th components of the two columns under consideration are non-zeros, is found (assuming again that no cancellation takes place). It is also clear that the diagonal elements of matrix B are non-zeros and, thus, no investigation concerning these elements is needed. Finally, the symmetry of matrix B should be taken into account; this means that only the sparsity pattern of the strictly upper triangular part of matrix B is to be searched. Matrix B may be rather dense even if matrix A is very sparse. ∎

To find a permutation matrix P means to find symmetric pivotal interchanges by which the sparsity of matrix B (when this matrix is sparse) is preserved as well as possible. The fact that B is symmetric has to be exploited in this part of the computational work. There are different ways to carry out symmetric pivotal interchanges. **The minimum degree algorithm** is very popular. This method is a modification of the well-known Markowitz pivotal strategy (see **Chapter 4**), which is often used in the Gaussian elimination for unsymmetric matrices. Also **the nested dissection algorithm** is often used in practice. Different pivotal strategies for symmetric and positive matrices are discussed in detail in [115]. In the context of this section it is important to emphasize that the pivotal interchanges carried out on matrix B by the use of the permutation matrix P correspond to column interchanges in the original matrix A. There exists software (see [115,117]) for the determination of matrix P. ∎

The third step can also be carried out by using subroutines for performing symbolic Cholesky decomposition of a symmetric and positive definite matrix (as, for example, the subroutines from [115,117]). In the particular algorithm which will be described here, the **George-Heath** algorithm, it is important to find a subroutine which determines a row-oriented sparsity pattern of the Cholesky factor L of matrix $P^T BP$ (because of the properties of the remaining part of the **George-Heath** algorithm, which will be described in the second part of this section). However, the algorithm could easily be modified for the case where the non-zero elements of L are ordered by columns. ■

The result of the three preliminary steps is the determination of the sparsity pattern of the upper triangular matrix R (as stated above the sparsity patterns of R and L are the same). The next question is: **how to determine the numerical values of the non-zero elements of the upper triangular matrix R without storing the intermediate fill-ins?**

Consider the **dense** case. Assume that the upper triangular matrix R has to be stored in an array with the same name and that the elements of vector $Q^T b$ are to be accumulated in array C, while matrix A and vector b are stored in arrays A and B respectively. Assume also that a subroutine, which will be called here **ROTATE(M,N,I,J,A,R,B,C)**, performs a plane rotation involving the i'th row of matrix A and the j'th row of matrix R ($j \leq i$) that annihilates element a_{ij}. The formulae given in **Section 12.2** are to be modified as follows in order to perform the plane rotation according to the notation adopted here:

$$(13.1) \quad \begin{vmatrix} \gamma & \delta \\ -\delta & \gamma \end{vmatrix} \begin{vmatrix} r_{j,j} & r_{j,j+1} & \cdots & r_{j,n} & c_j \\ a_{i,j} & a_{i,j+1} & \cdots & a_{i,n} & b_i \end{vmatrix}$$

with

$$(13.2) \quad \gamma = r_{jj}/(r_{jj}^2 + a_{ij}^2)^{\frac{1}{2}} \quad \wedge \quad \delta = a_{ij}/(r_{jj}^2 + a_{ij}^2)^{\frac{1}{2}} .$$

Subroutine **ROTATE** performs the multiplication shown by (13.1). The introduction of c_j and b_i is made to show that also the right hand side vector b is updated by **ROTATE**. Having such a subroutine, one can calculate both R and $c = Q^T b$ by using the piece of code given in **Fig. 13.1**.

```
        DO 20 I=1,M
          DO 10 J=1,MIN(I,N)
            CALL ROTATE(M,N,I,J,A,R,B,C)
10        CONTINUE
20      CONTINUE
```

Figure 13.1
Calculating matrix R and vector $c = Q^T b$ by using
the classical Givens plane rotations for a dense matrix A.

It is assumed that the contents of arrays R and C are equal to zero before the start of the loop in **Fig. 13.1**. The plane rotations are performed **by rows** when this algorithm is used. After the last call of **ROTATE** in the inner loop all plane rotations involving row i of matrix A are calculated. Therefore row i of matrix A is only needed when the i'th sweep of the outer loop is performed. This shows that matrix A may be held out of core (written, for example, on a file). If this is done, then one row at a time is read just before the performance of each inner loop.

In fact, if i≤n, then the i'th row of matrix A must be copied in the i'th row of matrix R during the last call of **ROTATE** in the inner loop. This means that no plane rotation is needed in this case and this can be exploited to reduce the number of rotations in the code given in **Fig. 13.1**.

Let I and J be fixed. Assume that a_{ij} (or A(I,J) in the notation used in **Fig. 13.1**) is a non-zero element. Assume also that r_{jj} (or R(J,J) in the notation used in **Fig. 13.1**) is a zero element. Then row i from matrix A and row j from matrix R should simply be exchanged. If $a_{ij}=0$, then no plane rotation is needed. These two facts could also be taken into account when subroutine **ROTATE** is prepared. ∎

The algorithm for dense matrices that has been described above can easily be modified for the case where **sparse** matrices are to be handled. The modifications can be described as follows (but note that this is not the only way to implement the algorithm). Assume that three conditions are satisfied.

(i) Matrix A is held in an array A ordered by rows. Moreover, the columns of matrix A are permuted by using the permutation matrix P obtained during the second preliminary step. The column numbers of the non-zero elements are held in an array CNA, so that if a_{ij} is stored in A(K), then CNA(K)=j. There are two **INTEGER** arrays, LNGTHA and STARTA, of length at least equal to M, where information about row starts and the length of the rows (the number of the non-zero elements in the rows) is stored.

(ii) SCATTER(N,I,WORK,CNR,LNGTHR(I),STARTR(I),A,CNA,LNGTHA(I),STARTA(I)) is a subroutine, which scatters the non-zero elements of the i'th row of matrix A in a **REAL** array WORK. Array WORK is of length at least equal to N and if $a_{ij}\neq0$, then this element is stored in the j'th location of array WORK The non-zero elements in the i'th row of matrix A are stored in an **REAL** array with the same name from position STARTA(I) to position STARTA(I)+LNGTHA(I)-1. Finally, the column numbers of the non-zero elements of row i in matrix A, which are stored in array CNA (in the same positions), are copied in array CNR (from position STARTR(I) to position STARTR(I)+LNGTHA(I)-1. Both CNA and CNR are **INTEGER** arrays.

(iii) ROTATE(M,N,I,J,WORK,R,CNR,B,C,STARTR(I),STARTR(J),LNGTHR(I),LNGTHR(J)) is a subroutine that performs a sparse plane rotation involving the i'th row of matrix A and the j'th row of matrix R. During the computations with the subroutine ROTATE fill-ins are created both in the j'th row of matrix R and in the i'th row of matrix A (stored in array WORK). Therefore both the contents of LNGTHR(J) and LNGTHR(I) (the numbers of non-zero elements in the two rows involved) are to be updated. Moreover, an appropriate storage scheme, which allows us to store the fill-ins created, should be used. The main ideas on which such a storage scheme is based are described in **Section 2.11**. It should be noted that the fill-ins in the i'th row of matrix A (which is scattered in array WORK) are the intermediate fill-ins. Thus, the intermediate fill-ins appear only in array WORK. On the other hand, one zero per call of subroutine ROTATE is created in array WORK. Hence, at the end of the inner loop all locations of array WORK will contain zeros, and the array is prepared for scattering the next row of matrix A.

Assume that the conditions stated above are satisfied and the pointers (row starts and row lengths) are stored in arrays STARTA, STARTR and LNGTHA before the start of the orthogonalization (by an algorithm based on the ideas sketched in **Section 2.11**). Then the classical Givens rotations may be performed by the following piece of code (assuming here that the contents of array WORK were set to zero before the start of the loop 20).

```
      DO 20 I=1,M
         CALL SCATTER(N,I,WORK,CNR,LNGTHA(I),STARTA(I),
     *               A,CNA,LNGTHR(I),STARTR(I))
         DO 10 J=1,MIN(I,N)
            IF(WORK(J).NE.0.0DO) THEN
               CALL ROTATE(M,N,I,J,WORK,R,CNR,B,C,STARTR(I),
     *                     STARTR(J),LNGTHR(I),LNGTHR(J))
            END IF
   10    CONTINUE
   20 CONTINUE
```

Figure 13.2
Calculating matrix R and vector $c=Q^T b$ by the classical Givens plane rotations of a sparse matrix A.

Subroutine ROTATE is called only if a_{ij} is a non-zero (when WORK(J) is not zero). The element r_{jj} can be zero. If this is so, then ROTATE should only exchange the contents of row j of matrix R with the contents of row i of matrix A (kept in array WORK).

The algorithm for sparse matrices outlined in this section has the following properties:

(a)	The intermediate fill-ins are not stored.
(b)	The computations with the intermediate fill-ins are not avoided.
(c)	The order in which the rows of matrix A are rotated in R is not essential (in connection with the numbers of non-zeros in R).
(d)	Row interchanges must be applied in an attempt to reduce the number of intermediate fill-ins and, thus, the computational work.

Different row pivotal strategies for sparse Givens transformations have been studied in three papers of **George et al. [118-120]**. Some even more efficient row pivotal strategies for the classical Givens plane rotations are discussed in **[166]**.

An algorithm (which will be called the **George-Heath algorithm**) that is very similar to that discussed in this section is implemented in a code described in [112]. The possibility of treating the case where matrix A is kept on an auxiliary storage device ("out of core") is discussed in **[114]**. Some comparisons of the algorithm implemented in the code from [112] with

other algorithms (the direct use of the normal equations via package
SPARSPAK-A **[115,117]** and an algorithm proposed in **[28]**) are presented and
discussed in **[113]**. An efficient modification of the **George-Heath algorithm**
is proposed in **[166]**.

The main advantage of the **George-Heath algorithm** is that the storage
needed is practically the same as the storage for the solution of least square
problems by the use of the normal equations. At the same time the danger for
obtaining unstable results when the normal equations are used (due to the
presence of the square of the condition number of matrix A in the error
bounds) is avoided because the stable Givens rotations are used.

Another advantage of this algorithm is that it can be implemented by the
use of a simple static storage scheme (see **Section 2.11**). The sparse matrix
computations can be carried out in a rather efficient way when static storage
schemes are in use.

The computational work with the fill-ins (both fill-ins in matrix R and
intermediate fill-ins) is not avoided. This is a potential **drawback** of the
George-Heath algorithm: it is not very easy to see how some "small" fill-ins
can be neglected when this algorithm is applied.

The main conclusion is that the **George-Heath algorithm** (and also other
algorithms based on the same principles) will often be the best choice in the
solution of problems where the coefficient matrix remains very sparse during
the computations (the number of fill-ins produced during the orthogonalization
process is small) or in the solution of very ill-conditioned problems (where
dropping "small" elements is not successful, because the iterative method
used to improve the solution found will not converge or will converge very
slowly). If this is not the case (if the number of fill-ins is large and the
matrix is not very ill-conditioned), then it will normally be better to switch
to some method in which removal of "small" elements during the decomposition
is allowed. Such an algorithm will be described in the following section.

13.2. Use of a dynamic storage scheme in connection with the Gentleman version of the Givens plane rotations

The application of the orthogonalization methods with a dynamic storage
scheme is another possibility when large and sparse linear least square
problems are to be treated. The application of a dynamic storage scheme alone
is normally not so efficient as the application of a static storage scheme as
in the algorithm described in the previous section. However, the use of a
dynamic storage scheme can be combined with the removal of "small" non-zero
elements in the course of the orthogonalization process. Of course, the
solution obtained in this way may be inaccurate, but the accuracy lost during
the orthogonalization (because some "small" elements are neglected) can
often be regained if some iterative process is applied in order to improve the
accuracy of the starting solution. The application of these ideas (the
computer-oriented manner of exploiting sparsity in connection with orthogonalization methods; see **Chapter 1**) will be studied in this section and in the
next chapters. The particular method that will be discussed is the Gentleman-
Givens algorithm from **Section 12.3**, but the same ideas can also be applied
in connection with other orthogonalization methods.

Let $m \in \mathbf{N}$, $n \in \mathbf{N}$, $b \in \mathbf{R}^{mxn}$ and $A \in \mathbf{R}^{mxn}$ be given. Assume that:

(a)	$m \geq n$,
(b)	$rank(A) = n$,
(c)	n is large (say, $n \geq 100$),
(d)	A is sparse (i.e. many of its elements are zeros).

Assume also that both row and column interchanges are carried out during the **Gentleman-Givens plane rotations** and, moreover, that the permutations are denoted by P_1 (for the row interchanges) and by P_2 (for the column interchanges). Then the decomposition can be written as

$$(13.3) \qquad QDR = P_1AP_2 + E,$$

where $P_1 \in R^{m \times m}$, $P_2 \in R^{n \times n}$, $E \in R^{m \times n}$ (E is a perturbation matrix), $Q \in R^{m \times n}$ (Q is a matrix with orthonormal columns; $Q^TQ = I_n \in R^{n \times n}$), $D \in R^{n \times n}$ (D is diagonal)   and $R \in R^{n \times n}$ (R is triangular).

Denote $A_1 = A$ and assume for the moment that the diagonal matrix $D_1 \in R^{m \times m}$ is given. The decomposition   (13.3)   is often carried out in   n **major steps**. During the k'th   major step, $k=1,2,\ldots,n$, the product D_kA_k (where $D_k \in R^{m \times m}$ is a diagonal matrix and $A_k \in R^{m \times n}$ contains only zero elements under the diagonal in its first $k-1$ columns for $k > 1$)   is transformed into the product $A_{k+1}D_{k+1}$ (where D_{k+1} is a diagonal matrix and $A_{k+1} \in R^{m \times n}$ contains only zero elements under the diagonal in its first k columns)   by the following algorithm:

(1) Choose a pivotal column j, $k \leq j \leq n$, and interchange column j and column k if $k \neq j$. Denote the matrix so found by A_k^*.

(2) Assume that the k'th column of A_k^* has σ_{k+1} non-zeros on and/or under the main diagonal.   Then σ_k multiplications with   elementary orthogonal matrices are to be performed when $\sigma_k > 0$,   so that one zero is produced in the pivotal column (on or under the main diagonal)   after each multiplication.

(3) Let the i'th   row, $k \leq i \leq n$,   be the only row which   contains   a   non-zero in   the pivotal column   after   the   last   multiplication with   an elementary orthogonal matrix during major step k. Interchange rows i and k if $k \neq i$.

When the computations in the n'th major step are completed, the product $D_{k+1}A_{k+1}$ will be available. It is clear that the first   n   rows of   matrix A_{k+1} form the upper triangilar matrix R,   while matrix D from (13.3) is the upper left n-by-n submatrix of D_{n+1}.

The multiplication with an elementary orthogonal matrix (or with a plane rotation; see **Section 12.2**) is often called a **minor step**.   In the algorithm described above the order in which the minor steps are performed is arbitrary.

This is exploited in the choice of the row interchanges (the pivotal interchanges will be discussed in detail in the following chapter). The performance of a minor step is discussed in **Section 10.3** and will not be repeated here.

Normally the straightforward implementation of the algorithm based on (1)-(3) will not be so efficient as the **George-Heath algorithm** (see the previous section). More precisely, the storage requirements of the algorithm based on (1)-(3) will as a rule be considerably greater than those of the **George-Heath algorithm**, while the computational work will be, roughly speaking, of the same order, but this will depend strongly of the particular implementation of the algorithm based on (1)-(3). The storage requirements can be reduced by a careful choice of the row pivotal interchanges. As mentioned before the order of the row pivotal interchanges is quite arbitrary. Therefore an idea proposed by **Gentleman [111]** can successfully be implemented and experiments show that the storage requirements (and the computing time also) can very often be reduced considerably when such an implementation is in use; this will be discussed in some more detail in the next chapter. However, it is fair to mention here that the **George-Heath algorithm** can also be modified so that the **Gentleman** idea can be exploited and, in fact, such a modification is proposed by **Liu [166]**. Therefore the improvement of the pivotal strategy does not give an advantage of the algorithm described in this section. The main disadvantage of this algorithm (in comparison with the **George-Heath algorithm**) is that intermediate fill-ins are to be stored when this algorithm is in use. This disadvantage can sometimes be compensated by introducing a rule for removing small elements during the orthogonalization process. Therefore, in the following discussion it will always be assumed that the algorithm sketched in this section is applied together with a device for removing "small" elements and together with some iterative process. If the algorithm is implemented in this way and if many "small" elements are removed during the orthogonal decomposition, then

(i)	the storage needed is reduced considerably,
(ii)	the computing time is also reduced,
(iii)	if the iterative process applied is convergent, then a sufficiently accurate solution is computed even when single precision is in use,
(iv)	a reliable error estimation can be calculated in a cheap way.

This shows that if the algorithm is combined with a device for removing "small" elements and with an iterative process, then in some cases it will perform more efficiently than the George-Heath algorithm.

13.3. Numerical results

Several experiments were carried out to compare the **George-Heath algorithm** (implemented in **SPARSPAK-B**) with the algorithm from **Section 13.2** (implemented in package **LLSS01-LLSS02**). **SPARSPAK-B** is run with default values of all parameters. The second code is run in three modes: (i) **LLSS01-DS**,

direct solution (with drop-tolerance T=10^{-25}), **(ii)** **LLSS01-IR**, iterative
refinement, and **(iii)** **LLSS02**, conjugate gradients. In the last two cases
the initial value T is T=2^{-4}, and T is varied as in **Chapter 11**. The Harwell-
Boeing matrices ([71,72]) used are listed in **Table 13.1**. Some results are
given in **Table 13.2** - **Table 13.3**. An **ALLIANT** computer is used in the runs.

Matrix	Rows	Columns	Number of non-zeros
well1033	1033	320	4732
well1850	1850	712	8758
illc1033	1033	320	4732
illc1032	1850	712	8758
abb313	313	176	1557
ash319	219	85	438
ash331	331	104	662
ash608	608	188	1216
ash958	958	292	1916

Table 13.1
The Harwell-Boeing matrices used in the runs

Matrix	SPARSPAK-B	LLSS01-DS	LLSS01-IR	LLSS02
well1033	12.1	2.2	Failed	2.1(13)
well1850	24.8	11.3	Failed	7.3(11)
illc1033	12.2	2.1	4.0(5)	4.1(3)
illc1032	24.9	11.1	10.9(16)	10.2(7)
abb313	6.1	1.9	2.1(24)	1.2(10)
ash319	3.7	0.6	0.3(7)	0.3(4)
ash331	4.6	1.2	0.4(6)	0.4(4)
ash608	7.1	4.3	0.8(7)	0.8(5)
ash958	10.8	10.0	1.2(5)	1.2(4)

Table 13.2
Computing times obtained when the Harwell-Boeing matrices are run.
The number of iterations are given in brackets.

r	SPARSPAK-B	LLSS01-DS	LLSS01-IR	LLSS02
10	36.3	72.1	1.4(19)	1.1(8)
20	61.1	106.9	1.5(9)	1.4(7)
30	73.3	143.8	2.0(13)	1.8(8)
40	100.6	180.7	2.2(22)	1.6(8)
50	128.3	207.7	2.1(14)	1.8(8)
60	136.2	214.2	3.6(34)	2.4(10)
70	141.0	256.9	Failed	6.6(11)
80	141.3	274.1	Failed	7.8(12)
90	149.8	282.0	Failed	9.6(13)
100	159.2	267.6	Failed	9.5(14)

Table 13.3
Computing times obtained when matrices F2(500,250,100,r,32.0)   are run.
The numbers of iterations are given in brackets

It is surprising that **LLSS01-DS** performs better than **SPARSPAK-B** when Harwell-Boeing matrices are run. These matrices are very sparse and stay sparse during the computations. This is probably the reason for the good performance of **LLSS01** (there are only a few changes in the dynamic scheme when the matrix stays sparse). On the other hand, the additional work carried out in **SPARSPAK-B** to prepare the static scheme is considerably large (compared with the total work) when the matrix is sparse and stays sparse. The situation is different when matrices of class **F2(m,n,c,r,α)** are used. These matrices produce many fill-ins (and, thus, many changes in the dynamic scheme are to be performed to find free locations for them). The additional work in **SPARSPAK-B** is small for these matrices (compared again with the total work needed to solve the problem). This explain why **SPARSPAK-B** performs better than **LLSS01-DS** in this case.

LLSS02 performs in general better than the other methods (and sometimes much better). **LLSS01-IR** is often comparable with **LLSS02**, but for many matrices it does not converge or converges very slowly when the largest value of the drop-tolerance that yields convergence of **LLSS02** is used. This is declared as a failure in **Table 13.2** and **Table 13.3**, but **LLSS01-IR** will certainly succeed with a smaller value of the drop-tolerance. **LLSS01** and **LLSS02** will be studied in more detail in the following chapters.

13.4. Concluding remarks and references concerning the implementation of orthogonalization methods in the solution of linear least square problems

Until 1980 it was commonly accepted that the use of orthogonalization methods in the solution of linear least square problems is too expensive (both in regard to the computing time used and in regard to the storage needed). Other methods are recommended in the well-known survey paper [75] published in 1976 (see also [23]). However, the advances made after 1980 in the development of new techniques and devices for orthogonal methods are very great and orthogonal methods with the new techniques and devices can successfully be applied in the solution of large and sparse linear least squares problems.

Only two implementations were discussed in this section. However, these implementations are based on very promising ideas and new algorithms based on these ideas are still appearing.

The first implementation was described in its classical form in 1980 in [112]. However, several improvements have been proposed after 1980. The improvement due to **Liu [166]** is probably the most important one.

The second implementation was just sketched here. It is based on more general ideas that can be applied with many other methods for solving algebraic problems of the type $x = A^{\dagger}b$ (see **Chapter 3**). Different details in connection with this implementation are given in [296,320,322]. The second implementation is further discussed in the following chapters.

CHAPTER 14

PIVOTAL STRATEGIES FOR GIVENS PLANE ROTATIONS

The general discussion of the orthogonalization methods (see **Chapter 12**) indicated that the Givens plane rotations are the most suitable method for handling large and sparse linear least squares problems. Two implementations of the Givens plane rotations for large and sparse linear least squares problems were discussed in the previous chapter. In the present chapter some pivotal strategies that can successfully be used with the second implementation will be described and compared. The same rule is used in the selection of pivotal columns in these pivotal strategies. The differences arise because the row interchanges are not the same.

An idea, proposed originally by **Gentleman** [111], is exploited in the construction of a pivotal strategy which will be called **Strategy 1**. This strategy often performs better than another pivotal strategy, **Strategy 2**, which has been advocated in several papers ([23,62,75]). **Strategy 1** is more difficult than **Strategy 2** from the implementation point of view. However, once implemented in a correct way, the strategy performs rather well. This is especially true for the important case where many fill-ins are created during the process of the orthogonal decomposition. The great efficiency in the latter case is demonstrated in this chapter by many numerical examples.

The implementation of an efficient pivotal strategy is much easier when a dynamic storage scheme is in use (when the second implementation from the previous chapter is studied). This is so because both the pivotal row and the pivotal column at any major step j are chosen immediately before the beginning of the major step j. The situation is not so simple when a static storage scheme is applied. In the latter case the pivotal columns and rows are to be determined in some way before the start of the actual computations needed in the orthogonalization process. Therefore, it is not a surprise that more complicated algorithms, often based on devices and results known from the graph theory, are to be applied in the design of a pivotal strategy in connection with a static storage scheme.

14.1. Definition of two pivotal strategies

Let us assume that an orthogonal decomposition,

$$(14.1) \quad QDR = P_1AP_2 + E,$$

is to be calculated in order to solve the linear least squares problem:

$$(14.2) \quad Ax = b - r \qquad (\text{with} \quad A^Tr = 0 \quad \text{also satisfied}).$$

The quantities $A \in \mathbf{R}^{m \times n}$, $b \in \mathbf{R}^{m \times 1}$, $m \in \mathbf{N}$ and $n \in \mathbf{N}$ are given data. It is assumed that $m \geq n$ and $rank(A) = n$. $P_1 \in \mathbf{R}^{m \times m}$ and $P_2 \in \mathbf{R}^{n \times n}$ are permutation matrices. $E \in \mathbf{R}^{m \times n}$ is a perturbation matrix (caused because the computations are carried out with rounding errors and, when the computer-oriented manner of exploiting sparsity is in use, because "small" elements are neglected). $Q \in \mathbf{R}^{m \times n}$ is a matrix with orthonormal columns (i.e. $Q^TQ = I_n$ if the rounding errors are neglected). $D \in \mathbf{R}^{n \times n}$ is a diagonal matrix. Finally, $R \in \mathbf{R}^{n \times n}$ is an upper triangular matrix.

In this chapter it will be assumed that the decomposition   QDR   is obtained by the **Gentleman-Givens plane rotations**. However, the pivotal strategies discussed could also be used, without any modification, in the case where the **classical Givens plane rotations** are used during the orthogonalization (assuming $D = I_n$ in the latter case). The pivotal strategies could be modified for other orthogonalization methods. For the Householder orthogonalization, for example, the selection of pivotal columns can be the same as that for the Givens plane rotations, while the computations could be carried out with no row interchanges $(P_1=I)$.

When the matrix is sparse, it is important to choose the permutation matrices P_1 and P_2 so that the sparsity of A is preserved as well as possible. An attempt to achieve this could be carried out as follows.

Definition 14.1. The set

(14.3) $\overline{C}_{kj} = \{a_{ij} \, / \, i=k,k+1,\dots,m\}$

is called the active part of column   j   (j=k,k+1,...,n)   at major step   k   (k=1,2,...,n).   ∎

Rule 14.1. At every major step   k   (k=1,2,...,n)   the column which contains a minimal number of non-zero elements in its active part (or one of these columns if there are several) is chosen as a pivotal column.   ∎

Definition 14.2. The set

(14.4) $\overline{R}_{ki} = \{a_{ij} \, / \, j=k,k+1,\dots,n\}$

is called the active part of row   i   (i=k,k+1,...,m)   at major step   k   (k=1,2,...,m).   ∎

Definition 14.3. Consider major step   k   (k=1,2,...,n).   The number of non-zero elements in the active part of row   i   (i=k,k+1,...,m)   before the minor step σ $(\sigma=1,2,\dots,\sigma_k)$ is denoted by $\rho(k,\sigma,i)$.   ∎

Rule 14.2. Consider major step   k   (k=1,2,...,n).   Choose row   i   for which the following three conditions are satisfied:

      (a) $k \le i \le m$,

      (b) $a_{ik} \neq 0$,

      (c)   if μ is any row for which $k \le \mu \le m$ and
         $a_{\mu k} \neq 0$,   then $\rho(k,1,i) \le \rho(k,1,\mu)$.

Consider any two rows ν and τ $(k \le \nu \le m, k \le \tau \le m, \nu \neq i, \tau \neq i)$ for which $a_{\nu k} \neq 0$ and $a_{\tau k} \neq 0$. The plane rotations $Q_{i\nu}$ and $Q_{i\tau}$ are used to transform $a_{\nu k}$ and $a_{\tau k}$ into zeros. Moreover, if $\rho(k,1,\nu) < \rho(k,1,\tau)$, then $a_{\nu k}$ will be transformed into zero before $a_{\tau k}$ (the multiplication with $Q_{i\nu}$ will be performed before the multiplication with $Q_{i\tau}$).   ∎

Rule 14.3. Consider major step   k   (k=1,2,...,n).   Before any minor step σ $(\sigma=1,2,\dots,\sigma_k)$ choose two rows μ and ν for which the following three conditions are satisfied:

    (a) $k \leq \mu \leq m$ $\wedge$ $k \leq \nu \leq m$,

    (b) $a_{\mu k} \neq 0$ $\wedge$ $a_{\nu k} \neq 0$,

    (c) if τ is any row for which $k \leq \tau \leq m$, $\tau \neq \mu$, $\tau \neq \nu$ and $a_{\tau k} \neq 0$,   then $max(\rho(k,\sigma,\mu),\rho(k,\sigma,\nu)) \leq \rho(k,\sigma,\tau)$.

Produce a zero element in one of these two rows (multiplying by the plane rotation matrix $Q_{\mu\nu}$) according to the algorithm in **Section 12.3**.   ■

Definition 14.4. The pivotal strategy based on **Rule 14.1** and **Rule 14.3** will be called **Strategy 1**. ■

Definition 14.5. The pivotal strategy based on **Rule 14.1** and **Rule 14.2** will be called **Strategy 2**. ■

The column with minimal number of non-zero elements in its active part, or one of the columns with minimal number of non-zero elements if there are several such columns, is chosen as a pivotal column at major step k (k=1,2,...n) both in **Strategy 1** and **Strategy 2**.

The row with a non-zero in the pivotal column and with a minimal number of non-zero elements in its active part (or one of the rows that satisfy these two conditions if there are several such rows) is used as a **fixed** pivotal row during the computations in major step k (k=1,2,...,n) in **Strategy 2**. For this reason **Strategy 2** is often called **a fixed pivotal row strategy**. The other rows that have non-zero elements in the pivotal column, **the target rows**, are ordered in an increasing number of the non-zero elements in their active parts and the minor steps in **Strategy 2** are carried out according to this order (see Rule 14.2).

In **Strategy 1** two rows, which have non-zero elements in the pivotal column and which have minimal numbers of non-zero elements in their active parts, are picked out (before each minor step of major step k), and a zero element is produced in one of them by a multiplication with plane rotation matrix. This explains why **Strategy 1** is called **a variable pivotal row strategy**. **Strategy 1** is based on an idea proposed originally by **Gentleman** [111]. A sparse matrix code in which **Strategy 1** is implemented is discussed in detail in [296,320-322].

A comparison of the performance of the two pivotal strategies will be carried out in the following sections. The main results of the comparison can be summarized as follows.

    (i)   If **Strategy 1** and **Strategy 2** are used with the same column interchanges, then the two pivotal strategies produce the same number of non-zero elements in matrix R. It has been mentioned in the previous two chapters that this should be so. However, it must also be emphsized that neither the computational work done nor the storage needed are the same for the two strategies in all situations. This is demonstrated in the next sections.

    (ii) If the number of non-zero elements in the active part of the pivotal column is less than four for each k

(k=1,2,...,n)   and if the same column interchanges are
applied in the two pivotal strategies, then the results
will be very similar.

(iii) If the matrix is sufficiently large (say, n≥100) and if
the matrix is not very sparse (say, the average number
of non-zero elements per column is greater than three),
then Strategy 1 will often give better results than
Strategy 2 with regard both to the storage used and
the computing time needed. The efficiency of Strategy
1 under these two conditions will be illustrated by
many numerical experiments.

Since the same rule is used in the selection of the pivotal columns in
the two strategies, it may happen that the calculations are carried out with
the same permutation matrix P_2 for both strategies. Therefore it is
interesting to answer first the question: what should be expected when the two
pivotal strategies are applied with the same column interchanges (with the
same permutation matrix P_2)? The answer to this question is given in the
next section.

14.2. Computations with the same column interchanges

The following definitions and theorems are needed before the discussion
of the influence of column interchanges on the computations during the
orthogonal decomposition of a sparse matrix by the Gentleman-Givens method.
The results could also be extended to the case where the classical Givens
method is applied.

Definition 14.6. Assume that major step k (k=1,2,...,n) is to be
carried out. Assume also that a pivotal column is chosen according to the
Rule 14.1, and that the column interchanges implied by this choice are
performed. Denote the matrix found after the column interchanges at major step
k by $\overline{A}_k$. The set

(14.5) $B_k = \{a_{ij} \in \overline{A}_k \ / \ k{\le}i{\le}m \ \wedge \ k{\le}j{\le}n\}$

is called   the active part of matrix $\overline{A}_k$.

Definition 14.7. The set

(14.6) $L_{ki} = \{j \ / \ a_{ij} \neq 0 \ \wedge \ a_{ij} \in B_k\}$

is called the sparsity pattern of the active part of row i at major step
k (i=k,k+1,..,m, k=1,2,...,n). ∎

Definition 14.8. The set

(14.7) $M_{kj} = \{i \ / \ a_{ij} \neq 0 \ \wedge \ a_{ij} \in B_k\}$

is called the sparsity pattern of the active part of column j at major step
k (j=k,k+1,..,n, k=1,2,...,n). ∎

Definition 14.9. The set

$$(14.8) \quad N_k = \{j \ / \ a_{ij} \neq 0 \quad \wedge \quad a_{ij} \in B_k \quad \wedge \quad i \in M_{kk}\}$$

is called **the union of the sparsity patterns of the active parts of rows involved in the computations at major step k** $(k=1,2,\ldots,n)$. ∎

Remark 14.1. It is clear that

$$(14.9) \quad N_k = \bigcup_{i \in M_{kk}} L_{ki} \ . \qquad ∎$$

Definition 14.10. The set

$$(14.10) \quad N^*_{kj} = \{i \ / \ a_{i\mu} \neq 0 \quad \wedge \quad a_{i\mu} \in C_{kj} \quad \wedge \quad i \in M_{kj} \quad \wedge \quad k < n\},$$

where the submatrix C_{kj} is determined by

$$(14.11) \quad C_{kj} = \{a_{i\mu} \in B_k \ / \ j \leq \mu \leq n\},$$

is called **the union sparsity pattern of the rows of matrix** C_{kj} **which have non-zero elements in the active part of column** j $(j=k+1,k+2,\ldots,n)$. ∎

The following lemma is well-known. Moreover, it can be proved in a very simple and elegant way. The proof given below is not simple. It is needed, however, in order to facilatate the proof of **Theorem 14.1.** In fact, the assertion of **Theorem 14.1** follows immediately from the proof of **Lemma 14.1.**

Lemma 14.1. **The number of non-zero elements in matrix R found by plane rotations does not depend on the permutation matrix P_1 (on the row interchanges) when the permutation matrix P_2 is kept fixed (when the same column interchanges are used).**

Proof. Assume that the multiplications with a fixed permutation matrix P_2 have been performed before the beginning of the computations; i.e. $A_1=AP_2$ and $A_k=\overline{A}_k$ $(k=1,2,\ldots,n)$. This assumption is not a restriction.

Consider the k'th major step $(k=1,2,\ldots,n)$. After the multiplication with any plane rotation matrix $Q_{\mu\nu}$ (where $\mu \in M_{kk} \ \wedge \ \nu \in M_{kk}$) the sparsity patterns of the active parts of the rows involved in the computations are:

$$(14.12) \quad L_{k\mu} \bigcup L_{k\nu} \qquad and \qquad (L_{k\mu} \bigcup L_{k\nu}) \setminus \{k\},$$

where the second sparsity pattern is for the row in which a zero element in the pivotal column is produced. The expressions in (14.12) show that after the last minor step within the k'th major step the sparsity patterns of the active parts of the involved rows are given by

$$(14.13) \quad \bigcup_{i \in M_{kk}} L_{ki} = N_k \qquad and \qquad (\bigcup_{i \in M_{kk}} L_{ki}) \setminus \{k\} \ .$$

The expressions in (14.13) do not depend on the order in which zero elements are produced.

Consider the sets N^*_{kj} $(j=k+1,k+2,\ldots,n, \ k<n)$. These sets are invariant during the computations at major step k $(k=1,2,\ldots,n-1)$. Indeed, new non-

zero elements (fill-ins) may be produced in some rows i ($i \in M_{kk}$), but this happens only in positions where there are non-zero elements in one or several of the other rows belonging to set M_{kk}. Therefore

(14.14) $N_{k,k+1}^{*} = N_k$ and $N_{kj}^{*} = N_{k+1,j}^{*}$ ($j=k+2,k+3,\ldots,n$, $k+1 < n$).

Thus the lemma is proved because:

(i)   an arbitrary major step k has been considered,

(ii)   the number, N_k, of the non-zero elements in the
        k'th row of matrix R does not depend on the
        order, in which the zero elements in the pivotal
        column are produced,

(iii)   the union sparsity pattern of the rows, which are
        involved in the computations during the next
        step, step k+1, does not depend on the computa-
        tions during the k'th major step. ■

The following results can easily be proved by using **Lemma 14.1.**

   Corollary 14.1. Assume that the same column interchanges have been used both with Strategy 1 and Strategy 2. Then the number of non-zero elements in matrix R will be the same for these two strategies.   ■

   Theorem 14.1. Assume that

(i)   if at any major step there are several columns which
        have a minimal number of non-zeros in their active
        parts, then the same column will be chosen as pivotal
        with both Strategy 1 and Strategy 2,

(ii)   after the column interchanges the number of non-zeros
        in the active part of column k is less than four
        for any k ($k=1,2,\ldots,n-1$).

Then the numbers of non-zero elements in the sets B_k ($k=1,2,\ldots,n-1$) found by Strategy 1 and Strategy 2 are the same for $\forall k$.   ■

   Denote by $W_1(k)$ the number of floating point operations needed to perform the orthogonal transformations at major step k ($k=1,2,\ldots,n$) by **Strategy 1.** Let $W_2(k)$ be the corresponding number for **Strategy 2.** Then the following result can easily be obtained from **Theorem 14.1.**

   Corollary 14.2. Assume that the conditions imposed in Theorem 14.1 are satisfied. Then

(14.15) $W_1(k) = W_2(k)$ for $\forall k \in \{1,2,\ldots,n\}$.   ■

The definition given below is introduced to facilitate the references to matrices for which the conditions imposed in **Theorem 14.1** are satisfied.

   Definition 14.11. If the conditions of **Theorem 14.1** are satisfied for a particular matrix A, then A is called **essentially sparse.** ■

The results formulated in this section show that the storage used in the orthogonal decomposition of an essentially sparse matrix A   by the Givens plane rotations is the same for the two pivotal strategies (**Theorem 14.1**). The computing time needed in the orthogonalization process for an essentially sparse matrix is nearly the same for the two pivotal strategies (**Corollary 14.2**). Some differences arise because the organization of the pivotal search depends on the pivotal strategy chosen.

One does not know in advance whether the matrix under consideration is essentially sparse or not. The requirement that the matrix is **very sparse** is only a necessary condition for essential sparsity of matrix A (the original matrix A is called **very sparse** here if the average number of non-zero per column is less than four). Although this condition is not sufficient (examples where **very sparse** matrices produce many fill-ins can easily be constructed), it can be used as an indication for essential sparsity. Therefore one should expect the two strategies to produce similar results for some **very sparse** matrices. This has been confirmed in many runs with **very sparse** matrices.

Some heuristics, which show clearly that one should expect **Strategy 1** to perform better than **Strategy 2** when the second condition in **Theorem 14.1** is not satisfied, can be formulated as follows. Assume that:

(i)   The conditions of Theorem 14.1 are satisfied for
$\forall \mu \in \{1,2,\ldots,k-1\}$.

(ii)   The same column is chosen as a pivotal column at
major step k in both Strategy 1 and Strategy 2.

(iii)   The number of non-zeros in the active part of the
pivotal column at major step k is greater than
three (say, q>3).

(iv)   The rows involved in the computations during major
step k are $i_1, i_2, \ldots, i_q$ with $\rho(k,1,i_\mu) \leq \rho(k,1,i_\nu)$
when $\mu < \nu$, $1 \leq \mu \leq q$ and $1 \leq \nu \leq q$.

If   **Strategy 2** is in use, then row i_1 will be chosen as a **fixed** pivotal row. At the end of computations during major step k the sparsity patterns of the active part of rows $i_1, i_2, \ldots, i_q$ are given by

(14.16)

$$
\begin{array}{l}
i_1: \quad \displaystyle\bigcup_{j=i_1,i_2,\ldots i_q} L_{kj} \\[2em]
i_2: \quad \displaystyle\bigcup_{j=i_1,i_2} L_{kj} \setminus \{k\} \\[2em]
i_3: \quad \displaystyle\bigcup_{j=i_1,i_2,i_3} L_{kj} \setminus \{k\} \\[2em]
\cdots\cdots\cdots\cdots\cdots\cdots\cdots\cdots\cdots\cdots \\[1em]
i_q: \quad \displaystyle\bigcup_{j=i_1,i_2,\ldots i_q} L_{kj} \setminus \{k\}
\end{array}
$$

From (14.16) it is easily seen that the number of non-zero elements in
the active part of the **fixed** pivotal row i_1 **may** increase quickly in the
course of the computations during major step k. Moreover, the increase of
the number of non-zero elements in the active part of row i_1 will often lead
to a considerable increase of the number of non-zero elements in the active
parts of the last rows involved in the orthogonal transformations in major
step k. Therefore the use of **Strategy 1**, where the last rows can be
transformed independently of row i_1 seems to be much more attractive when
q is large. An example is given below which demonstrates that **Strategy 1**
may be more efficient than **Strategy 2**.

Figure 14.1
The sparsity pattern of the active parts of the rows involved
in the computations during major step k (rows i_1, i_2, ..., i_{16}).
The non-zero elements are denoted by ■. The first column is
the pivotal column k. The next columns are j_1, j_2, ..., j_{16}.

Example 14.1. Let conditions **(i)-(iv)** stated above be satisfied.
Assume that q=16 and that the non-zero elements in the rows that will be
involved in the computations at major step k are the 16 elements in the
pivotal column k plus 16 other non-zero elements:

(14.17) a_{vk} for $v=i_1, i_2, \ldots, i_{16}$ **and** $a_{\mu v}$ for $(\mu, v)=(i_s, j_s)$ s=1,2,...,16.

The submatrix, where the non-zero elements involved in the transformations during the major step k are located, is given in **Fig. 14.1**. If
column k has been chosen as pivotal according to **Rule 14.1**, then the
columns j_1, j_2, ..., j_q must necessarily contain more non-zero elements in
their active parts than those shown in **Fig. 14.1**. However, these elements
are not involved in the computations at major step k and, for the sake of
simplicity, they are not shown in the figure.

The computations when **Strategy 2** is in use are to be carried out by
multiplying matrix A successively by the orthogonal matrices $Q_{\mu v}$ ($\mu=i_1$,
$v=i_2, i_3, \ldots, i_q$). If **Strategy 1** is applied, then matrix A is successively
multiplied by the plane rotation matrices given in (14.18). The multiplications are carried out successively from the left to the right until the
multiplication with the last matrix in a row in (14.18) is performed; then

the same process is repeated for the next row in (14.18). **The index i is omitted in (14.18); this means that, for example,** $Q_{9,11}$ **is the orthogonal matrix with indices** i_9 **and** i_{11}.

(14.18)

$$
\begin{array}{llllllll}
Q_{1,2} & Q_{3,4} & Q_{5,6} & Q_{7,8} & Q_{9,10} & Q_{11,12} & Q_{13,14} & Q_{15,16} \\
Q_{1,3} & & Q_{5,7} & & Q_{9,11} & & Q_{13,15} \\
Q_{1,5} & & & & Q_{9,13} \\
Q_{1,9}
\end{array}
$$

It is easy to verify that the number of new non-zero elements (fill-ins) produced by **Strategy 2** during major step k is 135 for this particular example, while the corresponding number for **Strategy 1** is only 64. ■

The above example is created in a very artificial way. Nevertheless, this example shows that **Strategy 1** **may** perform much better than **Strategy 2** when the second condition in **Theorem 14.1** is violated at some stage k. Of course, this will not happen always. Moreover, some examples, where **Strategy 2** will be efficient even when the second condition of **Theorem 14.1** is not satisfied, can be constructed. However, (14.16) shows that this is not typical. In general, one should expect that **Strategy 1** is more efficient when the matrix is not essentially sparse; see **Definition 14.11**. Several hundreds of experiments have been carried out with the two strategies. **Strategy 1** performed better for all examples in which matrix A was not very sparse (i.e. the average number of non-zeros per column was greater than three; it is obvious that A is not essentially sparse when it is not very sparse). Some of the numerical results will be presented in the next sections.

14.3. Choice of test-matrices

The results (especially the storage needed and the computing time used) depend on the preservation of the sparsity during the orthogonalization process. The preservation of the sparsity depends:

(i) on the dimensions of matrix A,
(ii) on the distribution of non-zeros within matrix A,
(iii) on the number of non-zeros in the beginning of the orthogonalization process,
(iv) on the magnitude of the non-zero elements.

A systematic investigation of the performance of the two pivotal strategies can easily be carried out if the matrix generator **MATRF2** that produces matrices F2(m,n,c,r,α) is used (see **Section 3.7**). The dimensions of the matrix can be varied by varying m and n. The sparsity pattern is varied by varying c. The number of non-zeros in the matrix generated is NZ=rm+110 (and can be varied by varying r even if m is fixed). Parameter

α is such that $\max(|a_{ij}|)/\min(|a_{ij}|) = \max[(rm-m)\alpha, 10\alpha^2]$, where non-zero elements only are considered. This shows that the magnitude of the matrix elements can be varied by varying α.

Matrices of class F2(m,n,c,r,α) are very suitable for studying the ability of the pivotal strategies to preserve the sparsity and are used below. All five parameters are varied in quite large intervals.

14.4. Comparison of the storage required by the two strategies

It is not very easy to compare the storage required by the two pivotal strategies, because the number of fill-ins produced during the orthogonal decomposition is not known in advance. This means that the optimal length of some arrays is not known when the orthogonalization is started. A similar problem has been discussed in connection with GE in the previous chapters.

The number of non-zero elements in matrix R should not be used in the comparison. This is so because if the column interchanges happen to be the same, then the same number of non-zero elements in R will be produced by both strategies. However, this does not mean that the same computational work will be needed and, what is more important in this section, that the same number of intermediate fill-ins will be created; see **Theorem 14.1**.

Assume that the non-zeros of A are kept in a **REAL** array AQDR and the transformations needed are carried out in this array, so that at the end of the computations the non-zeros of R are also stored in AQDR. Denote by COUNT1 and COUNT2 the maximal numbers of non-zeros kept in AQDR during the orthogonalization when **Strategy 1** and **Strategy 2** are used. COUNT1 and COUNT2 do not show the actual storage needed when the two strategies are used. Many other locations must be used, for example, as pointers and as working space (**Chapter 2**). However, the most important fact is that COUNT1 < COUNT2 shows that **Strategy 1** is more efficient, with regard to the storage requirements, than **Strategy 2**, while COUNT2 < COUNT1 shows the opposite.

| | S t r a t e g y 1 | | | | S t r a t e g y 2 | | | |
m	COUNT	Time	Iterations	Accuracy	COUNT	Time	Iterations	Accuracy
100	210	0.58	5	4.2E-26	210	0.60	5	1.9E-26
110	220	0.73	7	1.1E-27	227	0.82	9	2.9E-27
120	230	0.62	5	3.8E-27	230	0.65	5	3.5E-27
130	240	0.64	5	3.7E-27	240	0.64	5	8.5E-27
140	250	0.68	6	2.9E-27	250	0.74	7	2.6E-27
150	260	0.65	5	3.7E-27	260	0.71	6	6.2E-27
160	270	0.77	7	1.5E-27	270	0.67	5	6.0E-27
170	280	0.71	6	2.9E-27	280	0.84	8	1.8E-27
180	290	0.72	6	3.1E-27	290	0.67	5	5.4E-27
190	300	0.85	7	1.1E-27	300	0.96	9	2.9E-27
200	310	0.83	6	3.7E-27	310	0.97	8	1.7E-27

Table 14.1
Numerical results obtained in the solution of linear least square problems whose matrices are A=F2(m,100,11,1,10). NZ=m+110 in this experiment. The computer used is a CDC Cyber 173 and the times are given in seconds.

Denote

(14.19) E_s = COUNT1/COUNT2 .

The quantity E_s will be called the efficiency factor with regard to the storage requirements. By the use of E_s the following conclusions can be drawn for the two pivotal strategies from the experiments performed with matrices $F2(m,n,c,r,\alpha)$, which were run with many different values of the parameters.

(1) If the matrix is very sparse (the average number of non-zeros per column is smaller than three), then $E_s \approx 1$ is normally observed; compare the values of COUNT1 and COUNT2 in Table 14.1 and Table 14.2 (the notation COUNT is used for both strategies in the tables). This result should be expected, because in this situation it is very likely that the conditions of Theorem 14.1 are satisfied.

(2) If the matrix is not very sparse (if the average number of non-zeros per column is greater than three), then as a rule $E_s < 1$. This means that Strategy 1 performs better than Strategy 2 with regard to the storage requirements; see Table 14.3. This will also be observed in the next chapter.

(3) Many experiments carried out with different values of the parameters m, n, c and α show that the performance of the two pivotal strategies (with regard to the storage requirements) does not depend on these parameters. Only parameter r is important in this situation. If r is large, then the matrix is not very sparse and, therefore, Strategy 1 performs considerably better than Strategy 2. One should expect this to be true not only for the matrices $F2(m,n,c,r,\alpha)$ and runs with matrices from the Harwell-Boeing set, [71,72], confirm that Strategy 1 performs in general better than Strategy 2 for matrices that are not very sparse.

14.5. Comparison of the computing times

Denote by t_1 the computing time used when **Strategy 1** is applied and by t_2 the computing time used when **Strategy 2** is applied. The number $E_t = t_1/t_2$ will be called the efficiency factor with regard to the computing time. Conclusions (1)-(3) from the previous paragraph hold also when E_s is replaced with E_t. This means that if the matrix decomposed is not very sparse, then the computing time will often be reduced when **Strategy 1** is used instead of **Strategy 2**. This is illustrated in **Table 14.3**, while **Table 14.1** and **Table 14.2** show that the results obtained by the two pivotal strategies are similar when the matrix is very sparse. Numerical results obtained by using matrices from the Boeing-Harwell set, [71,72], also confirm such a conclusion.

m	Strategy 1				Strategy 2			
	COUNT	Time	Iterations	Accuracy	COUNT	Time	Iterations	Accuracy
100	878	1.42	5	4.5E-26	878	1.60	7	3.0E-26
110	559	1.23	7	9.6E-26	568	1.45	10	5.0E-27
120	522	1.18	7	2.9E-27	526	1.27	8	2.2E-27
130	536	1.20	7	1.9E-27	560	1.48	10	6.9E-27
140	611	1.15	5	2.0E-26	617	1.33	7	8.8E-27
150	677	1.54	8	9.6E-28	653	1.50	8	6.8E-27
160	638	1.42	7	1.6E-26	687	1.58	8	6.2E-27
170	717	1.75	9	2.3E-27	711	1.56	7	6.4E-27
180	776	1.73	7	9.6E-28	783	1.80	8	1.0E-27
190	811	1.84	7	1.3E-26	812	2.00	9	1.6E-26
200	817	1.73	5	3.0E-26	858	2.10	9	9.7E-26

Table 14.2
Numerical results obtained in the solution of linear least square problems
whose matrices are A=F2(m,100,11,2,10). NZ=2m+110 in this experiment.
The computer used is a CDC Cyber 173 and the times are given in seconds.

m	Strategy 1				Strategy 2			
	COUNT	Time	Iterations	Accuracy	COUNT	Time	Iterations	Accuracy
100	1997	5.26	15	2.7E-27	2076	4.70	10	1.6E-27
110	2217	4.64	7	3.9E-27	2439	5.76	7	2.7E-27
120	2331	4.89	5	1.3E-27	2095	5.09	8	2.0E-27
130	2336	5.96	9	7.6E-26	2378	6.50	9	2.0E-27
140	2287	5.94	7	5.6E-27	2476	6.71	8	9.6E-28
150	2896	8.35	7	2.1E-26	2504	6.66	5	1.3E-26
160	2644	8.82	9	2.3E-27	3052	11.48	10	8.6E-28
170	2676	8.73	7	4.0E-28	3198	13.21	11	1.4E-27
180	3361	14.14	8	1.9E-27	3595	16.93	5	9.0E-26
190	3334	14.46	7	3.1E-27	3435	16.33	5	1.3E-26
200	3690	16.83	7	3.3E-27	4143	22.64	8	2.6E-27

Table 14.3
Numerical results obtained in the solution of linear least square problems
whose matrices are A=F2(m,100,11,3,10). NZ=3m+110 in this experiment.
The computer used is a CDC Cyber 173 and the times are given in seconds.

14.6. Comparison of the accuracy achieved by the two strategies

The numerical results presented in this section are obtained by the use
of IR to improve the accuracy of the first solution (obtained by solving
$Rx=D^{-1}Q^Tb$,   where   QDR   is the orthogonal decomposition of matrix   A
calculated by the Gentleman-Givens plane rotations). The IR process applied
in connection with the orthogonalization method will be discussed in detail
in the next chapter. It is sufficient here to point out that all matrix-vector
multiplications are accumulated and stored in double precision. This means
that the vectors involved in the computations are declared as double precision
arrays, while all matrices (or, more precisely, the non-zero elements of all

matrices) are stored as single precision arrays. Assume that the mantissa of the real numbers in single precision contains t_1 digits, while the mantissa of the real numbers in double precision contains t_2 digits. Assume also that $t_2 \geq 2t_1$. Then one should expect to get about $2t_1$ correct digits in the calculated solution vector. The mantissa of the real numbers on the **CDC Cyber** computers is about 14 decimal digits in single precision and about 28 decimal digits in double precision. The corresponding figures for **IBM** computers are about 6 and 16 digits. Therefore the assumption $t_2 \geq 2t_1$ is satisfied for these computers. The numerical results show that the expected accuracy (about $2t_1$ correct digits) is achieved, for both strategies, when the iterative refinement converges; see **Table 14.1 - Table 14.3**.

α	Strategy 1			Strategy 2		
	m=100	m=150	m=200	m=100	m=150	m=200
1.0E+01	4.2E-26	2.9E-27	1.1E-27	2.0E-26	6.2E-27	1.7E-27
1.0E+02	4.4E-25	3.3E-26	2.3E-26	9.1E-26	3.1E-26	3.5E-26
1.0E+03	2.2E-24	2.2E-25	2.0E-24	2.0E-24	5.5E-25	2.4E-25
1.0E+04	1.7E-23	1.5E-24	4.6E-24	1.7E-23	1.2E-24	7.0E-25
1.0E+05	4.9E-23	2.3E-23	8.4E-24	3.1E-23	3.8E-23	2.1E-23
1.0E+06	1.1E-21	2.0E-22	1.1E-22	3.8E-22	2.6E-22	3.7E-22
1.0E+07	3.6E-21	1.2E-21	1.3E-21	2.6E-21	1.5E-21	2.2E-21
1.0E+08	1.6E-19	3.4E-20	1.1E-19	5.8E-20	1.2E-19	4.6E-19
1.0E+09	3.8E-17	4.5E-17	1.5E-17	8.4E-17	4.8E-17	1.8E-03
1.0E+10	2.4E-13	7.2E-16	2.7E-16	3.7E-15	1.1E-13	2.9E-00

Table 14.4

Numerical results obtained in the solution of linear least square problems whose matrices are $A=F2(m,100,11,3,\alpha)$. $NZ=3m+110$ in this experiment. The computer used is a **CDC Cyber 173**.

Of course, the expectation to get $2t_1$ correct digits is based on at least two assumptions:

> (i) the input data can be stored in a sufficiently accurate way in the computer memory,
>
> (ii) the matrix that is decomposed is not very ill-conditioned.

The matrices $F2(m,n,c,r,\alpha)$ become more and more ill-conditioned when parameter α is increased. Therefore, one should not expect about $2t_1$ digits in the computed solution to be correct for large values of α. This is demonstrated on **Table 14.4**. For some matrices the results obtained by **Strategy 1** are much better than those obtained by **Strategy 2** when α is large.

14.7. Robustness of the computations

Let $a^{(0)}$ be the maximal (in absolute value) element in matrix A before the start of the orthogonalization. Let $a^{(i)}$ be the maximal (in

absolute value) element found either among the elements of the original matrix
A or among the transformed elements during the orthogonal decomposition at
any major step j, j=1,2,...,i. This means that the relationship

(14.20) $a^{(i)} \geq a^{(i-1)} \geq \ldots \geq a^{(1)} \geq a^{(0)}$

holds for $\forall i \in \{1,2,\ldots,n\}$.   The quantity GROW determined by

(14.21) $GROW = a^{(n)}/a^{(0)}$

will be called **the growth factor** (a similar parameter has also been used in
connection with the Gaussian elimination; see **Chapter 4**).

	Strategy 1		Strategy 2	
m	Growth factor	Smallest element	Growth factor	Smallest element
100	6.50	6.2E-2	2.9E+15	3.0E-32
110	4.69	8.9E-3	1.1E+11	4.1E-37
120	2.36	1.7E-2	2.3E+17	7.1E-26
130	1.46	1.3E-1	5.9E+08	1.7E-23
140	2.07	3.3E-2	2.4E+12	2.3E-27
150	1.12	2.0E-2	1.5E+14	5.1E-30
160	2.41	3.9E-2	1.0E+09	2.2E-22
170	4.42	7.0E-3	1.8E+11	6.3E-27
180	4.50	1.6E-3	1.6E+08	1.6E-19
190	6.52	5.5E-4	4.1E+12	1.7E-27
200	4.14	2.7E-3	1.1E+08	1.0E-19

Table 14.5
Numerical results obtained in the solution of linear least square problems
whose matrices are $A=F2(m,100,11,3,\alpha)$. NZ=3m+110 in this experiment.
The values of parameter GROW are given under "Growth factor".
The values of the smallest elements in the diagonal matrices D^2 $(D=D_{n+1})$
in the orthogonal decompositions QDR are given under "Smallest element".
The computer used is a CDC Cyber 173.

     Assume that the linear least squares problem solved is not weighted. Then
one starts **the Gentleman-Givens algorithm** with a diagonal matrix $D_1=I$. During
the orthogonalization this matrix is modified; by using (12.27) and by
taking into account **Remark 12.2**. The algorithm described by (12.30)-(12.34)
can be applied when the first pivotal strategy is in use (because it is
allowed to use variable pivotal rows in **Strategy 1**). An attempt to reduce the
decrease of the diagonal elements of the matrices D_i^2 (i=1,2,...,n) is made
when (12.30)-(12.34) are in use; see **Remark 12.3**. From **Table 14.5** it is
seen that this attempt is very successful; the values of the growth factors
are considerably greater and the values of the smallest elements in matrix
$D^2 = D_{n+1}^2$ are considerably smaller when **Strategy 2** is used than the
corresponding values obtained when **Strategy 1** is in use.

     For some problems the values of the growth factors obtained when
Strategy 2 is applied are even greater than those shown in **Table 14.5**. If
the linear least-squares problem with matrix $A = F2(110,100,11,2,10)$ is
solved, then the growth factor is of magnitude $O(10^{140})$. On **CDC Cyber 173**
this does not cause trouble, but on many other computers such a large growth

factor will lead to overflow. Of course, there is no guarantee against overflows when **Strategy 1** is in use; see **Remark 12.3**. However, the results given in **Table 14.5** show clearly that **Strategy 1** is much more robust than **Strategy 2**.

If no requirement to fix the pivotal row before the beginning of the computations at major step k (k=1,2,...,n) is imposed, then the algorithm (10.30)-(10.34) can be applied to a strategy based on the same ideas as **Strategy 2**; the minor steps are performed in the same order as in **Strategy 2**, but at each minor step a zero element is produced in the row indicated by (10.30)-(10.34). Let us call the strategy so found **Strategy 2***. It is clear that both the magnitude of the elements of D_{n+1} and the magnitude of the growth factors found by **Strategy 2*** will be comparable to those found by **Strategy 1**. It is also clear that the storage needed when **Strategy 2** and **Strategy 2*** are used will be the same, while the computing time for **Strategy 2*** will be increased; extra work is needed to found out which non-zero element has to be transformed to zero at each minor step.

The experiments indicate that the fact that **Strategy 2** produces very small elements in D_{n+1} and very large elements in the matrices A_k (k=1,2,...n) does not have a great influence on the accuracy of the approximations computed by the use of this pivotal strategy. This is demonstrated by the results shown in **Table 14.1 - Table 14.4**, but the conclusion is also drawn by using many other tests. It is seen that the accuracy achieved when **Strategy 2** is used is of the same order of magnitude as that achieved when **Strategy 1** is applied; excepting the cases where parameter α is very large. This is a surprising result, but the relatively high accuracy of the approximations obtained by the Gentleman-Givens method is also discussed in **Golub and Van Loan [133]**. It should also be noted here that the use of iterative refinement contributes essentially to the accuracy of the computed solutions.

14.8. Concluding remarks and references concerning the pivotal strategies in sparse codes based on orthogonalization

Sparse pivotal strategies in connection with orthogonalization methods were discussed by **Tewarson** in 1969; see, for example, [243]. The idea of using a variable pivotal row has been proposed first by **Gentleman [111]** (his paper was published in 1976, but the idea was known by the specialists even before this year; both from his report and his presentations at different conferences). **Duff** published in 1974 ([62]) many numerical tests which indicated that the pivotal strategies based on the use of a variable pivotal row **do not** perform better than the pivotal strategies based on the use of a fixed pivotal row. Many of these results were also published in a survey paper of **Duff and Read, [75]**. The conclusion drawn in [62,75] was that pivotal strategies based on the use of a fixed pivotal row are more preferable than pivotal strategies based on a variable pivotal row because:

(i) **the numerical results obtained by the two strategies are very similar (both with regard to storage and computing time),**

(ii) **it is much more difficult to implement a strategy with a variable pivotal row than a strategy with a fixed pivotal row.**

After the publication of the two papers [62,75] pivotal strategies with a variable pivotal row were in practice not used in sparse codes implementing orthogonalization methods. Also until 1980 the orthogonalization methods were not very popular.

The important **Theorem 14.1** and **Corollary 14.2** have been proved in [296]. They show that the conclusions **(i)-(ii)** are necessarily true only when the matrix is essentially sparse (see **Definition 14.11**). One could also expect **(i)-(ii)** to hold when the matrix is close in some sense to an essentially sparse matrix (the matrix is very sparse and stays very sparse during the whole orthogonal decomposition).

A careful study of the set of test matrices used in [62-75] reveals that these matrices are extremely sparse. **There are matrices in which the number of non-zero elements is less than the number of rows.** Therefore it was interesting to investigate the case where the conditions of **Theorem 14.1** are not satisfied or, in other words, the case where at some major steps during the orthogonalization the number of non-zeros in the active part of the pivotal column is greater than three. This often happens in practice. Moreover, many matrices that do occur in practical problems are not very sparse (and, thus, the conditions of **Theorem 14.1** are violated from the very beginning of the computations). The numerical results show that:

(a) **The conclusions (i)-(ii) drawn in [62,75] are guaranteed for essentially sparse matrices only.**

(b) **If the matrix is not very sparse, then the performance of the strategy based on a variable pivotal row is usually considerably better than the performance of a strategy based on a fixed pivotal row (this being true both when the computing time and the storage needed are compared).**

Moreover, the pivotal strategies with a variable row can be implemented so that:

(c) **the code becomes much more robust with regard to overflows and underflows,**

(d) **the code gives more accurate results in the case where the matrix decomposed is very ill-conditioned.**

Application problems for pivotal strategies with a variable pivotal row were first studied in [296]; see also [153,166].

If a static storage scheme is to be applied, then the pivotal interchanges are to be determined before the actual start of the orthogonal decomposition. This fact puts a restriction on the computational process: one can determine the permutation matrices corresponding to the pivotal interchanges in advance, only if one has decided to keep all non-zero elements (i.e. the classical manner of exploiting sparsity is to be used). Thus, the computer oriented manner of exploiting sparsity can not be applied; not in an easy way, at least. It is difficult to see how an inaccurate, but easily computable, orthogonal decomposition can be obtained and used as a preconditioner when static storage schemes are applied. Therefore pivotal strategies for static storage schemes will not be studied here; such strategies are studied in [115,118-120,166-167].

CHAPTER 15

ITERATIVE REFINEMENT AFTER THE PLANE ROTATIONS

All numerical results presented in the previous chapter are obtained by the use of iterative refinement (**IR**) after the calculation of the first solution. However, the classical manner of exploiting sparsity (see **Chapter 1**) is in fact used in the calculations because the drop-tolerance used is so small ($T=10^{-25}$) that practically no non-zero elements are removed during the decomposition process.

The application of **IR** with a very small drop-tolerance (no "small" elements removed) leads to some **extra storage requirements** (several additional arrays of dimension m are needed). It must be emphasized here that the extra storage required is less than the extra storage that would be required when Gaussian elimination (GE) is used in the same way (with **IR**, but without dropping of "small" elements). A copy of matrix A is normally stored before the plane rotations (and used instead of matrix Q which should not stored because it is in general rather dense) even when no **IR** is to be applied. If **GE** is to be used without **IR** when no copy of matrix A is needed.

The application of **IR** also requires **extra computing time** when no "small" elements are removed (to carry out the iterations). However, the increase of the computing time is rather small, because normally a few iterations only are needed. The same effect has already been observed in connection with Gaussian elimination in **Chapter 5**.

What one expects to gain for the increase of the storage needed and the computing time spent is a much more accurate solution and a rather reliable error estimate. If one wishes to win even something more, then one should try to combine the use of **IR** with dropping some "small" elements during the process of the orthogonal decomposition. In other words, one should try to apply the computer-oriented manner of exploiting sparsity (**Chapter 1**). The ideas are very similar to those that have already been applied in connection with **GE** in **Chapter 5**. The approach is even more general and can successfully be applied to other methods also; see **Section 3.3**.

15.1. Statement of the problem

As in **Chapter 12** - **Chapter 14**, let us assume that $A \in \mathbb{R}^{m \times n}$, $b \in \mathbb{R}^{m \times 1}$, $m \in \mathbb{N}$ and $n \in \mathbb{N}$ are given. Assume also that $m \geq n$ and $rank(A)=n$. Consider the linear least square problem

(15.1) $x = A^{\dagger}b$,

where $A^{\dagger}$ is the **pseudo-inverse** of matrix A. Since $rank(A)=n$, the pseudo-inverse $A^{\dagger}$ can be written as

(15.2) $A^{\dagger} = (A^{T}A)^{-1}A^{T}$.

Assume that a matrix Q with orthonormal columns and such that

(15.3) $QDR = P_1AP_2 + E$

is calculated by the Gentleman-Givens method with pivotal interchanges
(defined by the two permutation matrices $P_1 \in R^{mxm} \wedge P_2 \in R^{nxn}$). As in the
previous chapter $D \in R^{nxn}$ is a diagonal matrix, while $R \in R^{nxn}$ is an
upper triangular matrix. Matrix $E \in R^{mxn}$ is a perturbation matrix. In the
classical theory the appearance of matrix E is due to rounding errors that
are practically unavoidable when an orthogonal decomposition is calculated on
a computer. In this chapter, however, matrix E is formed also by contribu-
tions, due to the fact that some "small" elements are removed.

Assume that the vector

(15.4) $c = Q^T b$

is calculated during the orthogonalization of matrix A. In **Section 12.9**
it has been shown that it is not necessary to keep matrix Q when vector
$c=Q^T b$ is available. An approximation x_1 to the solution x of the linear
least square problem (15.1) can be calculated by

(15.5) $x_1 = P_2 R^{-1} D^{-1} c$.

If the classical way of exploiting sparsity is applied, then the
approximation x_1 is normally sufficiently accurate and one often accepts it
as a solution of (15.1). If the computer-oriented manner of exploiting
sparsity is in use, then the factors R and D can be rather inaccurate when
many "small" elements have been removed. Therefore one should try to regain
the accuracy lost during the orthogonal decomposition by performing **IR** using
x_1 as a starting approximation. The **IR** applied in connection with the
Gentleman-Givens method will be discussed in this chapter.

The Gentleman-Givens method is used here, but all results could
immediately be extended to other orthogonalization methods. As an illustration
only it should be mentioned that the classical Givens method can be obtained
by simply setting $D=I_n$.

15.2. Introduction of the iterative refinement process

Assume that x_1 is calculated by (15.5). Then the **IR** process can be
carried out (for $i=1,2,\ldots,p$) by:

(15.6) $r_i = b - A x_i$,

(15.7) $r_i^* = A^T r_i$,

(15.8) $d_i^* = H r_i^*$,

(15.9) $x_{i+1} = x_i + d_i^*$,

where

$$\text{(15.10)} \quad H \overset{\text{def}}{=} P_2 R^{-1} D^{-2} (R^T)^{-1} P_2^T .$$

If several linear least square problems with the same matrix A but with
different right-hand side vectors b are to be solved and if the right hand
side for every problem after the first one can be calculated only after the

determination of the solution of the previous problem, then only a starting approximation for the first problem can be calculated by (15.5). Starting approximations for the other linear least squares problems can be calculated by the use of the formula:

(15.11) $x_1 = HA^Tb$.

One should be careful when the formulae (15.11) and (15.8) are in use, because the normal matrix A^TA is implicitly involved in the computations when these formulae are used (see also the Remark 12.7). The vectors b, r_i, r_i^* and x_i are stored in **DOUBLE PRECISION** arrays, while the non-zero elements of the matrices (A, D and R) are stored in **SINGLE PRECISION** arrays. In this way all inner product are both accumulated and stored in **DOUBLE PRECISION**. Therefore the accuracy achieved is normally twice as great as the accuracy of the computer under consideration in **SINGLE PRECISION**. Of course, this is so only if all input data can be represented exactly in the computer memory. The effect of this approach has already been demonstrated in the tables in the previous chapter (see also **Section 5.5**). It should be mentioned that in fact it is not necessary to store all vectors in **DOUBLE PRECISION** arrays (see again **Section 5.5**). However, this is convenient: in this way all vectors are stored in **DOUBLE PRECISION**, while all matrices are stored in **SINGLE PRECISION**.

15.3. Stopping criteria

Some stopping criteria are to be applied to terminate the iterative process when one of the following conditions is satisfied:

(i) **further calculations are not justified because the new corrections can not be represented on the computer used,**
(ii) **the iterative process is either not convergent or the rate of convergence is too slow,**
(iii) **it is judged (by the code) that the required accuracy is already achieved.**

The stopping criteria, by which the above three conditions are checked in the codes, are

(15.12) $\|x_i - x_{i-1}\| \le \epsilon \|x_i\|$,

(15.13) $\|d_i\| > \|d_{i-1}\| \ \wedge \ i>2$ *or* $\|d_i\| \le (1-RATE_i)\epsilon_{user} \ \wedge \ RATE_i < 1$,

(15.14) $i=MAXIT$,

where

(15.15) $RATE_i \overset{def}{=} \|d_i\| / \|d_{i-1}\|$, where ϵ_{user} as well as MAXIT are parameters prescribed by the user. By the choice of ϵ_{user} an attempt is made to stop the iterative process when the required accuracy is achieved. By the choice of MAXIT an attempt is made to stop the iterative process when it is convergent but the speed of convergence is very slow. An attempt is made to stop the iterative process when it does not converge by the use of the first condition in (15.13). Finally, if (15.12) is satisfied (with ϵ the smallest positive real number such that in the machine arithmetic on the computer used $1+\epsilon \neq 1$), then it is not worthwhile to continue the iterative process, because further corrections can not be represented in the computer arithmetic on the machine used.

These stopping criteria are similar to the stopping criteria used in connection with GE. The only difference is the criterion by which an attempt is made to stop the iterative process when the accuracy achieved is judged for acceptable. In fact, this criterion is still in an experimental stage, but the results obtained until now are quite good. Such a criterion will also be attached to **Y12M** in the near future. The results in the next sections are obtained with a rather small ϵ_{user}, so that the iterative process is stopped by the other stopping criteria. This is done only because the results in this chapter are to be compared with the results presented in the previous chapter. By setting a very small ϵ_{user} one can obtain the same order of accuracy as the accuracy obtained in the previous section. It will be shown that **Strategy 1** from **Chapter 13** performs better than **Strategy 2** not only when **IR** is used with a smal drop-tolerance, but also when a large drop-tolerance is applied.

15.4. Dropping "small" elements

The use of the iterative process (15.6)-(15.9) is to be combined with removing "small" elelements in the course of the orthogonal decomposition. It is clear that if many elements are removed, then both storage and computing time will be saved. In **Chapter 5** it has been demonstrated that this procedure could be very efficient for some classes of matrices. In **Chapter 3** it has been shown that the process of removing "small" elements can be applied in connection with a wide class of numerical algorithms united in a general scheme for solving algebraic problems of the type $x=A^{-1}b$. It will now be shown how the idea can be applied for orthogonalization methods.

One must have a criterion which can be used in the decision whether a non-zero element is "small" or not. In the codes, which will be studied here, a non-zero element is defined as a small element if it is smaller in absolute value than a prescribed in advance parameter; **the drop-tolerance**. This is an absolute drop-tolerance criterion; see (1.9) and **Section 5.2**. The fact that the columns of the matrix decomposed by orthogonal transformations have invariant two-norms could be exploited to introduce a relative (to the two-norms) drop-tolerance. Some experiments in this direction are still being carried out, but at present the drop-tolerance from (1.9) is used in the orthogonal codes. In general, the absolute drop-tolerance performs quite satisfactorily, especially when the columns of the matrix are scaled.

15.5. Numerical experiments

The test matrices are the same as the those in the previous chapter. The matrices are of class $F2(m,n,c,r,\alpha)$. All parameters are varied in quite wide ranges.

The results obtained by the two pivotal strategies will be compared with regard to: **(i)** the storage needed, **(ii)** the computing time spent, **(iii)** the accuracy achieved and **(iv)** the robustness of the computational process. Mainly the two pivotal strategies are compared in this chapter, but the effect of using a large drop-tolerance and IR can be seen either directly (**Table 15.1**) or by comparing the results with the corresponding results in **Chapter 14**.

15.6. Comparison of the storage required by the two pivotal strategies

Denote by COUNT1 and COUNT2 the maximal numbers of non-zeros kept in the main array AQDR. The abbreviation COUNT will be used both instead of COUNT1 and COUNT2 when there is no danger of misunderstanding. The efficiency factor with regard to the storage requirements, E_s, will be used also in this section. Results are presented in **Table 14.1 - Table 14.4**.

T	STRATEGY 1			STRATEGY 2		
	COUNT	Time	Iterations	COUNT	Time	Iterations
1.0E-25	2706(0.85)	8.91(0.85)	8.01(1.03)	2854	10.54	7.81
1.0E-02	1169(0.71)	3.72(0.75)	16.79(1.10)	1646	4.98	15.29
1.0E-01	816(0.65)	3.32(0.72)	22.21(1.02)	1259	4.58	21.71

Table 15.1
Numerical results that are obtained when linear least squares problems whose matrices are F2(m,100,11,3,10), m=100,110,...,200, are solved. NZ=3m+110 in this experiment. The average results (for the 11 problems solved for each value of the drop-tolerance T) are given in this table. The ratios of the characteristics obtained by Strategy 1 and the corresponding characteristics in Strategy 2 are given in brackets. The experiment was run an a CDC Cyber 173 computer.

n	STRATEGY 1				STRATEGY 2			
	COUNT	Time	Iterations	Accuracy	COUNT	Time	Iterations	Accuracy
100	4323	2.95	9	1.3E-14	4685	3.82	9	1.6E-14
110	4686	3.78	10	1.8E-14	4607	3.72	11	2.2E-14
120	3212	1.15	9	2.3E-14	4220	2.48	10	1.4E-14
130	3160	1.22	11	1.6E-14	4144	2.25	9	1.9E-14
140	2946	1.23	9	3.0E-14	4145	1.92	7	1.8E-14
150	3683	1.43	8	2.4E-14	4202	2.14	8	2.4E-14

Table 15.2
Numerical results obtained when linear least square problems whose matrices are F2(200,n,11,6,10) are solved. NZ=1310 in this experiment. The drop-tolerance is 1.0E-2. An IBM 3033 computer was used in this experiment.

	S T R A T E G Y 1				S T R A T E G Y 2			
c	COUNT	Time	Iterations	Accuracy	COUNT	Time	Iterations	Accuracy
20	4118	2.40	9	1.4E-14	4088	2.30	7	1.1E-14
25	3975	2.50	8	3.9E-14	4101	2.50	10	1.5E-14
30	3683	2.24	11	2.8E-14	4088	2.33	10	2.4E-14
35	4033	2.57	8	2.9E-14	4525	2.87	8	3.3E-14
40	3441	2.04	9	3.3E-14	4560	3.13	8	3.2E-14
45	2612	1.26	9	3.1E-14	3407	1.70	8	2.0E-14
50	3025	1.57	10	2.9E-14	3840	1.94	10	2.6E-14
55	3478	1.92	10	2.6E-14	4288	2.66	10	2.8E-14

Table 15.3

Numerical results obtained when linear least square problems whose matrices are   F2(200,100,c,6,10)   are solved.   NZ=1310   in this experiment. The drop-tolerance is   1.0E-2.   An   IBM 3033   computer was used in this experiment.

The same conclusions as those in Section 14.4 can be drawn:

(i)	If the matrix is very sparse (the average number of non-zeros per column is less than three), then the storage required by the two pivotal strategies is approximately the same; see the first row in Table 15.4.
(ii)	If the matrix is not very sparse (i.e. the average number of non-zero elements per column is greater than three), then some storage can be saved when Strategy 1 is used; see Table 15.1 - Table 15.4.
(iii)	The performance of the two pivotal strategies (with regard to the storage used) does not depend on the parameters m, n, c and α (some results that illustrate this statement are given in Table 15.2, Table 15.3 and Table 15.4).

Two additional conclusions can be drawn in the case where some "small" elements are dropped during the orthogonalization process.

(iv)	The storage needed when a positive drop-tolerance is used is smaller if the drop-tolerance is larger.
(v)	If no non-zero elements are dropped and if only one problem is to be solved, then it is not necessary to keep a copy of the original matrix. In such a case the direct solution of the problem (without attempt to improve the solution by IR) is the best choice when the storage requirements are taken into account.

15.7. Comparison of the computing time required by the two pivotal strategies

The comparison of the computing times for the two pivotal strategies in the case where a large drop-tolerance is used during the orthogonalization gave similar results as the comparison of the computing times for the two pivotal strategies in the case where no non-zero elements are dropped (see the previous chapter). The computing time is reduced considerably if the first starategy is used and if the matrix solved is not very sparse (this is demonstrated in **Table 15.4**), while the computing times for the two strategies are comparable when the matrix is very sparse (see the first row in **Table 15.4**). The case where $T=10^{-25}$ (see **Table 15.1**) can be considered as an attempt to prevent the dropping of "small" elements. Therefore, the results of the experiment presented in **Table 15.1** illustrate that the computing time can be reduced very considerably when a large drop-tolerance is in use. The reduction of the computing time when a large drop-tolerance is applied will be discussed in some more detail in the next chapter (here the main task is to compare the two pivotal strategies when some "small" elements are dropped).

	STRATEGY 1			STRATEGY 2		
r	COUNT	Time	Iterations	COUNT	Time	Iterations
2	427(0.96)	1.62(1.02)	16.71(1.06)	447	1.59	15.69
3	816(0.65)	3.32(0.72)	22.72(1.05)	1259	4.58	21.71
4	1346(0.68)	4.76(0.66)	20.09(1.00)	1989	7.16	20.19
5	1962(0.72)	7.96(0.71)	21.58(1.03)	2718	10.28	21.00
6	2144(0.70)	8.94(0.66)	21.30(1.22)	3044	13.47	14.39

Table 15.4

Numerical results that are obtained when linear least squares problems whose matrices are F2(m,100,11,r,10), m=100,110,...,200, are solved. NZ=rm+110 in this experiment. Average results (for the 11 problems solved for each value of the density parameter r) are given in this table. The ratios of the characteristics obtained by Strategy 1 and the corresponding characteristics in Strategy 2 are given in brackets. The value of the drop-tolerance used in all these runs was T=1.0E-1. The experiment was run an a CDC Cyber 173 computer.

15.8. Comparison of the accuracy achieved by the two pivotal strategies

The results obtained in the experiments with different values of the drop-tolerance show that the accuracy achieved is practically of the same order as the accuracy achieved when no non-zero is removed (assuming that the input data can be represented exactly in the computer and that the iterative process converges). Some illustrations of this are given in **Table 15.2** and in **Table 15.3**. The experiments described in these tables were run on an **IBM 3033** computer where the mantissa of the real numbers is approximately 6 digits in single precision and approximately 16 digits in double precision (which means that $t_1 \approx 6$, $t_2 \approx 16$ and, thus, the inequality stated in Section 14.6, $t_2 \geq 2t_1$, is clearly satisfied for this computer). Therefore one should expect that if the iterative process converges, then the number of accurate digits should be approximately equal to $2t_1$ (assuming again that the input data is exactly represented). It is seen from the tables that this is achieved for the matrices under consideration.

It must be emphasized that if A is ill-conditioned, then the drop-tolerance should be not very large. This has been discussed in **Chapter 3** (in a rather general context) and in **Chapter 5** (in connection with the use of iterative refinement in the solution of system of linear algebraic equations with Gaussian elimination). The problem of the choice of the drop-tolerance will be discussed in the following chapter.

15.9. Comparison of the robustness of the two pivotal strategies

Assume that the second pivotal strategy is applied. In **Section 14.7** it was shown that the elements of the matrix can grow very quickly during the orthogonalization with the Givens plane rotations when the second pivotal strategy is in use and when no "small" elements are removed. It has also been shown that the diagonal elements of the diagonal matrix D, produced when the Gentleman-Givens plane rotations are applied, decrease very quickly during the computational process. These two effects have been demonstrated in **Table 14.5**. Many experiments have also been carried out by the use of large values of the drop-tolerance T followed by **IR**. The behaviour is the same: again the elements of matrix A are normally growing rather quickly during the orthogonal decomposition and again the diagonal elements of the diagonal matrix D are normally decreasing rather quickly in the course of the computations. The situation is improved considerably when the first pivotal strategy is in use. The conclusion is that the first pivotal strategy, **Strategy 1**, is more robust than the second pivotal strategy, **Strategy 2**.

It must be emphasized, as in **Section 14.7**, that the experiments indicate that the fact that **Strategy 2** produces very small elements in the diagonal matrix $D = D_{n+1}$ and very large (in absolute value) elements in the matrices A_k ($k=1,2,...,n$) does not have a great influence on the accuracy of the approximations computed by the use of this pivotal strategy. As a rule, if the two strategies are applied with the same value of the drop-tolerance and if the **IR** process converges when **Strategy 1** is in use, then the **IR** process converges also when **Strategy 2** is applied. Moreover, if the **IR** process converges, then as a rule the results are very accurate (as mentioned in the previous section, $2t_1$ accurate digits are normally achieved when ϵ_{user} from **Section 15.3** is sufficiently small).

15.10. Concluding remarks and references concerning the iterative refinement process in the orthogonalization methods

The use of the **IR** process in an attempt to improve the accuracy of the first approximation of the solution of a linear least squares problem is a particular case of **the common approach for solving sparse problems** $x=A^{\dagger}b$ (see **Section 3.3**). The effects of applying this approach in connection with the orthogonalization methods are in principle the same as the effects of applying the **IR** in the solution of large and sparse systems of linear algebraic equations. The benefits of using **IR** in the latter case have been discussed in detail in **Chapter 5**. The numerical results presented in this chapter show clearly that the same benefits can be obtained when the **IR** is used together with the Gentleman-Givens method. More precisely, if a large

value of the drop-tolerance is specified and if the IR process defined in
Section 15.2 is applied together with the stopping criteria from Section
15.3, then the following advantages (compared with the case where the problems
are solved directly) can be achieved:

(i)   the storage needed can be reduced,

(ii)   the computing time can be reduced,

(iii)   the approximations obtained are normally very
        accurate (when the iterative process converges),

(iv)   a rather reliable error estimate can easily be
        obtained (practically with no extra computational
        efforts).

In this chapter the attention has been concentrated on the comparison of
the two pivotal strategies (that have been studied in the previous chapter)
in the new environments; in the situation where a large value of the drop-
tolerance is specified and IR is carried out. However, in the next chapter
more experiments will be presented, which show that the advantages (i)-(iv)
can be achieved for some classes of matrices.

The results presented in this chapter show clearly that Strategy 1
performs in general better than Strategy 2 also in the case where a large
drop-tolerance together with IR is in use. Therefore Strategy 1 should be
recommended as a general purpose strategy when the Gentleman-Givens plane
rotations are in use. Of course, for very sparse matrices Strategy 2 will
normally perform rather well and can also be used.

The results presented here are based on results given in [296]. Similar
conclusions (concerning the two pivotal strategies) are drawn in [153]. The
codes used to obtain the numerical results are described in [320-322].   In
these codes the general storage scheme discussed in Chapter 2 is used. The
particular implementation of the general storage scheme from Chapter 2 for
the case where the Gentleman-Givens method is applied in the solution of
sparse linear least square problems is discussed in detail in [320-322].

As pointed in Section 3.3 the IR is not the only possibility in the
efforts to regain the accuracy lost during the decomposition because some
"small" elements have been dropped. In connection with the orthogonalization
methods, the use of the conjugate gradient method is even more efficient.
The application of a preconditioned conjugate gradient algorithm in connection
with the Gentleman-Givens plane rotations will be described in the next
chapter.

CHAPTER 16

PRECONDITIONED CONJUGATE GRADIENTS

FOR GIVENS PLANE ROTATIONS

Assume, as in the previous four chapters, that $A \in \mathbf{R}^{m \times n}$ and $b \in \mathbf{R}^{m \times 1}$ are given and that $m \geq n$ as well as $rank(A)=n$ are satisfied. Assume also that an approximation to $x = A^{\dagger}b = (A^T A)^{-1} A^T b$ is to be calculated. In this chapter it will be shown that this problem can be transformed into an equivalent problem, which is a system of linear algebraic equations $Cy=d$ whose coefficient matrix C is symmetric and positive definite. Moreover, C can be written as $C = D^{-1}(R^T)^{-1} A^T A R^{-1} D^{-1}$, where D is a diagonal matrix and R is an upper triangular matrix. The conjugate gradient (CG) algorithm can be applied in the solution of the system of linear algebraic equations $Cy=d$ $(y=DRx, \quad d=D^{-1}(R^T)^{-1} A^T b)$. If $D=R=I$, then the CG algorithm is in fact applied to the system of normal equations $A^T A x = A^T b$ and the speed of convergence could be very slow (see also Chapter 3). If D and R are obtained by some kind of orthogonalization with some matrix $Q \in \mathbf{R}^{m \times n}$ with orthonormal columns (see Chapter 12) and if the calculations are performed without rounding errors, then $C=I$ and, thus, the CG algorithm converges in one iteration only. Even if the orthogonalization is carried out with rounding errors, the matrix C is normally close to the identity matrix I and the CG algorithm is quickly convergent. However, the orthogonal decomposition is an expensive process (both in regard to storage and in regard to computing time). Therefore it may be profitable to calculate an **incomplete orthogonal decomposition**. This is achieved in the same way as in the previous chapter: by using a special drop-tolerance parameter T (see also Chapter 3, Chapter 5 and Chapter 11). All elements that are smaller (in absolute value) than the drop-tolerance T are removed. Many numerical examples will be given to illustrate that the CG algorithm applied to the system of linear algebraic equations $Cy=d$ and used with incomplete factors D and R is very efficient for **some classes** of problems. Matrix C is never calculated explicitly. The whole computational work is carried out by the use of the matrices A, D, and R only.

16.1. The classical conjugate gradient algorithm

Let $A \in \mathbf{R}^{m \times n}$, $b \in \mathbf{R}^{m \times 1}$, $m \in \mathbf{N}$ and $n \in \mathbf{N}$ be given data. Assume that $m \geq n$ and $rank(A)=n$. Consider, as in Chapter 12 - Chapter 15, the problem of finding a vector $x \in \mathbf{R}^{n \times 1}$ that satisfies

(16.1) $\quad x = A^{\dagger}b,$

where $A^{\dagger}$ is the pseudo-inverse of matrix A (see, for example, [191]).

Since

(16.2) $\quad rank(A)=n \quad \Rightarrow \quad A^{\dagger} = (A^T A)^{-1} A^T,$

it is clear that (16.1) is equivalent to the system of the normal equations:

(16.3) $\quad A^T A x = A^T b$.

Denote

(16.4) $B = A^T A$ $\wedge$ $c = A^T b$.

Then vector x can be found by solving the system of linear algebraic equations:

(16.5) $Bx=c$.

Matrix B is symmetric and positive definite. Therefore the well-known conjugate gradient (**CG**) method (see [133,146,196]) could be written for the system (16.5) as in **Fig. 16.1**.

ALGORITHM 1	- Solving linear least squares problems by applying a conjugate gradient algorithm to the system of normal equations.
Step 1	- Calculate $r_0 = c - Bx_0 = A^T(b - Ax_0)$ (x_0 being an arbitrary starting approximation) and $\epsilon_0 = r_0^T r_0$.
Step 2	- Set $p_0 = r_0$ $\wedge$ $i = 0$.
Step 3	- Perform the i'th iteration: $\delta_i = \epsilon_i$, $s_i = Bp_i = A^T Ap_i$, $\gamma_i = p_i^T s_i$, $\alpha_i = \delta_i/\gamma_i$, $x_{i+1} = x_i + \alpha_i p_i$, $r_{i+1} = r_i - \alpha_i Bp_i = r_i - \alpha_i s_i$, $\epsilon_{i+1} = r_{i+1}^T r_{i+1}$, $\beta_i = \epsilon_i/\delta_i$, $p_{i+1} = r_{i+1} + \beta_i p_i$.

Step 4	- Carry out a termination check and **if**
	    (a) the process converges,
	    (b) the accuracy required is not achieved
	and
	    (c) the number of iterations performed, i is smaller than an upper bound prescribed by the user,
	then increase the value of parameter i by one and **go back to Step 3**.

Step 5	- Perform the output operations required by the user and **STOP** .

Figure 16.1
The algorithm of conjugate gradient applied in the solution of the system of linear algebraic equations (16.5).

Certain convergence and accuracy criteria are necessary in **Step 4** of the algorithm given in **Fig. 16.1**. These will be discussed in **Section 16.4**. It should be noted here that it is not necessary to compute explicitly matrix B and vector c when **Algorithm 1** is in use.

It is well-known that the speed of convergence of the CG algorithm is related to the **spectral condition number**, $\kappa(B) = \|B\|_2 \|B^{-1}\|_2$, of the coefficient matrix of the linear algebraic equations solved. The bound

(16.6) $\|x - x_i\|_B \leq \|x - x_0\|_B \{1 - [\kappa(B)]^{\frac{1}{2}}\}^{2i} / \{1 + [\kappa(B)]^{\frac{1}{2}}\}^{2i}$

is given, for example in [133]. The norm used in (16.6) is defined by

$$
(16.7) \quad \|w\|_B \overset{def}{=} w^T Bw \qquad (\ w \ \textit{being an arbitrary vector} \ ) .
$$

The quantities involved in (16.6) are the exact solution x of the system of linear algebraic equations (16.5) and the i'th iterate x_i (i=0,1,...) obtained by the **CG** algorithm.

The bound (16.6) is rather pessimistic but, nevertheless, this bound shows clearly that the speed of convergence may be very slow when the condition number of the original matrix A is large, or even only moderately large, because $\kappa(B)=[\kappa(A)]^2$. Therefore one must attempt to accelerate the speed of convergence. In this chapter it will be shown that this can be done by a kind of preconditioning of matrix B based on the use of factors of the original matrix A that are obtained by an orthogonalization process. The method will be applied in the numerical treatment of large and sparse linear least squares problems, where an incomplete orthogonal decomposition of A can efficiently be calculated by the use of a positive drop-tolerance (see **Chapter 3** and **Chapter 15**).

The content of the following sections in this chapter can be sketched as follows. In **Section 16.2** a transformation of the problem defined by (16.1) is described. A modification of **Algorithm 1** (see **Fig.16.1**) for the transformed problem is given in **Section 16.3**. Some stopping criteria, based on certain convergence and accuracy tests, are described in **Section 16.4**. Numerical results are presented in **Section 16.5**. In the last section, **Section 16.5**, some concluding remarks are given.

16.2. Transformation of the original problem

Assume that three matrices $Q \in \mathbf{R}^{mxn}$, $D \in \mathbf{R}^{nxn}$ and $R \in \mathbf{R}^{nxn}$ are available and such that

$$
(16.8) \quad A = QDR + E
$$

is satisfied with some perturbation matrix $E \in \mathbf{R}^{mxn}$. Assume also that the following relationships hold:

$$
(16.9) \quad Q^T Q = I_n \in \mathbf{R}^{nxn} \qquad (\ I_n \ \textit{being the identity matrix in} \ \mathbf{R}^{nxn} \ ),
$$

(16.10)   D   *is a diagonal matrix with*   rank(D)=n,

(16.11)   R   *is an upper triangular matrix with*   rank(R)=n.

Let

$$
(16.12) \quad C = D^{-1}(R^T)^{-1}A^T AR^{-1}D^{-1} ,
$$

$$
(16.13) \quad y = DRx,
$$

$$
(16.14) \quad d = D^{-1}(R^T)^{-1}A^T b
$$

and consider the system of n linear algebraic equations

(16.15)   Cy = d .

The following result can be proved by using (16.8)-(16.14).

Theorem 16.1. If **(16.8)-(16.14)** are satisfied and if $rank(A) = n$, then

    (i)   **the problems defined by (16.1) and (16.15) are equivalent,**

    (ii) **the matrix C in (16.15) is symmetric and positive definite,**

  (iii) **matrix C reduces to the identity matrix in $R^{n \times n}$ when E=0.**

Proof. **(i)** The problem defined by (16.15) can be obtained from the problem defined by (16.1) in the following way. Since $rank(A)=n$, (16.1) can be rewritten as (16.3); see **Section 16.1.** Since D^{-1} and R^{-1} exist, see the conditions stated by (16.10) and by (16.11), (16.3) can be rewriten as

(16.16) $D^{-1}(R^{T})^{-1}A^{T}A(R^{-1}D^{-1})(DR)x = D^{-1}(R^{T})^{-1}A^{T}b$

and (16.15) can easily be obtained by the use of (16.12)-(16.14). In a similar way (16.1) can be obtained from (16.15).

(ii) The second assertion of the theorem is nearly obvious because matrix C from (16.12) can be rewritten as

(16.17) $C = (AR^{-1}D^{-1})^{T}(AR^{-1}D^{-1})$.

(iii) It is clear that

(16.18) $A = QDR + E$ ⇒ $AR^{-1}D^{-1} = Q + ER^{-1}D^{-1}$

and

(16.19) $A^{T} = R^{T}DQ^{T} + E^{T}$ ⇒ $D^{-1}(R^{T})^{-1}A^{T} = Q^{T} + D^{-1}(R^{T})^{-1}E^{T}$.

Multiply the last equality in (16.19) with the last equality in (16.18). The result is

(16.20) $C = I_{n} + Q^{T}ER^{-1}D^{-1} + D^{-1}(R^{T})^{-1}E^{T}Q + D^{-1}(R^{T})^{-1}E^{T}ER^{-1}D^{-1}$.   ■

The statements formulated in the following remarks are corollaries of Theorem 16.1.

Remark 16.1. Since C is symmetric and positive definite, the method of conjugate gradients can be applied in connection with (16.15).   ■

Remark 16.2. The assertions of the theorem are true for any matrices Q, D and R that satisfy (16.9)-(16.11). In the following sections, however, it will be assumed that these matrices are obtained by the use of the Gentleman-Givens plane rotations (see **Chapter 12 - Chapter 15**). The code discussed in [296,320-322] will be used in the numerical demonstrations in Section 16.5. Pivotal interchanges, by which two permutation matrices P_1 and P_2 are induced, are to be applied in an attempt to preserve the sparsity of the original matrix during the orthogonal decomposition (Chapter 14 - Chapter 15). For the sake of simplicity, the permutation matrices are not included in the formulae given in this section. It is assumed that matrix A is multiplied by the two permutation matrices before the start of the computations and that the matrix so found is again called A. It is clear that this is not a restriction. ∎

Remark 16.3. If Q, D and R are obtained by the Gentleman-Givens method applied to the original matrix A, then it should be expected that matrix C is not very ill-conditioned, even when matrix B from (16.4) is, and that the CG method applied to the system of linear algebraic equation (16.15) will converge quickly. ∎

Remark 16.4. Matrix Q is not used in the transformed problem (16.15). This is an important fact, because Q is normally rather dense even when A is very sparse. In the code, which is used during the factorization of matrix A, this fact is exploited and the elements of matrix Q are never stored (see also Chapter 12 and Chapter 14). ∎

16.3. A conjugate gradient algorithm for the transformed problem

Assume that R and D are available. Then a CG algorithm for the problem defined by (16.15) can be defined as shown in Fig. 16.2.

The following remarks are useful in connection with the CG algorithm (which will be called Algorithm 2) defined in Fig. 16.2.

Remark 16.5. Algorithm 2 from Fig.16.2 is formulated in connection with (16.15) in the sense that the factors A, A^T, R, R^T and D of matrix C are used in the formulation. However, it is seen that one can calculate approximations, x_i, to the exact solution of (16.1) directly (see Step 3 of the algorithm in Fig. 16.2). ∎

Remark 16.6. Matrix A^TA is never calculated explicitly when Algorithm 2 from Fig. 16.2 is in use. ∎

Remark 16.7. It is probably better to compute some of the inner products in double precision (especially on computers such as an IBM where the mantissa of the real numbers in single precision is short). This has already been discussed in Chapter 12 - Chapter 15. In the numerical experiments, which will be presented in Section 16.5, the inner products are calculated in double precision (the same action was taken in the numerical experiments presented in the previous two chapters). ∎

Remark 16.8. The convergence and the accuracy criteria used to stop the process (see Step 4 in the algorithm described in Fig. 16.2) will be presented in the next section. ∎

| ALGORITHM 2 | - Solving the transformed problem (16.15) by the use of a preconditioned conjugate gradient algorithm. |

Step 1 — Calculate $\qquad r_0 = d-Cy_0 = D^{-1}(R^T)^{-1}A^T(b-Ax_0)$
(x_0 being an arbitrary starting approximation) and $\epsilon_0 = r_0^T r_0$.

Step 2 — Set $p_0 = r_0 \quad \wedge \quad i = 0$.

Step 3 — Perform the i'th iteration:
$$\delta_i = \epsilon_i, \qquad q_i = R^{-1}D^{-1}p_i, \quad s_i = Cp_i = D^{-1}(R^T)^{-1}A^TAq_i,$$
$$\gamma_i = p_i^T s_i, \quad \alpha_i = \delta_i/\gamma_i,$$
$$y_{i+1} = y_i + \alpha_i p_i \quad \Rightarrow \quad x_{i+1} = x_i + \alpha_i q_i,$$
$$r_{i+1} = r_i - \alpha_i Cp_i = r_i - \alpha_i s_i, \quad \epsilon_{i+1} = r_{i+1}^T r_{i+1},$$
$$\beta_i = \epsilon_{i+1}/\delta_i, \qquad p_{i+1} = r_{i+1} + \beta_i p_i.$$

Step 4 — Carry out a termination check and if
(a) the process converges,
(b) the accuracy required is not achieved
and
(c) the number, i, of iterations performed is smaller than an upper bound prescribed by the user,
then increase the value of parameter i by one and **go back to Step 3**.

Step 5 — Perform the output operations required by the user and **STOP** .

Figure 16.2
The preconditioned conjugate gradient algorithm
applied in the solution of the transformed linear least squares problem;
the system of linear algebraic equations (16.15).

16.4. Stopping criteria

The iterative processes defined by **Algorithm 1** and **Algorithm 2** (see **Section 16.1** and **Section 16.3**) have to be stopped when it is expected that one of the following conditions is satisfied:

(i)	the iterative process does not converge,
(ii)	the iterative process converges, but the speed of convergence is too slow,
(iii)	the accuracy that is required by the user is achieved.

The basic principles used in the stopping criteria applied in the codes in connection with **Algorithm 1** and **Algorithm 2** are discussed below.

16.4.1. Convergence control. Assume that all calculations in **Algorithm 1** and in **Algorithm 2** are carried out without rounding errors. Then

$$(16.21) \quad x_i \to x \quad as \quad i \to \infty \quad \Rightarrow \quad x = x_0 + \sum_{i=1}^{\infty} \alpha_i q_i$$

when **Algorithm 2** is in use (for **Algorithm 1**, q_i should be replaced by p_i). A necessary condition for the convergence of the series in (16.21) is

$$(16.22) \quad f_i \to 0 \quad as \quad i \to \infty \quad (f_i \overset{def}{=} \alpha_i \| q_i \|) .$$

Therefore the iterative processes are sometimes considered as convergent in computer codes if

$$(16.23) \quad RATE_i \overset{def}{=} f_i/f_{i-1} \leq RATE < 1 \quad (i=1,2,\ldots),$$

where   RATE   is a parameter that is to be prescribed by the user.

However, the iterative process could be convergent even if (14.23) is not satisfied for some values of i. This is illustrated in **Table 16.1**.

i	RATE
37	0.74
38	0.88
39	1.00
40	0.77
41	0.83
42	0.79
43	1.04
44	0.82
45	0.89
46	0.97
47	0.73
48	0.85
49	1.01
50	0.73
51	0.90
52	1.00
53	0.83
54	1.02

Table 16.1
The variation of parameter $RATE_i$ in the solution
of Problem 7 in the set of George, Heath amd Ng [113].
Algorithm 1 is used in this experiment. The values of
$RATE_i$ are given under "RATE" in this table.

It should be mentioned that the accuracy required in the solution of the problem shown in **Table 16.1** was 10^{-10} and this accuracy has been achieved in 98 iterations and 2.95 seconds of computing time. This is quite a good result and for this problem **Algorithm 1** is competitive with the other methods compared in **Section 16.5.**

The example discussed above, results being given in **Table 16.1**, shows clearly that the requirement imposed by (16.23) can be relaxed if the user is prepared to accept a larger number of iterations. This can be done in the following way.

Consider a variable $RMEAN_i$ defined by

$$(16.24) \quad RMEAN_i = 0.25 \sum_{j=0}^{3} RATE_{i-j} \qquad (\ RATE_k \overset{def}{=} RATE \quad for \quad k \leq 0 \ )$$

and assume that an integer variable NBAD is increased (at the i'th iteration):

$$
\begin{array}{llll}
\text{by} & 1 & \text{if} & RATE_i > RATE, \\
\text{by} & 2 & \text{if} & RATE_i > 1.0, \\
\text{by} & 3 & \text{if} & RATE_i > 1.5, \\
\text{by} & 4 & \text{if} & RATE_i > 2.0, \\
\text{by} & 5 & \text{if} & RATE_i > 3.0, \\
\text{by} & 6 & \text{if} & RATE_i > 4.0, \\
\text{by} & 7 & \text{if} & RATE_i > 5.0.
\end{array}
$$

NBAD is set to 0 if

$$(16.25) \quad RMAX_i \overset{def}{=} max(RMEAN_i, RMEAN_{i-1}, RMEAN_{i-2}, RMEAN_{i-3}) < RATE \ .$$

Consider also another integer variable NBAD1. If the value of NBAD is increased at the i'th iteration, then NBAD1 is increased by 1. NBAD1 is equal to 0 at the beginning of the iterative process and is reset to 0 at every iteration i at which NBAD is not increased.

The convergence criterion used in connection with Algorithm 1 and Algorithm 2 is defined by the use of $RMEAN_i$, NBAD and NBAD1.

Convergence criterion. The iterative processes in Algorithm 1 and Algorithm 2 are considered as convergent (for $i=1,2,\ldots$) until

$$(16.26) \quad (\ RMEAN_i < RATE) \lor (\ (\ NBAD \leq 40 \ ) \land (\ NBAD1 \leq 5 \ ) \ ) \ . \quad \blacksquare$$

16.4.2. Accuracy check. Assume that the convergence criterion (16.26) is satisfied at all steps $j \in \{1,2,\ldots,i\}$. Assume also that it is desirable to stop the iterative process if

$$(16.27) \quad \|x - x_i\| < EPS \ \|x\|,$$

where EPS<1 is an accuracy parameter prescribed by the user. Many experiments, and also the example given in **Table 16.1**, show that it is not very appropriate to replace the check (16.27) by

$$(16.28) \quad f_i < EPS \ \|x_i\| \ .$$

The latter check is often used when the iterative refinement processes (see the discussion in **Chapter 5** and in **Chapter 15**) are applied in the solution of the problem treated. A more stringent accuracy check is needed when **Algorithm 1** or **Algorithm 2** is in use because, among other reasons, the convergence control is relaxed; see the convergence criterion defined by (16.26). The following accuracy stopping criterion is used in the experiments both with **Algorithm 1** and **Algorithm 2**.

Accuracy criterion. The approximation, found at the i'th iteration either by **Algorithm 1** or by **Algorithm 2**, is declared as acceptable when

(16.29) $f_i < (1.0-RMAX_i)EPS \ \|x_i\|$ and $RMAX_i < RATE,$

and the iterative process is terminated. ∎

The test (16.29) is rather conservative when the iterative process converges slowly. If at the i'th iteration $RMAX_i \geq RATE$, then no attempt to evaluate the accuracy achieved is made. If RATE is close to 1, then the factor $1.0-RMAX_i$ could be rather small when $RMAX_i < RATE$ is satisfied, but $RMAX_i$ is close to RATE. Nevertheless, many numerical experiments indicate that this accuracy test performs rather well in practice. This will be illustrated in the next section.

16.4.3. Control of the speed of convergence. An attempt to stop the iterative process when the convergence is too slow seems to be desirable. In fact an attempt to also control the speed of convergence is made in the convergence criterion defined by (16.26). The user supplied parameter RATE involved in (16.26) can also be considered as a convergence speed parameter, because a requirement on the convergence speed is imposed by RATE. Small values of RATE correspond to a requirement for a high speed of convergence, while large values of RATE indicate that the user is not very concerned with the convergence speed and the code is allowed to perform many iterations (but not more than MAXIT, which is again prescibed by the user).

It should be emphasized that it is difficult to control the rate of convergence for **CG**-type methods. If the method converges slowly, then very often the error stays at a certain level during many iterations and after that it is reduced dramatically in two-three iterations. Sometimes such a behaviour can be observed several times. This phenomenon is not explained very well in the literature; especially when preconditioners of the type used here are applied. Extreme eigenvalues and/or Ritz values of the preconditioned matrix should be studied in order to explain the behaviour of the convergence (some results from [47,227,228] could probably be used to study the problem).

In the experiments, the results of which will be presented in the next section, RATE=0.9 was used in connection with both **Algorithm 1** and with **Algorithm 2**.

16.4.4. Other stopping criteria. Several other stopping criteria are to be used. The principles, on which these criteria are based, are very simple. The iterative process in **Algorithm 2** is terminated if any of the following conditions is satisfied:

(i) γ_i (see Fig. 16.2) is non-positive,

(ii) $\alpha_i \|p_i\|/\|x_i\|$ is very small,

(iii) $\alpha_i \|s_i\|/\|r_i\|$ is very small,

(iv) δ_i/ϵ_{i+1} is very small,

(v) $\text{RATE}_i > 2^{11}$.

Similar stopping criteria are used with **Algorithm 1**.

It is necessary to explain when a number is considered as very small in relation with another number. Assume that A and B are machine representations of $|a|$ and $|b| > 0$ respectively. It is said that $|a|$ is very small in comparison to $|b|$ if $10.0*B+A = 10.0*B$. A similar criterion, but without the factor 10.0, is used in **LINPACK**, [55].

It should be mentioned here that the relation used in **(v)** has been found experimentally (after many trials with badly scaled matrices). This relation could probably be improved.

16.4.5. Some remarks concerning the stopping criteria. Several remarks are needed in connection with the choice of stopping criteria.

Remark 16.9. All stopping criteria introduced in this chapter are based on heuristics, but a long sequence of experiments indicated that the stopping criteria selected work rather well on a wide range of test examples (some of the results will be presented in the next section).   ■

Remark 16.10. In the implementation of the two algorithms the last residual, r_{i+1}, is available on exit and the user could evaluate $\|r_{i+1}\|$. This norm could be used in the judgement of the accuracy achieved (as a supplement to the error evaluation calculated by (16.29). The experiments indicate that the estimates of the accuracy of the solution by the device described in **Section 16.4.2** tend to be pessimistic, while the estimates based on the residual norms tend to be optimistic. Thus, the exact error norm was often, in the experiments at least, between these two estimates.   ■

Remark 16.11. Some stopping criteria are based on the assumption that the computations are performed without rounding errors. Assume that EPS is not very stringent and that matrix A is not very ill-conditioned. Then one should expect the iterative process with the stopping criteria selected to be safely carried out because the truncation errors dominate over the rounding errors.   ■

16.5. Numerical results

Two codes based on the algorithms described in **Section 16.1** and **Section 16.3** as well as on the stopping criteria discussed in **Section 16.4** have been developed and compared with the code for solving linear least square problems by IR (see **Chapter 15**). Numerical results obtained in the comparison will be reported in this section. Some information about the implementation of **Algorithm 1** and **Algorithm 2** is needed before the presentation of the numerical results.

16.5.1. Implementation of conjugate gradient algorithms. Let NZ be the number of the non-zero elements in matrix A. Assume that the non-zero elements are stored in a **REAL** array AORIG(NZ) and their column and row numbers in two **INTEGER** arrays CNORIG(NZ) and RNORIG(NZ) respectively. The order of the non-zero elements of matrix A in array AORIG is not important. Any order will be accepted, but the following requirement is imposed:

$$(16.30) \quad AORIG(I) = a_{ij} \quad \Rightarrow \quad (CNORIG(I) = j \quad \wedge \quad RNORIG(I) = i)$$

$$for \quad \forall I \in \{1,2,\ldots,NZ\} .$$

This is the input storage scheme described in **Section 2.1**.

The implementation of **Algorithm 1** is straight forward when the storage scheme described above, and in **Section 2.1**, is in use. The code in which **Algorithm 1** is implemented will be referred to as **LLSSOX**. All operations involving vectors are carried out in double precision (and the vectors are stored in double precision). The non-zero elements of matrix A are stored in single precision. The total storage needed in **LLSSOX** is $3NZ + 2(4m+n)$ locations (assuming that reals and integers occupy the same storage in the computer memory, while reals in double precision occupy twice as much room as reals in single precision). This means that **LLSSOX** is the most efficient code (among the codes compared in this section) when the storage requirements are taken into account.

The implementation of **Algorithm 2** in a code is not so easy as the implementation of **Algorithm 1**. Arrays AORIG, CNORIG and RNORIG, with the same input as in **Algorithm 1**, are also used in **Algorithm 2**. The factors R and D, used in **Algorithm 2** (see **Section 16.3**), are calculated in code **LLSS01**. This is the same code as the code used to obtain the QDR decomposition in the previous chapter. Three extra arrays: a **REAL** array AQDR(NN) and two **INTEGER** arrays CNQDR(NN) and RNQDR(NN1), are needed. A. A non-negative drop-tolerance parameter $T \geq 0$ may be used during the calculations as described in the previous section.

The code first scans the non-zero elements of matrix A. All elements greater in absolute value than T are stored in array AQDR. Their column and row numbers are stored in the corresponding locations of arrays CNQDR and RNQDR so that

$$(16.31) \quad AQDR(I) = a_{ij} \quad \Rightarrow \quad (CNQDR(I) = j \quad \wedge \quad RNQDR(I) = i)$$

$$for \quad \forall I \in \{1,2,\ldots,NZ1\} ,$$

where NZ1 is the number of non-zeros in matrix A for which $|a_{ij}| > T$ holds. It is clear that $NZ1 \leq NZ$.

The second task is to reorder the non-zeros and their column numbers **by rows**: first the non-zeros (the column numbers of the non-zeros) of the first row, then the non-zeros (the column numbers of the non-zeros) of the second row, and so on. Some pointers are to be used and these are stored in an **INTEGER** array HA1(M,4); M being used in the code instead of the number of rows m in matrix A. Array HA1 is also used for other purposes (for example, to store information about the row interchanges).

The third task is to order the row numbers of the non-zeros **by columns**: first the row numbers of the non-zero elements of the first column, then the row numbers of the non-zero elements of the second column, and so on. The pointers needed in connection with this structure are stored in an **INTEGER** array HA2(N,4); N being used in the code instead of the number of columns in matrix A. Array HA2 is also used for other purposes (for example, to store information about the column interchanges).

After the reordering, performed in the three steps described above, the orthogonal decomposition can be started. The only important thing that should be stressed here is the fact that fill-ins will normally occur during the decomposition. Therefore both NN and NN1 have to be larger than NZ1 in order to ensure elbow room for the fill-ins; see **Chapter 2**. The optimal values of NN and NN1 are normally not known in advance. If NN and/or NN1 are not sufficiently large, then some extra computing time will be used to carry out the so-called **"garbage collections"**; see **Chapter 2** again. The experiments, some of which will be presented in this section, indicate that the choice

(16.32) $NN = NN1 \in [3NZ, 10NZ]$

is normally a good one when the drop-tolerance T is sufficiently large. Values of NN and NN1 from this interval are used in this chapter.

The code that uses the storage scheme sketched above is called **LLSS02**. It is obvious that **LLSS02** is more storage consuming then **LLSSOX**. The storage requirement for **LLSS02**, is given by

(16.33) $S_{LLSS02} = 3NZ + 2NN + NN1 + 7m + 15n,$

assuming again that reals and integers occupy the same storage in the computer memory, while reals in double precision occupy twice as much room as reals in single precision. It should be emphasized that the storage needed when **LLSS02** is used could be reduced very considerably by using an **EQUIVALENCE** statement (see **Remark 16.19**).

16.5.2. Organization of the experiments. Several common rules were used in all experiments on which the conclusions made are based (some results will be presented in this chapter). These rules (which are similar to the rules used in **Chapter 11**) are described below.

Column equilibration is performed if m > n, i.e. the non-zeros in every column are divided with the largest (in absolute value) element in the column or by one of the largest (in absolute value) elements when there are several such elements. Both column equilibration and row equilibration are carried out if m = n (the same procedure as above is also carried out for the rows). The column equilibration (or both the column equilibration and the row equilibration when m = n) is done only to facilitate the choice of the drop-tolerance. The equilibration time is negligible in comparison with the total computing time, but it is nevertheless added to the factorization time in all runs.

The starting value of the drop-tolerance is $T = 2^{-3}$ when m > n and **may** be $T = 2^{-5}$ when m = n (the latter situation occurs when the original matrix is such that the largest elements in the rows vary in a large range before the row equilibration). If the factorization fails (which may happen when too many elements are removed in the course of the orthogonaliza- tion process), then T is multiplied by 2^{-5} and a new attempt to perform the orthogonal process is carried out. Such a situation may occur several times. If the iterative process fails, then T is multiplied by 2^{-3} and a new orthogonal decomposition is performed to calculate more accurate factors D and R. This also may happen several times.

It is difficult to measure the storage needed, the parameter COUNT can used as an approximate measure. COUNT is the maximal number of locations occupied in arrays AQDR and CNQDR during the orthogonalization. The maximal number of locations occupied (in the course of the orthogonalization) in array RNQDR is not larger than COUNT (and it is normally smaller), while the storage due to the other arrays is fully determined by the parameters m, n and NZ, which are known in the beginning of the computations.

Unfortunately, COUNT is not known until the end of the orthogonaliza- tion process. Nevertheless, this parameter may be useful in two situations:

(i) If COUNT is much smaller than NN and NN1, then these parameters could be reduced in successive runs with the same matrix (but these two parameters have to be sufficiently larger than COUNT in order to avoid performing too many **"garbage collections"**).

(ii) Assume that two runs with drop-tolerances $T_1 = 0$ and $T_2 > 0$ are carried out. Then the value of COUNT corresponding to T_1 is normally larger and this indicates that the storage requirement with T_2 are normally smaller (in comparison with those for the run with T_1).   This will be demonstrated in the tables given in this section. This has also been demonstrated by many numerical results given in the previous chapters. Thus, NN and NN1 can as a rule be chosen smaller when T > 0.

If several failures (caused by large values of the drop-tolerance) are registered and if the drop-tolerance is reduced as explained above, then the factorization time given in the table under consideration is the sum of all factorization times and all iteration times (except the last one) used in the run. The value of COUNT obtained with the last (and the smallest) drop- tolerance is given in the tables.

The accuracy requirements are the same in all tests: an attempt to stop the iterative process when

(16.34) $\|x-x_p\|/\|x\| \le EPS$

is carried out with $EPS = 10^{-10}$. In order to facilitate the comparison of the accuracy actually obtained with the accuracy evaluated by the code, the test examples were constructed so that

(a) the linear least square problems solved are consistent (r=b-Ax=0),

(b) the right-hand side vectors b are chosen so that all components of the solution vectors are equal to **one** in all experiments.

The numerical results are obtained by the use of an **IBM 3081** computer.

The **FORTVS** compiler with **OPT=O3** is specified and the computing times are always measured in seconds. The non-zeros of the matrices (A, R and D) are stored in single precision. All inner product are accumulated and stored in double precision (see also **Remark 16.7**).

   16.5.3. The Waterloo set of test matrices. This set consists of **ten** rectangular matrices; see **Table 16.2**. Test matrices from this set have been used in [112,113,166,167]. All ten matrices were run with T = 0 (**Table 16.3**) and with T > 0 (**Table 16.4**). In the runs the **IR** code from the previous chapter is compared with the **CG** codes.

   It is clearly seen, in **Table 16.3**, that if T = 0 the **IR** code is better than the preconditioned **CG** code. The opposite is true when T > 0 is specified; see **Table 16.4**.

Problem No.	m	n	NZ
1	313	176	1557
2	1033	320	4732
3	1033	320	4719
4	1850	712	8755
5	1850	712	8636
6	784	255	3136
7	1444	400	5776
8	1512	402	7152
9	1488	784	7040
10	900	269	4208

Table 16.2
**Some characteristics of the Waterloo test-matrices
that are used in the experiments in this chapter.**

   The storage used could be reduced when T > 0 is chosen. Sometimes the reduction is very considerable; compare, for example, the values of COUNT for the seventh problem in **Table 16.3** and in **Table 16.4**. It is also seen that the value of COUNT could be even smaller than NZ when T > 0; compare the values of NZ and COUNT for the second problem in **Table 16.2** and **Table 16.4**.

   The iteration times are very small (compared with the factorization times) when T = 0; see **Table 16.3**. Therefore the total computing times obtained with the **IR** code and with T = 0 are close to the total computing times needed to solve the problem with the direct orthogonalization method. These total computing times are given in the second column in **Table 16.5**. The total computing times obtained by the preconditioned **CG** method (**Algorithm 2**) are given in the fifth column of the same table. The results obtained by the pure **CG** method (**Algorithm 1**) are given in the last three columns of **Table 16.5**.

   The "pure orthogonalization" (i.e. the use of **IR** with T = 0) performs better than the preconditioned **CG** with a starting tolerance $T = 2^{-3}$ in three examples: problems 2, 3 and 10. For problems 2 and 10 this is so because in these cases the runs with the starting drop-tolerance $T = 2^{-3}$ failed and the drop-tolerance was reduced (two and three times respectively).

The attempt with the starting drop-tolerance $T = 2^{-3}$ was succesful for the second problem, but (i) the number of the removed elements is not very large (compare the values of COUNT for this problem) and (ii) the number of iterations for the preconditioned CG algorithm (45) is much larger than that for the "pure orthogonalization" (3). However, the increase in the computing time is not large (for the three examples where the preconditioned CG does not perform better than the "pure orthogonalization").

No.	COUNT	Fact. Time	Iterative refinement				Preconditioned CG			
			Time	Iters	Exact	Eval.	Time	Iters	Exact	Eval.
1	3662	1.05	0.06	3	5.0E-16	4.5E-15	0.06	2	2.7E-16	4.3E-11
2	5282	1.54	0.12	3	7.2E-16	2.4E-13	0.14	2	2.0E-16	1.1E-14
3	5213	1.48	0.18	5	4.7E-14	9.3E-13	0.18	4	5.9E-15	2.9E-11
4	15835	7.31	0.25	3	1.6E-15	3.4E-12	0.30	3	1.8E-16	2.3E-13
5	16196	7.44	0.32	4	5.2E-14	5.2E-13	0.37	4	1.5E-15	8.3E-14
6	14663	9.38	0.13	3	4.7E-16	5.2E-13	0.15	3	2.0E-16	2.2E-14
7	40195	36.08	0.27	3	6.2E-16	8.9E-12	0.32	3	1.9E-16	2.3E-13
8	10619	5.30	0.17	3	6.7E-16	1.2E-12	0.20	3	2.2E-16	1.4E-14
9	14432	7.21	0.24	3	7.8E-16	1.5E-12	0.28	3	2.7E-16	2.1E-14
10	17827	11.64	0.19	4	4.7E-12	2.2E-11	0.22	4	9.8E-13	4.1E-12

Table 16.3

Results obtained with $T=0$ and $EPS=10^{-10}$. "COUNT" is the maximal number of non-zero elements kept in AQDR. "Fact. time" is the time needed to obtain the orthogonalization and the first solution. "Time" is the iteration time. "Iters" is the number of iterations. "Exact" is the exact error in the final solution. "Eval." is the error by the code.

No.	COUNT	Fact. Time	Iterative refinement				Preconditioned CG			
			Time	Iters	Exact	Eval.	Time	Iters	Exact	Eval.
1	2021	0.41	1.05	67	9.6E-12	7.6E-11	0.32	18	4.2E-12	3.4E-11
2	4065	0.65	does	not	converge		1.61	45	2.5E-14	4.9E-11
3	4727	1.43	0.39	11	1.8E-12	2.4E-11	0.25	6	2.2E-13	5.8E-11
4	8193	2.61	does	not	converge		2.11	28	9.8E-12	9.1E-11
5	11535	4.56	0.98	13	9.2E-12	6.7E-11	0.83	10	1.3E-13	8.5E-12
6	3210	0.86	0.72	30	1.2E-11	7.0E-11	0.39	15	1.0E-11	8.1E-11
7	6190	1.85	1.31	30	1.3E-11	7.4E-11	0.70	15	6.6E-12	5.9E-11
8	7564	1.85	1.44	29	2.2E-11	4.8E-11	0.80	15	1.0E-11	9.1E-11
9	8084	1.76	18.75	294	3.6E-12	9.8E-11	1.69	24	9.7E-12	3.4E-11
10	16496	14.11	0.50	11	5.2E-12	1.1E-11	0.39	8	4.1E-13	9.8E-13

Table 16.4

Results obtained with $T>0$ and $EPS=10^{-10}$. "COUNT" is the maximal number of non-zero elements kept in AQDR. "Fact. time" is the time needed to obtain the orthogonalization and the first solution. "Time" is the iteration time. "Iters" is the number of iterations. "Exact" is the exact error in the final solution. "Eval." is the error evaluated by the code.

No.	Pure orth.			Precond. CG			Pure CG		
	Time	Iters	COUNT	Time	Iters	COUNT	Time	Iters	COUNT
1	1.11	3	3682	0.73	18	2021	1.34	133	0
2	1.66	3	5282	2.26	45	4065	does not converge		
3	1.66	5	5213	1.78	6	4727	does not converge		
4	7.56	3	15835	4.72	28	8193	does not converge		
5	7.76	4	16196	5.39	10	11535	does not converge		
6	9.51	3	14663	1.25	15	3210	1.41	82	0
7	36.35	3	40195	2.55	15	6190	2.94	98	0
8	5.47	3	10619	2.65	15	7564	3.64	103	0
9	7.45	3	14432	3.45	24	8084	9.27	234	0
10	11.83	4	17827	14.50	8	16496	does not converge		

Table 16.5

Comparison of the performance of three options on the Waterloo test matrices.
"Pure orth." is the option where IR with T=0 is used.
"Precond. CG" is the option in which the preconditioned CG code
is used with a starting T= 2^{-3}.
"Pure CG" is the option where CG without any preconditioning is applied.

The preconditioned CG algorithm performs better than the "pure orthogonalization" for the other seven problems. For the fifth problem this is so in spite of the fact that the attempt to calculate an orthogonal factorization with $T = 2^{-3}$ failed and the orthogonalization was repeated with $T = 2^{-8}$. It is seen that the reduction in computing time can be very significant when the preconditioned CG algorithm is used. For example a reduction by a factor of about 15 is achieved for the seventh problem.

The results for the preconditioned CG algorithm in **Table 16.5** are not the best possible. The best computing time obtained in the solution of the tenth problem was 2.93 sec. (found when no equilibration is performed and when $T = 2.0$ is used). However, it is not very easy to find the best value of the drop-tolerance. Therefore a common strategy, **equilibration + starting drop-tolerance** $T = 2^{-3}$, was used in **all** experiments. Of course, if a simulation process, where many problems with the same coefficient matrix or similar coefficient matrices are to be handled, is carried out, then it may be profitable to attempt to determine an optimal (or nearly optimal) value of the drop-tolerance at the beginning of the simulation process and after that to solve the remaining problems using the value of the drop-tolerance found. This is the same principle as that applied for **ODE's** in **Chapter 8**.

The computing times obtained with the pure CG process (**Algorithm 1** implemented in code **LLSSOX**) are rather good when the process is convergent (or, more precisely, when the speed of convergence is not very slow). It should be pointed out here that the stringent accuracy requirement imposed in this experiment, EPS = 10^{-10}, is favourable for the other two algorithms (first and foremost, for the "pure orthogonalization", but also for the preconditioned CG algorithm). If EPS = 10^{-3} is used, then the results for the pure CG algorithm will become the best when the algorithm converges and the speed of convergence is not very slow. However, the pure CG algorithm is not convergent (or is slowly convergent) in 50% of the problems in the Waterloo set of test-examples.

The storage needed for the pure **CG** algorithm is smaller than the storage needed for the other two methods. In this case the factors D and R are not needed and, therefore, no orthogonalization is carried out (this fact is expressed in **Table 16.5** by setting COUNT = 0).

The storage needed when the preconditioned **CG** algorithm is specified is normally smaller than the storage needed when the pure orthogonalization method is applied (or, in other words, when T = 0, or a very small value of the drop-tolerance as in the previous two chapters, is used without any attempt to improve the first solution). The reduction can be very significant; see the values of COUNT obtained in the solution of the seventh problem with these two options.

The numerical results indicate that no method is best for any problem and for any accuracy requirement. It is also clear that the preconditioned **CG** algorithm leads to very significant reductions both in computing time and in storage for some problems (compared with the "pure orthogonalization" method).

16.5.4. The Harwell set of test matrices. This set consists of 37 test matrices, but only five of them are with rectangular matrices (the others being with square matrices). Some information about these test matrices is given in **Duff and Reid [76]**. Matrices of this set have been used in many studies: as, for example, [75,122,167,291]. Recently many other test-matrices have been added to the Harwell set: the new set is called the Harwell-Boeing set (see, [71,72] as well as **Chapter 10, Chapter 11** and **Chapter 13**).

All 37 matrices of the original Harwell set of sparse test matrices were run in a similar manner as the matrices from the Waterloo set. The results are presented in **Table 16.7 - Table 16.9**. Some characteristics of these matrices are given in **Table 16.6**.

It is seen from **Table 16.7** that even for T = 0 the use of the preconditioned **CG** algorithm is quite competitive with the **IR**. The same is also true for the tests of the Waterloo set (see the previous paragraph). However, for some tests from the Harwell set the preconditioned **CG** algorithm is slightly better than the **IR**, while for all test-examples from the Waterloo set the **IR** option is slightly better than the preconditioned **CG** algorithm.

For five test matrices from the Harwell set (SHL 0, SHL 200, SHL 400, STR 0 and BP 0) the codes determined the exact solution. This means that the first residual vectors are exact zero-vectors and the iterative processes (both the **IR** and the **CG**) are practically not performed; the calculations being terminated after the determination of the first residual vector.

The results for T > 0 are given in **Table 16.8**. The starting value of the drop-tolerance is $T = 2^{-3}$ if m>n or if $(m=n) \wedge max_{1 \leq j \leq n}(|a_{ij}|)/|a_{ii}| \leq 1$ for $\forall i) \wedge (a_{ii} \neq 0$ for $\forall i)$. If at least one of these conditions is not satisfied, then $T = 2^{-5}$ is used.

It is seen that the preconditioned **CG** algorithm performs normally better than the **IR** algorithm; for many test-examples the **IR** does not converge for the value of the drop-tolerance for which the **CG** algorithm is convergent.

No.	Name	m	n	NZ
1	LUND A	147	147	2249
2	LUND B	147	147	2241
3	ERIS1176	1176	1176	18552
4	GENT113	113	113	655
5	IBM 32	32	32	126
6	CURTIS54	54	54	291
7	WILL 57	57	57	281
8	WILL199	199	199	701
9	ASH 292	292	292	2208
10	ASH 85	85	85	523
11	ARC 130	130	130	1282
12	SHL 0	663	663	1687
13	SHL 200	663	663	1726
14	SHL 400	663	663	1712
15	STR 0	363	363	2454
16	STR 200	363	363	3068
17	STR 400	363	363	3157
18	STR 600	363	363	3279
19	BP 0	822	822	3276
20	BP 200	822	822	3802
21	BP 400	822	822	4028
22	BP 600	822	822	4172
23	BP 800	822	822	4534
24	BP1000	822	822	4661
25	BP1200	822	822	4726
26	BP1400	822	822	4790
27	BP1600	822	822	4841
28	ASH 219	219	85	431
29	ASH 958	958	292	1916
30	ASH 331	331	104	662
31	ASH 608	608	188	1216
32	ABB 313	313	176	1557
33	FS 541-1	541	541	4285
34	FS 541-2	541	541	4285
35	FS 541-3	541	541	4285
36	FS 541-4	541	541	4285
37	JPWH 991	991	991	6027

Table 16.6
Some characteristics of the test-matrices
from the Harwell set.

The total computing times, the numbers of iterations and the values of COUNT obtained with the IR and with $T = 0$ are given in the second, third and fourth columns of Table 16.9. The corresponding figures obtained with the preconditioned CG algorithm and with $T > 0$ are given in the fifth, sixth and seventh columns of Table 16.9. Finally, the values obtained with the pure CG algorithm are given in the last three columns of the table.

The "pure orthogonalization" (the IR with $T = 0$) performs better than the preconditioned CG algorithm for many test matrices from the Harwell set. In some cases this is so because the problems solved are very small

according to modern standards (IBM 32, WILL 57). In other cases (see the
results for the classes STR and BP) this is so because (1) the first
trial (with the starting value of the drop-tolerance) was not successful and
the computations were repeated with a smaller value of the drop-tolerance or
(2) because the number of iterations is large (even when the run with the
starting value of the drop-tolerance was not successful and the drop-tolerance
was reduced). It should be noted, however, that the increase of the computing
time for the examples where the preconditioned CG algorithm does not perform
better than the "pure orthogonalization" is normally not large.

The preconditioned CG algorithm performs better than the "pure
orthogonalization" for **all** five test-examples with rectangular matrices (i.e.
when least square problems are to be solved). The reduction both in computing
time and in storage is sometimes very significant. For problem ASH 608, for
example, the computing time is reduced by a factor of 6.57 and the value of
COUNT is reduced by a factor of 4.62 when the preconditioned CG algorithm
is used.

A very significant reduction is also achieved in the solution of the
problems of class FS, although the starting value of the drop-tolerance was
not successful for three of them. This means that the computing times for
these three examples are sums of two computing times: **the computing time with
the starting value of the drop-tolerance + the computing time with the reduced
value(s) of the drop-tolerance.**

A rather significant reduction in both computing time and storage is
achieved in the solution of problem JPWH 991, which is very difficult for
the "pure orthogonalization".

The results obtained with the preconditioned CG algorithm are not the
best possible. The best results, obtained with the optimal values of the drop-
tolerance are nearly always better than the corresponding results obtained
with "pure orthogonalization". However, it may sometimes be difficult to find
the optimal values. Therefore, the common and somewhat crude strategy for
specifying and varying the drop-tolerance, which is outlined in §16.5.2, is
used in connection with all examples in the previous and this paragraphs.
Also the stringent accuracy requirement (EPS=10^{-10}) is clearly favourable for
the "pure orthogonalization". If the accuracy requirement is not so
stringent, then the performance of the preconditioned CG algorithm will be
improved considerably for the examples for which the convergence is slow,
while the performance of the "pure orthogonalization" will remain
practically the same.

The pure CG algorithm is the best one, with regard to the computing
time, when its convergence is sufficiently quick. However, the convergence of
this algorithm is prohibitively slow for many examples from the Harwell set.

The same conclusion as in the previous paragraph can be drawn with regard
to the storage used: **the pure CG algorithm is best one (when it converges
quickly, of course), while the preconditioned CG is normally better than the
"pure orthogonalization" (the first two examples being an exception).** The
reduction of the storage when the preconditioned CG algorithm is used may
be very large; see the values of COUNT for the last five examples in **Table
16.9.**

Again, as in §16.5.3, the numerical results indicate that no method is best for any problem and for any accuracy requirement.

No.	COUNT	Fact. Time	Iterative refinement				Preconditioned CG			
			Time	Iters	Exact	Eval.	Time	Iters	Exact	Eval.
1	5662	1.76	0.10	3	3.7E-14	4.0E-11	0.11	3	3.0E-15	4.7E-12
2	5426	1.65	0.10	3	2.0E-14	3.8E-11	0.11	3	4.1E-15	3.1E-12
3	48202	18.46	0.79	3	3.7E-15	8.4E-15	0.93	3	6.8E-16	1.2E-14
4	1428	0.15	0.14	12	6.5E-12	3.3E-12	0.07	5	6.1E-12	1.3E-11
5	347	0.04	0.01	2	4.4E-16	4.2E-11	0.01	2	2.1E-16	3.0E-12
6	593	0.07	0.01	2	4.7E-16	4.4E-11	0.02	2	3.8E-16	2.9E-12
7	468	0.05	0.01	2	1.2E-15	2.1E-11	0.01	2	3.0E-16	1.6E-12
8	2881	0.42	0.07	4	1.0E-13	5.4E-12	0.09	4	1.6E-14	1.8E-14
9	9571	2.36	0.14	3	4.7E-16	6.9E-15	0.13	2	3.8E-16	2.7E-11
10	1339	0.22	0.03	3	6.9E-17	1.3E-15	0.03	2	3.0E-16	1.0E-11
11	2836	0.36	0.06	3	2.4E-15	1.9E-14	0.05	2	1.4E-15	2.1E-11
12	1687	0.30	0.01	0	0.0	0.0	0.01	0	0.0	0.0
13	1726	0.23	0.01	0	0.0	0.0	0.01	0	0.0	0.0
14	1712	0.26	0.01	0	0.0	0.0	0.01	0	0.0	0.0
15	2454	0.15	0.01	0	0.0	0.0	0.01	0	0.0	0.0
16	6442	0.81	0.18	4	5.4E-14	5.1E-12	0.20	4	8.2E-15	9.4E-16
17	7050	0.91	0.18	4	3.4E-14	1.2E-13	0.17	3	2.0E-15	8.9E-12
18	8735	1.31	0.31	6	2.8E-13	3.2E-12	0.24	4	3.2E-14	3.5E-12
19	3276	0.30	0.01	0	0.0	0.0	0.01	0	0.0	0.0
20	8447	0.83	0.19	3	4.4E-14	3.5E-12	0.24	3	7.4E-15	3.1E-13
21	28914	3.95	0.66	5	2.3E-13	1.2E-12	0.61	4	1.6E-14	2.5E-13
22	25774	3.71	0.61	5	1.3E-13	5.2E-13	0.57	4	3.5E-14	1.4E-12
23	30892	4.95	0.57	4	7.6E-14	5.3E-11	0.66	4	4.6E-15	1.0E-13
24	44482	9.04	0.93	5	7.4E-13	2.5E-12	0.86	4	9.1E-14	9.9E-13
25	49820	10.77	2.23	11	7.3E-11	4.7E-11	1.15	5	3.1E-13	1.5E-12
26	43314	8.49	0.91	5	5.8E-13	2.4E-12	0.84	4	5.5E-14	4.6E-12
27	35802	6.41	0.79	5	6.8E-14	4.5E-13	0.74	4	1.9E-14	1.3E-12
28	1176	0.25	0.02	3	4.2E-15	7.9E-15	0.02	2	1.3E-16	2.3E-11
29	10071	4.35	0.10	3	3.4E-14	8.7E-14	0.09	2	1.9E-15	8.4E-11
30	2252	0.56	0.04	3	5.6E-17	1.6E-14	0.03	2	1.8E-16	3.4E-11
31	5702	2.11	0.06	3	1.3E-14	4.9E-14	0.05	2	1.9E-15	8.3E-11
32	3579	0.87	0.06	3	5.0E-16	8.4E-15	0.05	2	4.9E-16	5.5E-11
33	21192	7.29	0.30	3	1.3E-15	2.3E-14	0.26	2	4.8E-16	9.9E-11
34	21161	7.19	0.40	4	3.6E-14	8.4E-14	0.36	3	3.5E-15	6.3E-12
35	23240	8.58	0.52	5	2.7E-13	6.9E-13	0.48	4	2.4E-14	1.4E-11
36	23632	9.68	0.41	4	4.3E-14	9.2E-12	0.37	3	4.9E-14	5.4E-11
37	233480	334.55	3.00	4	2.3E-15	3.2E-12	2.65	3	1.6E-15	9.4E-12

Table 16.7
Results obtained with T=0 and EPS=10^{-10}. "COUNT" is the maximal number of non-zero elements kept in AQDR. "Fact. time" is the time needed to obtain the orthogonalization and the first solution. "Time" is the iteration time. "Iters" is the number of iterations. "Exact" is the exact error in the final solution. "Eval." is the error evaluated by the code.
The names of the matrices (together with their dimensions and numbers of non-zeros) are given in Table 16.6.

No.	COUNT	Fact. Time	Iterative refinement				Preconditioned CG			
			Time	Iters	Exact	Eval.	Time	Iters	Exact	Eval.
1	5710	2.32	does	not	converge		1.26	40	1.7E-11	9.7E-11
2	6220	2.29	does	not	converge		1.71	52	5.0E-12	1.8E-11
3	24824	9.53	does	not	converge		4.12	22	8.4E-12	5.3E-11
4	1337	0.20	does	not	converge		0.15	12	7.6E-12	1.1E-11
5	176	0.02	0.03	10	4.2E-13	7.9E-12	0.03	7	7.9E-13	3.7E-11
6	346	0.04	0.05	9	4.3E-12	8.8E-11	0.04	7	2.3E-12	7.7E-11
7	289	0.03	0.04	9	7.3E-13	1.8E-11	0.04	7	9.3E-13	4.1E-11
8	1875	0.20	does	not	converge		1.01	62	5.4E-10	3.8E-11
9	4062	0.83	0.41	14	1.8E-12	1.7E-11	0.30	9	4.1E-12	9.9E-11
10	625	0.08	0.08	11	1.7E-12	2.7E-11	0.07	8	4.7E-12	8.6E-11
11	361	0.05	0.09	9	1.1E-12	3.3E-11	0.09	7	5.7E-13	4.7E-11
12	1437	0.14	4.12	178	6.7E-12	9.9E-11	0.10	3	9.9E-15	5.3E-11
13	1451	0.14	does	not	converge		0.13	4	1.0E-14	6.4E-11
14	1428	0.16	does	not	converge		0.13	4	3.2E-15	7.0E-11
15	1126	0.09	0.51	24	1.1E-11	6.7E-11	0.30	12	2.6E-12	4.5E-11
16	1406	0.33	does	not	converge		1.14	34	2.8E-12	6.3E-11
17	1426	0.31	does	not	converge		1.82	53	2.1E-12	7.6E-11
18	8033	2.13	does	not	converge		0.89	17	3.2E-12	1.9E-11
19	2895	0.20	7.59	203	5.6E-12	9.9E-11	0.68	15	1.6E-12	4.2E-11
20	6912	0.67	does	not	converge		5.68	90	1.6E-12	3.9E-11
21	24574	4.60	does	not	converge		1.76	14	3.4E-13	7.6E-12
22	23343	4.00	does	not	converge		2.07	17	3.0E-12	4.4E-11
23	21692	5.57	does	not	converge		4.86	42	1.2E-11	2.1E-11
24	26545	8.12	does	not	converge		7.92	60	1.8E-10	4.2E-11
25	41404	10.06	does	not	converge		6.41	33	1.3E-08	3.2E-11
26	36006	8.15	does	not	converge		4.64	28	7.3E-12	8.4E-12
27	34716	9.14	does	not	converge		3.25	20	2.5E-12	2.4E-11
28	379	0.05	0.10	16	6.3E-12	4.8E-11	0.07	10	2.3E-13	5.9E-12
29	1885	0.32	0.23	12	7.2E-12	7.2E-11	0.20	9	1.1E-12	3.2E-11
30	678	0.10	0.10	12	9.8E-12	9.5E-11	0.08	9	2.3E-12	3.8E-11
31	1234	0.20	0.15	12	7.0E-12	7.8E-11	0.13	9	2.0E-12	4.3E-11
32	1998	0.38	does	not	converge		0.34	20	9.5E-12	5.5E-11
33	759	0.11	0.33	10	3.1E-12	7.6E-11	0.29	7	6.9E-13	3.9E-11
34	9405	1.34	0.86	13	7.4E-12	4.2E-11	0.75	10	1.5E-12	1.7E-11
35	11838	1.86	does	not	converge		2.74	37	9.3E-12	4.3E-11
36	10159	1.40	1.40	22	5.8E-12	3.7E-11	0.92	13	1.8E-12	2.5E-11
37	43965	32.18	does	not	converge		25.45	131	5.0E-11	9.9E-11

Table 16.8

Results obtained with T>0 and EPS=10^{-10}. "COUNT" is the maximal number of non-zero elements kept in AQDR. "Fact. time" is the time needed to obtain the orthogonalization and the first solution. "Time" is the iteration time. "Iters" is the number of iterations. "Exact" is the exact error in the final solution. "Eval." is the error evaluated by the code.
The names of the matrices (together with their dimensions and numbers of non-zeros) are given in Table 16.6.

No.	Pure orth.			Precond. CG			Pure CG		
	Time	Iters	COUNT	Time	Iters	COUNT	Time	Iters	COUNT
1	1.86	3	5662	3.58	40	5710	does not converge		
2	1.75	3	5426	4.00	32	6220	does not converge		
3	19.25	3	48202	13.85	22	13252	4.33	48	0
4	0.29	12	1428	0.45	12	1337	does not converge		
5	0.05	2	347	0.06	7	176	0.03	11	0
6	0.08	2	593	0.08	7	346	0.05	14	0
7	0.06	2	468	0.07	7	289	0.05	14	0
8	0.49	4	2881	1.21	62	1875	does not converge		
9	2.50	3	9571	1.13	9	4062	0.28	19	0
10	0.25	3	1339	0.15	8	625	0.09	17	0
11	0.42	3	2836	0.14	7	361	0.25	30	0
12	0.31	0	1687	0.24	3	1437	does not converge		
13	0.24	0	1726	0.27	4	1461	does not converge		
14	0.27	0	1712	0.29	4	1428	does not converge		
15	0.16	0	2454	0.39	12	1126	does not converge		
16	0.98	4	6442	1.47	34	2819	does not converge		
17	1.07	4	7050	2.13	53	3010	does not converge		
18	1.62	6	8735	3.02	17	8003	does not converge		
19	0.31	0	3276	0.88	15	2895	does not converge		
20	1.02	3	8447	6.35	90	6912	does not converge		
21	4.61	5	28914	6.36	14	24574	does not converge		
22	4.32	5	25774	6.07	17	23343	does not converge		
23	5.52	4	30892	10.43	42	21692	does not converge		
24	9.97	5	44482	16.04	60	26545	does not converge		
25	13.00	11	49820	16.47	33	41404	does not converge		
26	9.40	5	43314	12.79	28	36006	does not converge		
27	7.20	5	35802	12.39	20	34716	does not converge		
28	0.27	3	1176	0.12	10	379	0.18	39	0
29	4.45	3	10071	0.52	10	1885	0.46	34	0
30	0.60	3	2252	0.18	9	678	0.17	29	0
31	2.17	3	5702	0.33	9	1234	0.32	34	0
32	0.93	3	3579	0.72	20	1998	0.78	78	0
33	7.59	3	21192	0.40	7	4727	0.67	24	0
34	7.59	4	21161	2.09	10	9405	does not converge		
35	9.10	5	23240	4.60	37	11838	does not converge		
36	10.09	4	23632	2.32	13	10159	does not converge		
37	337.55	4	233480	57.63	131	43965	does not converge		

Table 16.9
Comparison of the performance of three options
on the Harwell test matrices.
"Pure orth." is the option where IR with T=0 is used.
"Precond. CG" is the option in which the preconditioned CG code
is used with a starting $T=2^{-3}$ or $T=2^{-5}$.
"Pure CG" is the option where CG without any preconditioning is applied.
The names of the matrices (together with their dimensions
and numbers of non-zeros) are given in Table 16.6.

16.5.5. Tests with matrix generators. Several matrix generators were used in the experiments. Some results obtained with **MATRF2** (see **Chapter 3**) will be presented in this paragraph. This generator produces matrices dependent on five parameters: the number of rows m, the number of columns n, the identifier of the sparsity pattern c, the avarage number of non-zeros per row r and the identifier for the magnitude of the non-zero elements α. The dependency of the magnitude of the non-zero elements on parameter α is expressed by the relationship:

(16.34) $\displaystyle \max_{\substack{1 \le i \le m \\ 1 \le j \le n}} (|a_{ij}|) / \min_{\substack{1 \le i \le m \\ 1 \le j \le n \\ a_{ij} \ne 0}} (|a_{ij}|) = max(10\alpha^2, (rm-n)\alpha).$

It has already been pointed out (**Chapter 3**) that the number of matrices that can be produced by this generator is practically infinite and no huge files are needed to keep sets of these matrices (each matrix can easily be produced by a simple call of subroutine **MATRF2**. Moreover, many properties of the sparse codes tested can be efficiently studied in a systematic way by the use of matrices generated by **MATRF2**. One of the experiments, carried out in connection with the sparse codes for linear least square problems, is described in this paragraph as an illustration of the above statement.

The condition numbers of the matrices generated by **MATRF2** will in general be increased when larger values of parameter α are selected (keeping the other four parameters fixed), which is confirmed in **Chapter 9**; see also **Zlatev et al. [338]**. This property of the matrices generated by **MATRF2** has been exploited to compare in a systematic way the largest values of the drop-tolerance T that can be specified for the **IR** option and for the preconditioned **CG** algorithm.

In all runs in this section the first four parameters are fixed (m=500, n=400, c=150, r=12). This means that the number NZ of non-zero elements is the same for all experiments: NZ=rm+110=6110 (see again **Chapter 3**). Six values of parameter α were used: α=2^k, k=0,3,6,9,12,15.

First the six matrices generated by **MATRF2** with these values of the parameters were decomposed with $T = 0.0$ and the linear least square problems (created as in the previous two paragraphs) were solved by using **IR** and preconditioned **CG**. The results are given in **Table 16.10**. The factorization times and the storage requirements are the same in this part of the experiment (the orthogonal decomposition is carried out with the same subroutines when both the **IR** option is used and the preconditioned **CG** algorithm is chosen). The solution times are slightly different; being in general smaller when the **IR** process is in use.

During the second part of the experiment, $T = 2^{-2}$ was used as a starting drop-tolerance in the decomposition of each matrix. If the accuracy requirement, prescribed by $EPS = 10^{-10}$, was not satisfied (either because the iterative process under consideration was not convergent or because the speed of convergence was judged, by the code, to be very slow), then the drop-tolerance was reduced by a factor of two and a new decomposition was computed. Sometimes several reductions were performed. In this way for each value of α and for each of the two iterative processes the largest value of the drop-tolerance T was found, that allows us to calculate a sufficiently accurate solution. The results are given in **Table 16.11**. In general larger values of the drop-tolerance could be used with the CG algorithm, leading to savings in both computing time and storage. Even if the two iterative processes could be run with the same value of the drop-tolerance T, which is the case for well-conditioned matrices, the CG algorithm leads to a reduction of the numbers of iterations and, thus, to savings in computing time; see the first two examples in **Table 16.11**. Of course, the same effect could be traced by studying the results presented in the previous two paragraphs, but here the relationship between the largest drop-tolerance that can be applied and the condition number of the matrix decomposed is much more transparent.

k	Iterative refinement				Preconditioned CG				
	T	COUNT	Fact.	Sol.	Iters	T	COUNT	Fact.	Sol.	Iters
0	0.0	62459	46.46	0.61	3	0.0	62459	46.46	0.72	3
3	0.0	60979	43.40	0.61	3	0.0	60979	43.40	0.72	3
6	0.0	61998	45.39	0.61	3	0.0	61998	45.39	0.72	3
9	0.0	62284	45.45	0.81	4	0.0	62284	45.45	0.72	3
12	0.0	62886	47.05	0.84	4	0.0	62886	47.05	0.94	4
15	0.0	61991	45.50	1.20	6	0.0	61991	45.50	1.12	5

Table 16.10

Results obtained by using matrices produced by MATRF2 with $m=500$, $n=400$, $c=150$, $r=12$ (NZ=6110) and six different values of α ($\alpha=2^k$, $k=0,3,6,9,12,15$). The accuracy requirement was determined by EPS=10^{-10} and has been satisfied for all runs. Under "T", "COUNT", "Fact.", "Sol." and "Iters" the drop-tolerances used, the numbers of non-zeros in AQDR, the factorization times, the solution times and the numbers of iterations are given.

k	Iterative refinement				Preconditioned CG				
	T	COUNT	Fact.	Sol.	Iters	T	COUNT	Fact.	Sol.	Iters
0	2.E-02	4231	0.22	1.24	28	2.E-2	4231	0.22	0.92	19
3	2.E-02	3722	0.20	20.75	469	2.E-2	3722	0.20	0.98	20
6	2.E-04	5307	0.67	1.03	20	2.E-2	3590	0.19	1.12	23
9	2.E-07	6502	1.34	0.89	16	2.E-2	5017	0.53	2.15	40
12	2.E-10	9885	2.30	3.36	51	2.E-4	5284	0.75	2.07	38
15	2.E-11	12963	3.38	29.97	396	2.E-8	7098	1.45	1.77	29

Table 16.11

Results obtained by using matrices produced by MATRF2 with $m=500$, $n=400$, $c=150$, $r=12$ (NZ=6110) and six different values of α ($\alpha=2^k$, $k=0,3,6,9,12,15$). The accuracy requirement was determined by EPS=10^{-10} and has been satisfied for all runs. Under "T", "COUNT", "Fact.", "Sol." and "Iters" the drop-tolerances used, the numbers of non-zeros in AQDR, the factorization times, the solution times and the numbers of iterations are given.

If the condition number of the matrix is large, then the largest value
of the drop-tolerance by which the decomposition can be successfully carried
out and the iterative method used converges is normally small (but being
considerably larger when the preconditioned CG algorithm is in use). This
is not a surprise: the relationship between the condition number and the
largest drop-tolerance that yields convergence is studied in Zlatev
[299,301], see also Section 3.5 and Section 3.8.

Comparing the results in Table 16.10 and Table 16.11, it is seen that
the use of a large drop-tolerance may be profitable even if the speed of
convergence is very slow; compare the results obtained by the IR for the
matrices generated by $\alpha=2^3$ and α^{15} in Table 16.10 and Table 16.11.

The pure CG algorithm was also tested. It did not converge for the
examples used in this experiment.

16.6. Concluding remarks and references concerning the conjugate gradient algorithm for orthogonalization methods

The results presented in this chapter demonstrate clearly the fact that
methods of conjugate gradient-type can successfully be used in connection with
large and sparse algebraic problems (both systems of linear algebraic
equations and linear least squares problems). Several remarks concerning the
particular algorithms used and the possibilities of improving the performance
of the algorithms are given in this section. Also many references, concerning
some interesting works in this field, are quoted.

Remark 16.12. The method of conjugate gradient was originally proposed
as a direct method by Hestenes and Stiefel [146]. The interest to this
method became much greater after the proposal made by Reid [196] to use it
as an iterative method. The method is studied in many publications as, for
example, Axelsson and Barker [15], Golub and Van loan [133] and Lawson and
Hanson [165]. It is normally used with symmetric and positive definite
matrices, but also some attempts to apply the method to some classes of non-
symmetric matrices were carried out; see, for example, [13,16]. ∎

Remark 16.13. The matrix A^TA appears in the formulation of the CG
algorithm discussed in this chapter. Therefore it seems that the speed of
convergence will depend on $[\kappa(A)]^2$ (at least when the pure CG algorithm
is in use). The experiments carried out indicate that this was not the case.
In many examples the pure CG algorithm was convergent even when $\kappa(A)$ is
rather close to the reciprocal of the machine precision. Probably the reason
for this is that A^TA is never calculated explicitly. It should be noted here
that Michlin [171], p. 21, has derived (for m=n and where the CG is used
as a direct method) some error estimates that depend on $\kappa(A)$ and not on
$[\kappa(A)]^2$, but these error estimates are derived under an assumption that $\kappa(A)$
is much smaller than the reciprocal of the machine precision. ∎

Remark 16.14. The orthogonalization method used in this chapter is the
Gentleman-Givens plane rotations; [109,110,141]. Other methods were discussed
in Chapter 12 (see also [23,25,26,75,153,166,167,184]). The method proposed
by George and Heath [112], see also [113,114,121], should be noted. This
method is very efficient in the case where the storage is an important factor
(see also Chapter 13). Unfortunately, it is not very clear how a preconditio-

ner can efficiently be calculated when the latter method is in use. Surveys of the new sparse matrix techniques used in connection with linear least squares problems have recently been published by **Heath** [143], **Ikramov** [153] and **Zlatev** [306]. ∎

Remark 16.15. It is perhaps more natural to drop an element if it is smaller, in absolute value, than the product of the drop-tolerance and the minimum of the maximal elements in the active parts of its row and column (this possibility has also been discussed in the previous chapters). In other words, it is more natural to introduce a relative drop-tolerance. Unfortunately, the maximal elements in the active parts of the columns can not be found in an efficient way, because the non-zeros are stored by rows. Nevertheless, the work towards an efficient implementation of a relative drop-tolerance is continuing. ∎

Remark 16.16. The pivotal strategy is based on an idea originally proposed by **Gentleman** [111]. This means that the plane rotations are carried out for the pair with minimal number of non-zeros in the active parts of the rows involved (at each minor step). Thus, the pivotal row is not fixed (at the beginning of a major step), but variable. A variable pivotal row has recently been used also by **Liu** [166].

Remark 16.17. The particular versions of the **CG** algorithm described in **Algorithm 1** and **Algorithm 2** (see **Fig. 16.1** and **Fig. 16.2** respectively) are not the only possible versions. These particular versions are the same as the corresponding algorithms in **Dongarra et al.** [59]. Different modifications have been proposed by many authors (as, for example, in [16,45,46,80,87-89,168,170,181,207]). Of course, all authors claim their modifications are better than the classical version, but not always there is sufficient experimental evidence to support such claims when **large and sparse general matrices** are to be treated. A rather big set of large test matrices, the Harwell-Boeing matrices ([71,72]), is now easily available. This set should be used in the verification of advocated changes. The use of a standard set of test matrices will facilitate the comparison of different changes: one will not need to test all other versions if results obtained by using a standard set of test matrices are easily available (together with demonstration programs). ∎

Remark 16.18. The assumption $rank(A)=n$ was imposed in this chapter (and also in the previous chapters). In fact, even a stronger assumption is needed: **matrix A should not be too close to a matrix A^* with $rank(A^*)<n$ when a computer is in use.** For many practical problems it is desirable to determine the numerical rank of matrix A and to use special algorithms when it is smaller than n. ∎

Remark 16.19. The evaluation of the storage needed by (16.33) is a considerable overestimation. A careful study of the processes that are to be carried out shows that array RNQDR, whose length is NN1, is only used (as a working array) during the orthogonal decomposition. On the other hand, many of the working arrays of length m and n (most of them declared in double precision) that are used in the iterative processes are not needed during the orthogonal decomposition. This shows that the storage actually needed can be reduced substantially, compared with (16.33), when an **EQUIVALENCE** statement is introduced in an appropriate way in the main program. ∎

REFERENCES

1. *P. R. Amestoy and I. S. Duff:* "Vectorization of a multiprocessor multi-frontal code". *Report No. CSS 231. Computer Science and Systems Division, A.E.R.E., Harwell, Oxfordshire, England, 1988.*

2. *R. Alexander:* "Diagonally implicit Runge-Kutta methods or stiff ODE's". *SIAM J. Numer. Anal., 14(1977), 1006-1021.*

3. *E. Anderson:* "Parallel implementation of preconditioned conjugate gradient methods for solving sparse systems of linear equations". *Report No. 805. Center for Supercomputing Research and Development, University of Illinois at Urbana-Champaign, Urbana, Illinois, 1988.*

4. *E. Anderson and Y. Saad:* "Preconditioned conjugate gradient methods for general sparse matrices on shared memory machines". In: *"PARALLEL PROCESSING FOR SCIENTIFIC COMPUTING"* (G. Rodrigue, ed.), pp. 88-92. *Society for Industrial AND Applied Mathematics, Philadelphia, 1989.*

5. *N. Andrei and C. Rasturnoiu:* "Matrice rare et aplicatile lor". *Editura Technica, Bucuresti, 1983* (in Roumanian).

6. *M. Arioli, J. W. Demmel and I. S. Duff:* "Solving sparse linear systemes with sparse backward error". *Report No. CSS 214. Computer Science and Systems Division, A.E.R.E., Harwell, England, 1988.*

7. *M. Arioli and I. S. Duff:* "Experiments in tearing large sparse matrices". *Report No. CSS 217. Computer Science and Systems Division, A.E.R.E., Harwell, England, 1988.*

8. *M. Arioli, I. S. Duff, N. I. M. Gould and J. K. Reid:* "Use of the P^4 and P^5 algorithms for in-core factorization of sparse matrices". *SIAM J. Sci. Statist. Comput., 11(1990), 913-927.*

9. *M. Arioli, I. S. Duff and P. P. M. de Rijk:* "On the augmented system approach to sparse least-squares problems". *Numer. Math., 55(1989), 667-684.*

10. *W. E. Arnoldi:* "The principle of minimized iteration in the solution of the matrix eigenvalue problem". *Quart. Appl. Math., 9(1951), 17-29.*

11. *O. Axelsson:* "A generalized SSOR method". *BIT, 12(1972), 443-467.*

12. *O. Axelsson:* "On preconditioned conjugate gradients method". In: *"CONJUGATE GRADIENT METHODS AND SIMILAR TECHNIQUES"* (I.S.Duff, ed.), pp. 23-55. *Report No. R9636, A.E.R.E., Harwell, England, 1979.*

13. *O. Axelsson:* "Conjugate type methods for unsymmetric and inconsistent systems of linear equations". *Lin. Alg. Appl., 29(1980), 1-16.*

14. *O. Axelsson:* "A survey of preconditioned iterative methods for linear systems of algebraic equations". *BIT, 25(1985), 166-187.*

15. *O. Axelsson and V. A. Barker:* "Finite element solution of boundary value problems". *Academic Press, Orlando, 1984.*

16. O. Axelsson and I. Gustavsson: "A modified upwind scheme for convective transport equations and the use of a conjugate gradient method for the solution of non-symmetric systems of equations". J. Inst. Math. Applics., 23(1979), 321-337.

17. V. A. Barker: "Numerical solution of sparse singular systems of equations arising from ergodic Markov chains". Stochastic Models, 5(1989), 335-381.

18. C. H. Bischoff: "Incremental condition number". SIAM J. Matr. Anal. Appl., 11(1990), 312-322.

19. C. H. Bischoff and P. C. Hansen: "Structure-preserving and rank revealing QR-factorizations". Preprint No. MCS-P100-0989. Mathematics and Computer Division, Argonne National Laboratory, Argonne, Illinois, 1989, to appear in SIAM J. Sci. Statist. Comput.

20. C. H. Bischoff, J. G. Lewis and D. J. Pierce: "Incremental condition estimation for general matrices". SIAM J. Matr. Anal. Appl., 11(1990), 644-662.

21. Å. Björck: "Iterative refinement for linear least squares solutions. I". BIT, 7(1967), 257-278.

22. Å. Björck: "Iterative refinement for linear least squares solutions. II". BIT, 8(1968), 8-30.

23. Å. Björck: "Methods for sparse linear least squares problems". In: "SPARSE MATRIX COMPUTATIONS" (J. R. Bunch and D. J. Rose, eds.), pp. 177-199. Academic Press, New York, 1976.

24. Å. Björck: "Coments on the iterative refinement of least-square solutions". J. Amer. Statist. Assoc., 73(1978), 161-166.

25. Å. Björck: "Numerical algorithms for linear least-square problems". Report No. 2. Matematisk Institut, Universitetet i Trondheim, Trondheim, Norway, 1978.

26. Å. Björck: "Least squares methods". Working paper. Department of Mathematics, Linköping University, Linköping, Sweden, 1986. To appear in: "HANDBOOK ON NUMERICAL ANALYSIS", Vol.1 (P. G. Ciarlet and J. L. Lions eds). North Holland, Amsterdam, 1987.

27. Å. Björck: "Stability analysis of the method of seminormal equations for linear least squares problems". Lin. Alg. Appl., 88(1987), 31-48.

28. Å. Björck and I. S. Duff: "A direct method for the solution of sparse linear least squares problems". Lin. Alg. Appl., 34(1980), 43-67.

29. Å. Björck and T. Elfving: Accelerated projection methods for computing pseudo-inverse solutions of systems of linear equations". BIT, 19(1979), 145-163.

30. Å. Björck and G. H. Golub: "ALGOL programming, contribution No. 22: Iterative refinement of linear least squares solutions by House-holder transformations". BIT, 7(1967), 322-337.

31. Å. Björck and Z. Zlatev: "Exploiting the separability in the solution of linear ordinary differential equations". Comp. Math. Appl., 18(1989), 421-438.

32. I. D. L. Bogle: "A comparison of methods for solving large sparse sy-stems of nonlinear equations". Report. Department of Chemical Engineering and Chemical Technology, Imperial College, London, 1982.

33. R. K. Brayton, F. G. Gustavson and R. A. Willoughby: "Some results on sparse matrices". Math. Comp., 29(1970), 937-954.

34. A. M. Bruaset, A. Tveito and R. Winther: "On the stability of relaxed incomplete LU factorizations". Math. Comp., 54(1990), 701-719.

35. J. R. Bunch: "The weak and strong stability of algorithms in numerical linear algebra". Lin. Alg. Appl., 88(1987), 49-66.

36. K. Burrage, J. C. Butcher and F. N. Chipman: "STRIDE: Stable Runge-Kutte integrator for differential equations". Report Series No. 150 (Computational Mathematics No. 20). Department of Mathematics, University of Auckland, Auckland, New Zealand, 1979.

37. K. Burrage, J. C. Butcher and F. N. Chipman: "An implementation of singly implicit Runge-Kutta methods". BIT, 20(1980), 326-340.

38. J. C. Butcher: "The numerical analysis of ordinary differential equations: Runge-Kutta and general linear methods". Wiley, New York, 1987.

39. T. F. Chan: "Rank revealing QR factorization". Lin. Alg. Appl., 88(1987), 67-82.

40. T. F. Chan and P. C. Hansen: "Computing truncated singular value decomposition least squares solutions by rank revealing QR factorization". SIAM J. Sci. Statist. Comput., 11(1990), 519-530.

41. R. J. Clasen: "Techniques for automatic tolerance control in linear programming". Comm. ACM, 9(1966), 802-803.

42. A. K. Cline, A. R. Conn and C. F. Van Loan: "Generalizing the LINPACK condition number estimator". In: "NUMERICAL ANALYSIS" (J. P. Hennart, ed.), pp. 73-83. Lecture Notes in Mathematics, No. 909. Springer, Berlin, 1982.

43. A. K. Cline, C. B. Moler, G. W. Stewart and J. H. Wilkinson: "An esti-mate for the condition number of a matrix". SIAM J. Numer. Anal., 16(1979), 368-375.

44. **A. K. Cline and R. K. Rew:** "A set of counter-examples to three condition number estimators". *SIAM J. Sci. Statist. Comput.*, **4(1983)**, 602-611.

45. **P. Concus, G. H. Golub and G. Meurant:** "Block preconditioning for the conjugate gradient method". *SIAM J. Sci. Statist. Comput.*, **6(1985)**, 220-252.

46. **P. Concus, G. H. Golub and D. P. O'Leary:** "A generalized conjugate gradients methods for the solution of elliptic partial differential equations". In: **"SPARSE MATRIX COMPUTATIONS"** (J. R. Bunch and D. J. Rose, eds.), pp. 309-332. Academic Press, New York, **1976**.

47. **J. Cullum and R. A. Willoughby:** "Computing eigenvectors (and eigenvalues) of large, symmetric matrices using Lanczos tridiagonalization". In: **"NUMERICAL ANALYSIS DUNDEE 1979"** (G. A. Watson, ed.), pp. 46-6-3. Lecture Notes in Mathematics, No. 773. Springer, Berlin, **1980**.

48. **A. R. Curtis and J. K. Reid:** "The solution of large sparse unsymmetric systems of linear equations". *J. Inst. Math. Applics.*, **8(1971)**, 344-353.

49. **C. Daly and J. J. Du Croz:** "Performance of a subroutine library on vector-processing machines". *Computer Physics Communications*, **37(1985)**, 181-186.

50. **T. A. Davis and E. S. Davidson:** "Pairwise reduction for the direct, parallel solution of sparse unsymmetric sets of linear equations". *IEEE Trans. Comput.*, **37(1988)**, 1648-1654.

51. **T. A. Davis and P.-C. Yew:** "A nondeterministic parallel algorithm for general unsymmetric sparse LU factorization". *SIAM J. Matrix. Anal. Appl.*, **3(1990)**, 383-402.

52. **M. J. Dayde and I. S. Duff:** "Use of level 3 BLAS in LU factorization on CRAY-2, the ETA-10P, and the IBM 3090-200/VF". *Report No. CSS 229. Computer Science and Systems Division, A.E.R.E., Harwell, England, **1988**.

53. **D. S. Dodson and J. G. Lewis:** "Proposed sparse extensions to the basic linear algebra subprograms". *ACM SIGNUM Newsletter*, **20(1985)**, 22-25.

54. **J. J. Dongarra:** "Performance of various computers using standard linear equations software in a Fortran environment". *Technical Memorandum No. 23. Mathematics and Computer Science Division, Argonne National Laboratory, Argonne, Illinois, USA, **1986**.

55. **J. J. Dongarra, J. R. Bunch, C. B. Moler and G. W. Stewart:** "LINPACK: users' guide". *Society for Industrial and Applied Mathematics, Philadelphia, **1979**.

56. **J. J. Dongarra, J. J. Du Croz, S. Hammarling and R. J. Hanson:** "A proposal for an extended set of FORTRAN basic linear algebra subprograms". *ACM SIGNUM Newsletter*, **Vol. 20, No. 1, 1985**, 1-18.

57. **J. J. Dongarra and S. C. Eisenstat:** "Squeezing the most out of an algorithm in CRAY FORTRAN". ACM Trans. Math. Software, **10(1984)**, 219-230.

58. **J. J. Dongarra, F. G. Gustavson and A. Karp:** "Implementing linear algebra algorithms for dense matrices on a vector pipeline machine". SIAM Rev., **26(1984)**, 91-112.

59. **J. J. Dongarra, G. K. Leaf and M. Minkoff:** "A preconditioned conjugate gradient method for solving a class of non-symmetric linear systems". Report No. ANL-81-71. Mathematics and Computer Science Division, Argonne National Laboratory, Argonne, Illinois, USA, **1986**.

60. **J. J. Dongarra and A. H. Sameh:** "On some parallel banded system solvers". Parallel Computing, **1(1984)**, 223-235.

61. **J. J. Du Croz and J. Wasniewski:** "Basic linear algebra computations on the Sperry ISP". Supercomputer, **21(1987)**, 45-54.

62. **I. S. Duff:** "Pivot selection and row ordering in Givens reduction of sparse matrices". Computing, **13(1974)**, 239-248.

63. **I. S. Duff:** "MA28: a set of FORTRAN subroutines for sparse unsymmetric linear equations". Report No. R8730. Computer Science and Systems Division, A.E.R.E., Harwell, England, **1977**.

64. **I. S. Duff:** "Practical comparisons of codes for the solution of sparse linear systems". In: **"SPARSE MATRIX PROCEEDINGS 1978"** (I. S. Duff and and G. W. Stewart, eds.), pp. 107-134. Society for Industrial and Applied Mathematics, Philadelphia, **1979**.

65. **I. S. Duff:** "The solution of sparse linear equations on the CRAY1". Report No. CSS 125 (Revised). Computer Science and Systems Division, A.E.R.E., Harwell, Oxfordshire, England, **1983**.

66. **I. S. Duff:** "Data structures, algorithms and software for sparse matrices". In: **"SPARSITY AND ITS APPLICATIONS"** (D. J. Evans, ed.), pp. 1-29. Cambridge University Press, Cambridge-London, **1985**.

67. **I. S. Duff:** "The solution of large-scale least-squares problems on supercomputers". Report No. CSS 225. Computer Science and Systems Division, A.E.R.E., Harwell, Oxfordshire, England, **1988**.

68. **I. S. Duff:** "Experience with the HARWELL Subroutine Library on the CRAY-2". Report No. CSS 227. Computer Science and Systems Division, A.E.R.E., Harwell, England, **1988**.

69. **I. S. Duff, A. M. Erisman and J. K. Reid:** "Direct methods for sparse matrices". Oxford University Press, Oxford-London, **1986**.

70. **I. S. Duff, N. I. M. Gould, M. Lescrenier and J. K. Reid:** "The multifrontal method in a parallel environment". Report No. CSS 211. Computer Science and Systems Division, A.E.R.E., Harwell, England, **1987**.

71. I. S. Duff, R. G. Grimes and J. C. Lewis: *"Sparse matrix test problems".* ACM Trans. Math. Software, **15(1989)**, 1-14.

72. I. S. Duff, R. G. Grimes, J. C. Lewis and G. W. Poole, Jr.: *"Sparse matrix test problems".* SIGNUM Newsletter, **17**, 2 (1982), 22.

73. I. S. Duff and J. K. Reid: *"A comparison of sparsity orderings for obtaining a pivotal sequence in Gaussian elimination".* J. Inst. Math. Applics., **14(1974)**, 281-291.

74. I. S. Duff and J. K. Reid: *"On the reduction of sparse matrices to condensed form by similarity transformations".* J. Inst. Math. Applics., **15(1975)**, 217-224.

75. I. S. Duff and J. K. Reid: *"A comparison of some methods for the solution of sparse overdetermined systems of linear equations".* J. Inst. Math. Applics., **17(1976)**, 267-280.

76. I. S. Duff and J. K. Reid: *"Performance evaluation of codes for sparse matrices".* Report No. CSS 66. Computer Science and Systems Division, A.E.R.E., Harwell, England, **1978**.

77. I. S. Duff and J. K. Reid: *"Some design features of a sparse matrix code".* ACM Trans. Math. Software, **5(1979)**, 18-35.

78. I. S. Duff and J. K. Reid: *"The multifrontal solution of indefinite sparse symmetric linear systems".* Report No. CSS 122. Computer Science and Systems Division, A.E.R.E., Harwell, England, **1982**.

79. I. S. Duff, J. K. Reid and J. A. Scott: *"The use of profile reduction algorithms with a frontal code".* Report No. CSS 224. Computer Science and Systems Division, A.E.R.E., Harwell, England, **1988**.

80. S. C. Eisenstat: *"Efficient implementation of a class of preconditioned conjugate gradient methods".* SIAM J. Sci. Statist. Comput., **2(1981)**, 1-4.

81. S. C. Eisenstat, H. C. Elman and M. H. Schultz: *"Variational methods for nonsymmetric systems of linear equations".* SIAM J. Numer. Anal., **20(1983)**, 345-357.

82. S. C. Eisenstat, M. C. Gursky, M. H. Schultz and A. H. Sherman: *"The YALE sparse matrix package: The symmetric codes".* Internat. J. Numer. Meth. Engng., **18(1982)**, 1145-1151.

83. S. C. Eisenstat, M. C. Gursky, M. H. Schultz and A. H. Sherman: *"The YALE sparse matrix package: The non-symmetric codes".* Report No. 114. Department of Computer Science, Yale University, New Haven, Connecticut, **1977**.

84. S. C. Eisenstat, M. H. Schultz and A. H. Sherman: *"Software for sparse Gaussian elimination with limited core storage".* In: **"SPARSE MATRIX PROCEEDINGS 1978"** (I.S.Duff and G. W. Stewart, eds.), 135-153. Society for Industrial and Applied Mathematics, Philadelphia, **1979**.

85. S. C. Eisenstat, M. H. Schultz and A. H. Sherman: "Algorithms and data structure for sparse symmetric Gaussian elimination". SIAM J. Sci. Statist. Comput., 2(1981), 225-237.

86. T. Elfving: "A note on sparsity in Gauss and Givens methods". Report No. LITH-MAT-R-1976-5. Department of Mathematics, Linköping University. Linköping, Sweden, 1976.

87. H. C. Elman: "Iterative methods for large, sparse, nonsymmetric systems of linear equations". Ph. D. Thesis, Department of Computer Science, Yale University, New Haven, Connecticut, 1982.

88. H. C. Elman, Y. Saad and P. E. Saylor: "A hybrid Chebyshev Krylov subspace algorithm for solving nonsymmetric systems of linear equations". SIAM J. Sci. Statist. Comput., 7(1986), 840-855.

89. H. C. Elman and M. H. Schultz: "Preconditioning by fast direct methods for non-self-adjoint nonseparable elliptic equations". SIAM J. Numer. Anal., 23(1986), 44-57.

90. A. M. Erisman, R. G. Grimes, J. G. Lewis and G. W. Poole, Jr.: "A structurally stable modification of Hellerman-Raric's P^4 algorithm for reordering unsyymetric sparse matrices". SIAM J. Numer. Anal., 22(1985), 369-385.

91. A. M. Erisman, R. G. Grimes, J. G. Lewis, G. W. Poole, Jr. and H. D. Simon: "Evaluation of orderings for unsymmetric sparse matrices". SIAM J. Sci. Statist. Comput., 8(1987), 600-624.

92. A. M. Erisman and J. K. Reid: "Monitoring the stability of the triangular factorization of a sparse matrix". Numer. Math., 22(1974), 183-186.

93. D. J. Evans: "The use of preconditioning in iterative methods for solving linear equations with symmetric positive definite matrices". J. Inst. Math. Applics., 4(1968), 295-314.

94. D. J. Evans: "The analysis and application of sparse matrix algorithms in the finite element method". In: "THE MATHEMATICS OF FINITE ELEMENTS AND APPLICATIONS" (J. R. Whiteman, ed.), pp. 427-447. Academic Press, London, 1973.

95. D. J. Evans: "Iterative sparse matrix algorithms". In: "SOFTWARE FOR NUMERICAL MATHEMATICS" (D. J. Evans, ed.), pp. 49-83. Academic Press, London, 1974.

96. D. J. Evans: "Iterative methods for sparse matrices". In: "SPARSITY AND ITS APPLICATIONS" (D. J. Evans, ed.), pp. 45-111. Cambridge University Press, Cambridge-London, 1985.

97. D. K. Fadeev and V. N. Fadeeva: "Computational methods of linear algebra". Freeman, San Francisco, 1963.

98. R. Fletcher: "Expected conditioning". IMA J. Numer. Anal., 5(1985), 247-273.

99. *G. E. Forsythe, M. A. Malcolm and C. B. Moler:* "Computer methods for mathematical computations". Prentice-Hall, Englewood Cliffs, N. J., *1977*.

100. *G. E. Forsythe and C. B. Moler:* "Computer solution of linear algfebraic equations". Prentice-Hall, Englewood Cliffs, N. J., *1967*.

101. *R. Freund:* "On conjugate gradient type methods and polynomial preconditioners for a class of complex non-Hermitian matrices". Numer. Math. 57(1990), 285-312.

102. *K. A. Gallivan, W. Jalby and U. Meier:* "The use of BLAS3 in linear algebra on a parallel processor with hierarchical memory". SIAM J. Sci. Statist. Comput., *8(1987)*, 1079-1084.

103. *K. A. Gallivan, R. J. Plemmons and A. H. Sameh:* "Parallel algorithms for dense linear algebra computations". SIAM Rev., *32(1990)*, 54-135.

104. *K. A. Gallivan, A. H. Sameh and Z. Zlatev:* "Solving general sparse linear systems using conjugate gradient-type methods". In: *"PROCEEDINGS OF THE 1990 INTERNATIONAL CONFERENCE ON SUPER-COMPUTING, June 11-15 1990, Amsterdam, The Netherlands"*. ACM Press, New York, *1990*.

105. *K. A. Gallivan, A. H. Sameh and Z. Zlatev:* "A parallel hybrid sparse linear system solver". Computing Systems in Engineering, *1(1990)*, 183-195.

106. *C. W. Gear:* "Numerical initial value problems in ordinary differential equations". Prentice-Hall, Englewood Cliffs, N.J., USA, *1971*.

107. *C. W. Gear:* "Numerical error in sparse linear equations". Report No. UIUCDCS-F-75-885. Department of Computer Science, University of Illinois at Urbana-Champaign, Urbana, Illinois, *1975*.

108. *C. W. Gear and Y. Saad:* "Iterative solution of linear equations in ODE codes". SIAM J. Sci. Statist. Comput., *4(1983)*.

109. *W. M. Gentleman:* "Least squares computations by Givens transformations without square roots". J. Inst. Math. Applics., *12(1973)*, 329-336.

110. *W. M. Gentleman:* "Error analysis of QR by Givens transformations". Lin. Alg. Appl., *10(1975)*, 189-197.

111. *W. M. Gentleman:* "Row elimination for solving sparse linear systems and least squares problems". In: *"NUMERICAL ANALYSYS DUNDEE 1975"* (G. A. Watson, ed.), pp. 122-133. Lecture Notes in Mathematics, No. 506. Springer, Berlin, *1976*.

112. *J. A. George and M. T. Heath:* "Solution of sparse linear least squares problems using Givens rotations". Lin. Alg. Appl., *34(1980)*, 69-83.

113. *J. A. George, M. T. Heath and E. Ng:* "A comparison of some methods for solving sparse linear least squares problems". SIAM J. Sci. Statist. Comput., *4(1983)*, 177-187.

114. **J. A. George, M. T. Heath and R. J. Plemmons:** *"Solution of large-scale linear least squares problems using auxiliary storage".* SIAM J. Sci. Statist. Comput., *2(1981),* 416-429.

115. **J. A. George and J. W. Liu:** *"Computer solution of large sparse positive definite systems".* Prentice-Hall, Englewood Cliffs, N. J., *1981.*

116. **J. A. George and J. W. Liu:** *"Householder reflections versus Givens rotations in sparse orthogonal decomposition".* Lin. Alg. Appl., *88(1987),* 223-238.

117. **J. A. George, J. W. Liu and E. Ng:** *"User guide for SPARSPACK: Waterloo sparse linear equations package".* Report No. CS-78-30. Department of Computer Science, University of Waterloo, Waterloo, Ontario, Canada, *1980.*

118. **J. A. George, J. W. Liu and E. Ng:** *"Row ordering schemes for sparse Givens rotations: I. Bipartite graph model".* Lin. Alg. Appl., *61(1984),* 55-81.

119. **J. A. George, J. W. Liu and E. Ng:** *"Row ordering schemes for sparse Givens rotations: II. Implicit graph model".* Lin. Alg. Appl., *75(1986),* 203-223.

120. **J. A. George, J. W. Liu and E. Ng:** *"Row ordering schemes for sparse Givens rotations: III. Analyses for a model problem".* Lin. Alg. Appl., *75(1986),* 225-240.

121. **J. A. George, J. W. Liu and E. Ng:** *"A data structure for sparse QR and LU factors".* Report No. CS 85-16. Department of Computer Science, University of Waterloo. Waterloo, Ontario, Canada, *1985.*

122. **J. A. George and E. Ng:** *"An implementation of Gaussian elimination with partial pivoting for sparse systems".* SIAM J. Sci. Statist. Comput., *6(1985),* 390-405.

123. **J. A. George and E. Ng:** *"Symbolic factorization for sparse Gaussian elimintion with partial pivoting".* SIAM J. Sci. Statist. Comput., *8(1987),* 877-898.

124. **J. R. Gilbert:** *"An efficient parallel sparse partial pivoting algorithm".* Report No. 88/45052-1. Chr. Michelsen Institute, Department of Science and Technology, Centre for Computer Science, Fantoftvegen 38, N-5036 Fantoft, Bergen, Norway, *1988.*

125. **J. R. Gilbert and T. Peierls:** *"Sparse partial pivoting in time proportional to arithmetic operations".* SIAM J. Sci. Statist. Comput., *9(1988),* 862-874.

126. **Ph. E. Gill, W. Murray, M. A. Saunders and M. H. Wright:** *"Maintainning LU factors of a general sparse matrix".* Lin. Alg. Appl., *88(19878),* 239-270.

127. J. W. **Givens:** *"Numerical computation of the characteristic values of a real matrix". Report No. ORNL-1574. Oak Ridge National Laboratory, Oak Ridge, Tennessee,* **1954**.

128. J. W. **Givens:** *"Computation of plane unitary rotations transforming a general matrix to a triangular form". J. Soc. Ind. Appl. Math.,* **6(1958)**, *26-50*.

129. G. H. **Golub:** *"Numerical methods for solving linear least squares problems". Numer. Math.,* **7(1965)**, *206-216*.

130. G. H. **Golub and W. Kahan:** *"Calculating the singular values and pseudo inverse of a matrix". J. SIAM, Ser. B: Numer. Anal.,* **2(1965)**, *205-224*.

131. G. H. **Golub, P. Manneback and Ph. Toint:** *"A composition between some direct and iterative methods for large geodetic least squares problems". SIAM J. Sci. Statist. Comput.,* **7(1986)**, *799-816*.

132. G. H. **Golub and C. Reinsch:** *"Singular value decomposition and least squares solutions". Numer. Math.,* **14(1970)**, *402-420*.

133. G. H. **Golub and C. F. Van Loan:** *"Matrix computations". The John Hopkins University Press, Baltimore, Maryland, USA,* **1983**.

134. G. H. **Golub and J. H. Wilkinson:** *"Note on the iterative refinement of linear least squares solution". Numer. Math.,* **9(1966)**, *139-148*.

135. R. G. **Grimes and J. G. Lewis:** *"Condition number estimation for sparse matrices". SIAM J. Sci. Statist. Comput.,* **2(1981)**, *384-388*.

136. F. G. **Gustavson:** *"Some basic techniques for solving sparse systems of linear equations". In:* **"SPARSE MATRICES AND THEIR APPLICATIONS"** *(D. J. Rose and R. A. Willoughby, eds.), 41-52. Plenum Press, New York,* **1972**.

137. F. G. **Gustavson:** *"Two fast algorithms for sparse matrices: multiplication and permuted transposition". ACM Trans. Math. Software,* **4(1978)**, *250-269*.

138. G. D. **Hachtel:** *"The sparse tableau approach to finite element assembly". In:* **"SPARSE MATRIX COMPUTATION"** *(J. R. Bunch and D. J. Rose, eds.), pp. 349-363. Academic Press, New York,* **1976**.

139. A. L. **Hageman and D. M. Young:** *"Applied iterative methods". Academic Press, New York,* **1981**.

140. E. **Hairer, S. P. Nørsett and G. Wanner:** *"Solving ordinary differential equations: I. Nonstiff problems". Springer, Berlin,* **1987**.

141. S. **Hammarling:** *"A note on modifications to the Givens plane rotations". J. Inst. Math. Applics.,* **13(1974)**, *215-218*.

142. **Harwell Subroutine Library Bulletin:** *"Changes to the MA28 suite". (I. S. Duff, ed.), p. 2. A.E.R.E., Harwell, England,* **1983**.

143. **M. T. Heath**: "Numerical methods for large sparse linear least squares problems". *SIAM J. Sci. Statist. Comput.*, *5(1984)*, 497-513.

144. **E. Hellerman and D. C. Rarick**: "Reinversion with the preassigned pivot procedure". *Programming, 1(1971)*, 195-216.

145. **E. Hellerman and D. C. Rarick**: "The partitioned preasigned pivot procedure (P⁴)". In: *"SPARSE MATRICES AND THEIR APPLICATIONS"* (D. J. Rose and R. A. Willoughby, eds.), pp. 67-76. Plenum Press, New York, *1972*.

146. **M. R. Hestenes and E. Stiefel**: "Methods of conjugate gradients for solving linear systems". *J. Res. Nat. Bur. Stand.*, *49(1952)*, 409-436.

147. **N. J. Higham**: "A survey of condition number estimation for triangular matrices". *SIAM Rev.*, *29(1987)*, 575-596.

148. **N. J. Higham**: "Fortran codes for estimating the one-norm of a real or complex matrix with application to condition estimation. *Numerical Analysis Report No. 135. Department of Mathematics, University of Manchester, Manchester M13 9PL, ENGLAND, 1987.*

149. **A. C. Hindmarsh**: "LSODE and LSODI, two new solvers of initial value ordinary differential equations". *ACM SIGNUM Newsletter, 15(1980)*, 10-11.

150. **A. S. Householder**: "Unitary triangularization of a nonsymmetric matrix". *J. Assoc. Comput. Mach.*, *5(1958)*, 339-342.

151. **Ø. Hov, Z. Zlatev, R. Berkowicz, A. Eliassen and L. P. Prahm**: "Comparison of numerical techniques for use in air pollution models with nonlinear chemical reactions". *Atmospheric Environment, 23(1989)*, 967-983.

152. **A. S. Hunding**: "Dissipative structures in reaction-diffusion systems: Numerical determination of bufurcations in the sphere". *J. Chem. Phys.*, *72(1980)*, 5241-5248.

153. **H. D. Ikramov**: "Sparse linear least squares problems". In: *"ADVANCES IN SCIENCES AND TECHNOLOGY: MATHEMATICAL ANALYSIS"* (R. V. Gamkrelidze, ed.), Vol. 23, pp. 219-285. Academy of Sciences of USSR, Moscow, *1985* (in Russian).

154. **M. Jankowski and H. Wozniakowski**: "Iterative refinement implies numerical stability". *BIT, 17(1977)*, 303-311.

155. **A. Jennings**: "Matrix computations for engineers and scientists". Wiley, New York, *1977*.

156. **A. Jennings and G. M. Malik**: "Partial elimination". *J. Inst. Math. Aplics.*, *20(1977)*, 307-316.

157. **W. D. Joubert and D. M. Young:** *"Necessary and sufficient conditions for the simplification of generalized conjugate-gradient algorithms".* Lin. Alg. Appl., **88(1987)**, 449-485.

158. **E. F. Kaasshieter:** *"The solution of nonsymmetric systems by biconjugate gradients and conjugate gradients squared".* Report No. 86/21. Department of Mathematics and Informatics, Delft University of Technology, Delft, Netherlands, **1986**.

159. **D. Kincaid, T. Oppe, J. Respess and D. Young:** *"ITPACKV 2C: User's guide".* Report No. CNA-191. Center for Numerical Analysis, University of Texas at Austin, Austin, Texas, **1984**.

160. **D. Kincaid, J. Respess, D. Young and R. Grimes:** *"Algorithm 586, ITPACK 2C: A FORTRAN package for solving large sparse systems by adaptive accelerated iterative methods".* ACM Trans. Math. Software, **8(1982)**, 302-322.

161. **D. J. Kuck, E. S. Davidson, D. H. Lawrie and A. H. Sameh:** *"Parallel surepcomputing today and the Cedar approach".* Science, **231(1986)**, 967-974.

162. **J. D. Lambert:** *"Computational methods in ordinary differential equations".* Wiley, London, **1973**.

163. **L. Lapidus and J. H. Seinfeld:** *"Numerical solution of ordinary differential equations".* Academic Press, New York, **1971**.

164. **C. L. Lawson and R. J. Hanson:** *"Solving least squares problems".* Prentice-Hall, Englewood Cliffs, N. J., **1974**.

165. **C. L. Lawson, R. J. Hanson, O. R. Kincaid and F. T. Krogh:** *"Basic linear algebra subprograms for Fortran usage".* ACM Trans. Math. Software, **5(1979)**, 308-323.

166. **J. W. H. Liu:** *"On general row merging schemes for sparse Givens rotations".* SIAM J. Sci. Statist. Comput., **7(1986)**, 1190-1211.

167. **P. Manneback:** *"On some numerical methods for solving sparse linear least squares problems".* Ph D Thesis. Department of Mathematics, University of Namur. Namur, Belgium, **1985**.

168. **P. E. Manneback, Ch. Murigande and Ph. L. Toint:** *"A modification of an algorithm by Golub and Plemmons for large least squares problems in the context of Dopler positioning".* IMA J. Numer. Anal., **5(1985)**, 221-233.

169. **H. M. Markowitz:** *"The elimination form of the inverse and its applications to linear programming".* Management Sci., **3(1957)**, 255-267.

170. **J. A. Meijerink and H. A. van der Vorst:** *"An iterative solution method for linear systems of which the coefficient matrix is a symmetric M-matrix".* Math. Comp., **31(1977)**, 148-162.

171. **S. G. Michlin:** "On the rounding error in the conjugate gradients method". Vestnik of the Leningrad State University, *19(1984)*, 15-23 (in Russian).

172. **C. B. Moler:** "Iterative refinement in floating point". J. Assoc. Comput. Mach., *14(1967)*, 316-321.

173. **C. B. Moler:** "Three research problems in numerical linear algebra". In: *"NUMERICAL ANALYSIS"* (G. H. Golub and J. Oliger, eds.), pp. 1-18. American Mathematical Society, Providence, Rhode Island, *1978*.

174. **E. B. Moore:** "On the reciprocal of the general algebraic matrix". Bull. Amer. Math. Soc., *26(1920)*, 394-395.

175. **N. Munksgaard:** "Fortran subroutines for direct solution of sets of sparse and symmetric linear equations". Report No. NI-77-05. Institute for Numerical Analysis, Technical University of Denmark. Lyngby, Denmark, *1977*.

176. **N. Munksgaard:** "Solving sparse symmetric sets of linear equations by preconditioned conjugate gradients". ACM Trans. Math. Software, *6(1980)*, 206-219.

177. **Ch. Murigande, P. Paquet and Ph. L. Toint:** "Solution of large scale least squares in geodetic surveying computations". Report No. 85-11. Department de Mathematique, Facultes Universitaires de Namur, Namur, Belgium, *1985*.

178. **NAG LIBRARY FOTRAN MANUAL:** "Mark 7, Vol. 3-4". Numerical Algorithms Group, Banbury Road 7, Oxford, England, *1979*.

179. **H. B. Nielsen:** "Iterative refinement". Report No. NI-76-02. Institute for Numerical Analysis, Technical University of Denmark, Lyngby, Denmark, *1976*.

180. **D. P. O'Leary:** "Estimating matrix condition numbers". SIAM J. Sci. Statist. Comput., *1(1980)*, 205-209.

181. **D. P. O'Leary:** "The block conjugate gradient algorithm and related methods". Lin. Alg. Appl., *29(1980)*, 293-322.

182. **D. P. O'Leary and J. A. Simmons:** "A bidiagonalization-regularization procedure for large scale discretizations of ill-posed problems". SIAM J. Sci. Statist. Comput., *2(1981)*, 474-489.

183. **T. C. Oppe, W. D. Joubert and D. R. Kincaid:** "NSPCG User's guide, Version 1.0: A package for solving large sparse linear systems by various iterative methods". Report CNA-216. Center for Numerical Analysis, University of Texas at Austin, Austin, Texas, *1988*.

184. **E. E. Osborne:** "On least squares solutions of linear equations". J. Assoc. Comput. Mach., *8(1961)*, 628-636.

185. *C. C. Paige and M. A. Saunders:* "LSQR: an algorithm for sparse linear equations and sparse leasts squares". ACM Trans. Math. Software, *8(1982)*, 43-71.

186. *C. C. Paige and M. A. Saunders:* "Algorithm 583, LSQR: Sparse linear equations and leasts squares". ACM Trans. Math. Software, *8(1982)*, 195-209.

187. *B. N. Parlett:* "The symmetric eigenvalue problem". Prentice-Hall, Englewood Cliffs, N. J., *1980.*

188. *B. N. Parlett:* "The software scene in the extraction of eigenvalues from sparse matrices". SIAM J. Sci. Statist. Comput., *5(1984)*, 590-604.

189. *B. N. Parlett, D. R. Taylor and Z. A. Liu:* "A look-ahead Lanczos algorithm for unsymmetric matrices". Math. Comp., *44(1985)*, 105-124.

190. *G. V. Paulini and G.Radicati di Brozolo:* "Data structures to vectorize CG algorithms for general sparse paterns." BIT, *29(1989)*, 703-718.

191. *R. Penrose:* "A generalized inverse for matrices". Proc. Cambridge Phil. Soc., *51(1955)*, 406-413.

192. *F. J. Peters:* "Parallel pivoting algorithms for sparse symmetric matrices". Parallel Computing, *1(1984)*, 99-110.

193. *G. Peters and J. H. Wilkinson:* "The least squares problem and pseudo-inverses". Computer J., *13(1970)*, 309-316.

194. *S. Pissanetzky:* "Sparse matrix technology". Academic Press, New York, *1984.*

195. *R. J. Plemmons:* "Linear least squares by elimination and MGS". J. Assoc. Comput. Mach., *21(1974)*, 581-585.

196. *J. K. Reid:* "On the method of conjugate gradients for the solution of large systems of linear equations". In: *"LARGE SPARSE SETS OF LINEAR EQUATIONS"* (J. K. Reid, ed.), pp. 231-254. Academic Press, New York, *1971.*

197. *J. K. Reid:* "A note on the stability of Gaussian elimination". J. Inst. Math. Applics., *8(1971)*, 374-375.

198. *J. K. Reid:* "Fortran subroutines for handling sparse linear programming bases". Report No. R8269. A.E.R.E., Harwell, England, *1976.*

199. *J. K. Reid:* "Solution of linear systems of equations: direct methods general)". In: *"SPARSE MATRIX TECHNIQUES"* (V. A. Barker, ed.), pp. 102-109. Lecture Notes In Mathematics, No. 572. Springer, Berlin, *1977.*

200. *J. K. Reid:* "Sparse matrices".In: *"THE STATE OF THE ART IN NUMERICAL ANALYSIS"* (D. A. H. Jacobs, ed.), pp. 85-146. Academic Press, London, *1977.*

201. J. K. Reid: "TRESOLVE, a package for solving large sets of linear finite-element equations". In: **"PDE SOFTWARE: MODULES, INTERFACES AND SYSTEMS"** (B. Engquist and T. Smedsaas, eds.), pp. 1-17. North-Holland, Amsterdam, **1984**.

202. J. K. Reid: "Sparse matrices". Report No. CSS-201. Computer Science and Systems Division, Harwell Laboratory. Harwell, Oxfordshire, OX11 0RA, England, **1986**.

203. J. R. Rice: "Experiments on Gram-Schmidt orthogonalization". Math. Comp., **20(1966)**, 325-328.

204. J. R. Rice and R. F. Boisvert: "Solving elliptic problems using ELLPACK". Springer, New York, **1984**.

205. Å. Ruhe: "SOR methods for the eigenvalue problem with large sparse matrices". Math. Comp., **28(1974)**, 695-710.

206. Å. Ruhe: "Implementation aspects of band Lanczos algorithm for computation of eigenvalues of large sparse symmetric matrices". Math. Comp., **33(1979)**, 680-687.

207. Å. Ruhe and T. Wiberg: "The method of conjugate gradients used in inverse iteration". BIT, **12(1972)**, 543-554.

208. Y. Saad: "Variations of Arnoldi's method for computing eigenelements of large unsymmetric matrices". Lin. Alg. Appl., **34(1980)**, 269-295.

209. Y. Saad: "Krylov subspace methods for solving large unsymmetric linear systems". Math. Comp., **37(1981)**, 105-126.

210. Y. Saad and M. H. Schultz: "CMRES: A generalized minimal residual algorithm for solving nonsymmetric linear systems". SIAM J. Sci. Statist. Comput., **7(1986)**, 856-869.

211. K. Schaumburg and J. Wasniewski: "Use of a semiexplicit Runge-Kutta integration algorithm in a spectroscopic problem". Comput. Chem., **2(1978)**, 19-24.

212. K. Schaumburg, J. Wasniewski and Z. Zlatev: "Solution of ordinary differential equations with time-dependent coefficients. Development of a semiexplicit Runge-Kutta algorithm and application to a spectroscopic problem". Comput. Chem., **3(1979)**, 57-63.

213. K. Schaumburg, J. Wasniewski and Z. Zlatev: "The use of sparse matrix techniques in the numerical integration of stiff systems of linear ordinary differential equations". Comput. Chem., **4(1980)**, 1-12.

214. K. Schaumburg, J. Wasniewski and Z. Zlatev: "Classification of the systems of ordinary differential equations and practical aspects in the numerical integration of large systems". Comput. Chem., **4(1980)**, 13-18.

215. D. S. Scott: "Solving sparse symmetric generalized eigenvalue problems without factorization". SIAM J. Numer. Anal., **18(1981)**, 102-110.

216. **D. S. Scott and R. C. Ward:** *"Solving quadratic λ-matrix problems without factorization". SIAM J. Sci. Statist. Comput., 3(1982), 58-67.*

217. **L. F. Shampine and M. K. Gordon:** *"Computer solution of ordinary differential equations: The initial value problem", Freeman, San Fransisco, California, 1975.*

218. **A. H. Sherman:** *"On efficient solution of sparse systems of linear and non-linear equations". Report No. 46 (Ph D Thesis). Department of Computer Science, Yale University, New Haven, Connectucutt, 1975.*

219. **A. H. Sherman:** *"Algorithms for sparse Gaussian elimination with partial pivoting". ACM Trans. Math. Software, 4(1978), 330-338.*

220. **G. M. Shroff and C. H. Bischoff:** *"Adaptive condition estimation for rank-one updates of QR factorization". Preprint No. MCS-P166-0790. Mathematics and Computer Division, Argonne National Laboratory, Argonne, Illinois, 1990.*

221. **R. D. Skeel:** *"Scaling for numerical stability in Gaussian elimination". J. Assoc. Comput. Mach., 26(1979), 494-526.*

222. **R. D. Skeel:** *"Iterative refinement implies numerical stability for Gaussian elimination". Math. Comp., 35(1980), 817-832.*

223. **R. D. Skeel:** *"Effect of equilibration on residual size for partial pivoting". SIAM J. Numer. Anal., 18(1981), 449-454.*

224. **A. van der Sluis:** *"Condition numbers and equilibration matrices". Numer. Math., 14(1969), 14-23.*

225. **A. van der Sluis:** *"Condition, equilibration and pivoting in linear algebraic systems". Numer. Math., 15(1970), 74-86.*

226. **A. van der Sluis:** *"Stability of the solutions of linear least squares problems". Numer. Math., 23(1975), 241-254.*

227. **A. van der Sluis and H. A. van der Vorst:** *"The rate of convergence of conjugate gradients". Numer. Math., 48(1986), 543-560.*

228. **A. van der Sluis and H. A. van der Vorst:** *"The convergence behaviour of Ritz values in the presence of close eigenvalues". Report No 86-08. Department of Mathematics and Informatics, Delft University of Technology, Julianalaan 132, 2628 BL Delft, Netherlands, 1986.*

229. **P. Sonneveld:** *"CGS, a fast Lanzos-type solver for nonsymmetric linear systems". SIAM J. Sci. Statist. Comput., 10(1989), 36-52.*

230. **H. J. Stetter:** *"Analysis of discretization methods for ordinary differential equations". Springer, Berlin, 1973.*

231. **G. W. Stewart:** *"Introduction to matrix computations". Academic Press, New York, 1973.*

232. **G. W. Stewart:** *"The economical storage of plane rotations".* Numer. Math., *25(1976),* 137-138.

233. **G. W. Stewart:** *"On the perturbation of pseudo-inverses, projections and linear least squares problems".* SIAM Review, *19(1977),* 634-662.

234. **H. S. Stone:** *"An efficient parallel algorithm for the solution of a tridiagonal system of equations".* J. Assoc. Comput. Mach., *20(1973),* 27-38.

235. **H. S. Stone:** *"Parallel tridiagonal equations solvers".* ACM Trans. Math. Software, *1(1975),* 289-307.

236. **G. Strang:** *"Linear algebra and its applications".* Academic Press, New York, *1976.*

237. **R. P. Tewarson:** *"On the product form of inverses of sparse matrices"* SIAM Review, *8(1966),* 336-342.

238. **R. P. Tewarson:** *"The product form of inverses of sparse matrices and graph theory".* SIAM Review, *9(1967),* 91-99.

239. **R. P. Tewarson:** *"Row-column permutation of sparse matrices".* Computer J., *10(1967),* 300-305.

240. **R. P. Tewarson:** *"Solution of a system of simultaneous linear equations with a sparse coefficient matrices by elimination methods".* BIT, *7(1967),* 226-239.

241. **R. P. Tewarson:** *"On the orthogonalization of sparse matrices".* Computing, *3(1968),* 268-279.

242. **R. P. Tewarson:** *"The Crout reduction for sparse matrices".* Computer J., *12(1969),* 158-159.

243. **R. P. Tewarson:** *"The elimination and the orthogonalization methods for the inversion of sparse matrices".* In: **"ADVANCING FRONTIERS IN OPERATIONAL RESEARCH"**, pp. 315-334. Hindustan Publishing Corporation, Delhi, India, *1969.*

244. **R. P. Tewarson:** *"Sparse matrices".* Academic Press, New York, *1973.*

245. **R. P. Tewarson:** *"On the solution of sparse non-linear equations and some applications".* In: **"SPARSITY AND ITS APPLICATIONS"** (D. J. Evans, ed.), pp. 137-152. Cambridge University Press, Cambridge-London, *1985.*

246. **W. F. Tinney and W. S. Meyer:** *"Solution of large sparse systems by ordered triangular factorization".* IEEE Trans. Automatic Control, *18(1973),* 333-346.

247. **W. F. Tinney and J. W. Walker:** *"Direct solution of sparse network equations by optimally ordered triangular factorization".* Proc. IEEE, *55(1967),* 1801-1809.

248. **L. B. Tosovic:** *"Some experiments on sparse sets of linear equations".* *SIAM J. Appl. Math.*, **25(1973)**, *142-148.*

249. **Ll. N. Trefethen:** *"Three mysteries of Gaussian elimination".* *ACM SIGNUM Newsletter,* **Vol. 20, No. 4** *(1983), 2-5.*

250. **Ll. N. Trefethen and R. S. Schreiber:** *"Avarage-case stability of Gaussian elimination".* *SIAM J. Matr. Anal. Appl.*, **11(1990)**, *335-360.*

251. **A. D. Tuff and A. Jennings:** *"An iterative method for large systems of linear structural equations".* *Internat. J. Numer. Methods Engng.*, **7(1973)**, *175-183.*

252. **R. Underwood:** *"An iterative block Lanczos method for the solution of large symmetric eigenproblems".* *Report No. Stan-CS-75-496. Department of Computer Science, Stanford University, Stanford, California, 1975.*

253. **C. F. Van Loan:** *"Generalized singular values with algorithms and applications".* *Ph D Thesis. University of Michigan, Ann Arbor, Michigan,1973.*

254. **C. F. Van Loan:** *"Generalizing the singular value decomposition".* *SIAM J. Numer. Anal.*, **13(1976)**, *76-83.*

255. **J. M. Varah:** *"On the solution of block-triangular systems arising from certain finite elements equations".* *Math. Comp.*, **26(1972)**, *859-862.*

256. **J. M. Varah:** *"On the numerical solution of ill-conditioned linear systems with applications to ill-posed problems".* *SIAM J. Numer. Anal.*, **10(1973)**, *257-267.*

257. **R. S. Varga:** *"Matrix iterative analysis".* *Prentice-Hall, Englewood Cliffs, N. J.*, **1970.**

258. **P. K. W. Vinsome:** *"Orthomin, an iterative method for solving sparse sets of simultaneous linear equations".* *In:* **"PROCEEDINGS OF THE FOURTH SYMPOSIUM ON RESERVOIR SIMULATION"**, *pp. 140-159. Society of Petroleum Engineers of AIME,* **1976.**

259. **V. V. Voevodin:** *"Computational bases of the linear algebra".* *Nauka, Moscow,* **1977** *(in Russian).*

260. **H. van der Vorst:** *"Experience with parallel vector computers for sparse linear systems".* *Supercomputer,* **37(1990)**, *28-35.*

262. **R. C. Ward and L. J. Gray:** *"Eigensystem computation for skew-symmetric and a class of symmetric matrices".* *ACM Trans. Math. Software,* **4(1978)**, *278-285.*

261. **E. L. Washpress:** *"Iterative solution of elliptic systems".* *PrenticeHall, Englewood Cliffs, N. J.*, **1966.**

263. J. Wasniewski, K. Schaumburg and Z. Zlatev: "Vectorizing codes for solving systems of linear ODE's". Supercomputer, 18(1987), 40-51.

264. J. Wasniewski, Z. Zlatev and K. Schaumburg: "A method for reduction of the storage requirements by the use of special computer facilities; application to linear systems of algebraic equations". Comput. Chem., 4(1982), 181-192.

265. P. A. Wedin: "Perturbation bounds in connection with singular value decomposition". BIT, 12(1972), 99-111.

266. P. A. Wedin: "Perturbation theory for pseudo-inverses". BIT, 13(1973), 217-232.

267. P. A. Wedin: "On the almost rank-deficient case of the least squares problem". BIT, 13(1973), 344-354.

268. P. Wesseling: "Theoretical and practical aspects of a multi-grid method". SIAM J. Sci. Statist. Comput., 3(1982), 387-407.

269. O. Widlund: "A Lanczos method for a class of nonsymmetric systems of linear equations". SIAM J. Numer. Anal., 15(1978), 801-812.

270. J. H. Wilkinson: "Error analysis of direct methods of matrix inversion". J. Assoc. Comput. Mach., 8(1961), 281-330.

271. J. H. Wilkinson: "Rounding errors in algebraic processes". Notes in Applied Sciences, No. 32, HMSO, London, 1963.

272. J. H. Wilkinson: "The algebraic eigenvalue problem". Oxford University Press, Oxford-London, 1965.

273. J. H. Wilkinson: "Modern error analysis". SIAM Review, 13(1971), 548-568.

274. J. H. Wilkinson: "Note on matrices with a very ill-conditioned eigen-problem". Numer. Math., 19(1972), 176-178.

275. J. H. Wilkinson: "Some recent advances in numerical linear algebra". In: "THE STATE OF THE ART IN NUMERICAL ANALYSIS" (D. A. H. Jacobs, ed.), pp. 3-51. Academic Press, London, 1977.

276. J. H. Wilkinson: "Singular value decomposition-basic aspects". In: "NUMERICAL SOFTWARE-NEEDS AND AVAILABILITY" D. A. H. Jacobs, ed.), pp. 109-135. Academic Press, London, 1978.

277. J. H. Wilkinson and C. Reinsch: "Handbook for automatic computation: Vol. 2 Linear Algebra". Springer, Berlin, 1971.

278. P. Wolfe: "Error in the solution of linear programmig problems". In: "ERROR IN DIGITAL COMPUTATION" (L. B. Rall, ed.), Vol. 2, pp. 271-284. Wiley, New York, 1965.

279. **G. C. Wright and G. A. Miles:** "An economical method for determining the smallest eigenvalue of large linear systems". Internat. J. Numer. Methods Engng., **3(1971)**, 25-33.

280. **G-C. Yang:** "Paraspice: A parallel direct circuit simulator for shared-memory multiprocessors". Ph D Thesis. Department of Computer Science, University of Illinois at Urbana-Champaign, Urbana, Illinois, USA, **1990**.

281. **D. M. Young:** "Iterative methods for solving partial differential equations of elliptic type". Trans. Amer. Math. Soc., **75(1954)**, 92-111.

282. **D. M. Young:** "Convergence properties of the symmetric and unsymmetric over-relaxation methods". Math. Comp., **24(1970)**, 793-807.

283. **D. M. Young:** "Iterative solution of large linear systems". Academic Press, New York, **1971**.

284. **D. M. Young:** "Generalization of Property A and consistent ordering". SIAM J. Numer. Anal., **9(1972)**, 454-463.

285. **D. M. Young and K. C. Jea:** "Generalized conjugate gradient acceleration of non-symmetric iterative methods". Lin. Alg. Appl., **34(1980)**, 159-194.

286. **D. M. Young and D. R. Kincaid:** "The ITPACK package for large sparse linear systems". In: **"ELLIPTIC PROBLEM SOLVERS"** (M. Schultz, ed.), pp. 163-185. Academic Press, New York, **1981**.

287. **D. M. Young and D. R. Kincaid:** "The ITPACK software package". In: **"PDE SOFTWARE: MODULES, INTERFACES AND SYSTEMS"** (B. Engquist and T. Smedsaas, eds.), pp. 193-206. North Holland, Amsterdam, **1984**.

288. **G. Zielke:** "Inversion of modified symmetric matrices". J. Assoc. Comput. Mach., **15(1968)**, 402-408.

289. **G. Zielke:** "Report on test matrices for generalized inverses". Computing, **36(1986)**, 105-162.

290. **Z. Zlatev:** "Stability properties of variable stepsize variable formula methods". Numer. Math., **31(1978)**, 175-182.

291. **Z. Zlatev:** "On some pivotal strategies in Gaussian elimination by sparse technique". SIAM J. Numer. Anal., **17(1980)**, 18-30.

292. **Z. Zlatev:** "On solving some large linear problems by direct methods". Report No. DAIMI PB-111, Department of Computer Science, Aarhus University. Aarhus, Denmark, **1980**.

293. **Z. Zlatev:** "Zero-stability properties of the three-ordinate variable stepsize variable formula methods". Numer. Math., **37(1981)**, 157-166.

294. **Z. Zlatev:** *"Modified diagonally implicit Runge-Kutta methods". SIAM J. Sci. Statist. Comput., 2(1981), 321-334.*

295. **Z. Zlatev:** *"Use of iterative refinement in the solution of sparse linear systems". SIAM J. Numer. Anal., 19(1982), 381-399.*

296. **Z. Zlatev:** *"Comparison of two pivotal strategies in sparse plane rotations". Comput. Math. Appl., 8(1982), 119-135.*

297. **Z. Zlatev:** *"Consistency and convergence of general linear multistep variable stepsize variable formula methods". Computing, 31(1983), 47-67.*

298. **Z. Zlatev:** *"Application of predictor-corrector schemes in solving air pollution problems". BIT, 24(1984), 700-714.*

299. **Z. Zlatev:** *"General scheme for solving linear algebraic problems by direct methods". Applied Numerical Mathematics, 1(1985), 176-186.*

300. **Z. Zlatev:** *"Variable stepsize variable formula methods based on predictor-corrector schemes". Applied Numerical Mathematics, 1(1985), 395-416.*

301. **Z. Zlatev:** *"Sparse matrix technique for general matrices: pivotal strategies, decompositions and applications in ODE software". In: "SPARSITY AND ITS APPLICATIONS" (D. J. Evans, ed.), pp. 185-228. Cambridge University Press, Cambridge-London, 1985.*

302. **Z. Zlatev:** *"Mathematical model for studying the sulphur pollution over Europe". J. Comput. Appl. Math., 12(1985), 651-666.*

303. **Z. Zlatev:** *"Solving large systems of linear algebraic equations by the use of package Y12M". In: "STRUCTURAL ANALYSIS SYSTEMS" (A. Niku-Lari, ed.), Vol. 2, pp. 152-160. Pergamon Press, New York, 1985.*

304. **Z. Zlatev:** *"Numerical treatment of some mathematical problems describing long-range transport of air pollution". In: "MATHEMATICAL ANALYSIS AND ITS APPLICATIONS" (S. M. Mazhar, A. Hamoui and N. S. Faour, eds.), pp. 367-383. Pergamon Press, New York, 1986.*

305. **Z. Zlatev:** *"Transition to variable stepsize variable formula methods for solving ordinary differential equations". In: "COMPUTATIONAL MATHE-MATICS II" (S. O. Fatunla, ed.). Boole Press, Dublin, 1987.*

306. **Z. Zlatev:** *"Survey of the advances of exploiting the sparsity in the solution of large problems". J. Comput. Appl. Math., 20(1987), 83-105.*

307. **Z. Zlatev:** *"Treatment of some mathematical models describing long- range transport of air pollutants on vector processors". Parallel Computing, 6(1988), 87-98.*

308. **Z. Zlatev:** "Advances in the theory of variable stepsize variable formula ordinary differential equations". Applied Mathematics and Computation, **31(1989)**, 209-249.

309. **Z. Zlatev:** "Computations with large and band matrices on vector processors". In: **"ADVANCES IN PARALLEL COMPUTING"** (D. J. Evans, ed.), Vol. 1, pp. 7-37. JAI Press, London-Cambridge, England, **1990**.

310. **Z. Zlatev:** "Solving band systems of linear equations by iterative methods on vector processors". In: **"NUMERICAL LINEAR ALGEBRA,- DIGITAL SIGNAL PROCESSING AND PARALLEL COMPUTING"** (G. Golub and P. Van Dooren, eds.), pp. 713-725. Springer-Verlag, Berlin, **1990**.

311. **Z. Zlatev and V. A. Barker:** "Logical procedure SSLEST-an Algol W procedure for solving sparse systems of linear equations". Report No. NI-76-13. Institute for Numerical Analysis, Technical University of Denmark, Lyngby, Denmark, **1976**.

312. **Z. Zlatev, V. A. Barker and P. G. Thomsen:** "SSLEST: a FORTRAN IV subroutine for solving sparse systems of linear equations (Users' guide)". Report No. NI-78-01. Institute for Numerical Analysis, Technical University of Denmark, Lyngby, Denmark, **1978**.

313. **Z. Zlatev, R. Berkowicz and L. P. Prahm:** "Stability restrictions on time-step size for numerical integration of first-order partial differential equations". J. Comput. Phys., **51(1983)**, 1-27.

314. **Z. Zlatev, R. Berkowicz and L. P. Prahm:** "Testing subroutines solving advection-diffusion equations in atmospheric environments". Comput. Fluids., **11(1983)**, 12-38.

315. **Z. Zlatev, R. Berkowicz and L. P. Prahm:** "Three-dimensional advection-diffusion modeling for regional scale". Atmos. Environ., **17(1983)**, 491-499.

316. **Z. Zlatev, R. Berkowicz and L. P. Prahm:** "Implementation of a variable stepsize variable formula method in the time-integration part of a code for long-range transport of air pollutants". J. Comput. Phys., **55(1984)**, 279-301.

317. **Z. Zlatev, J. Christensen, J. Moth and J. Wasniewski:** "Vectorizing codes for studying long-range transport of air pollutants". Comp. Math. Appl., **20(1991)**, to appear.

318. **Z. Zlatev and H. B. Nielsen:** "Preservation of sparsity in connection with iterative refinement". Report No. NI-77-12. Institute for Numerical Analysis, Technical University of Denmark, Lyngby, Denmark, **1977**.

319. **Z. Zlatev and H. B. Nielsen:** "SIRSM: a package for the solution of sparse systems by iterative refinement". Report No. NI-77-13. Institute for Numerical Analysis, Technical University of Denmark, Lyngby, Denmark, **1977**.

320. **Z. Zlatev and H. B. Nielsen:** "LLSS01: a FORTRAN subroutine for solving linear least squares problems (Users'guide)". Report No. NI-79-07. Institute for Numerical Analysis, Technical University of Denmark, Lyngby, Denmark, **1979**.

321. **Z. Zlatev and H. B. Nielsen:** "Least squares solution of large linear problems". In: **"SYMPOSIUM I ANVENDT STATISTIK 1980"** (A. Hoeskuldsson, K. Konradsen, B. Sloth Jensen and K. Esbensen, eds.), pp. 17-52. NEUCC, Lyngby, Denmark, **1980**.

322. **Z. Zlatev and H. B. Nielsen:** "Solving large and sparse linear least-squares problems by conjugate gradients algorithms". Comput. Math. Appl., **15(1988)**, 185-202.

323. **Z. Zlatev, K. Schaumburg and J. Wasniewski:** "Implementation of an iterative refinement option in a code for large and sparse systems". Comput. Chem., **4(1980)**, 87-99.

324. **Z. Zlatev and P. G. Thomsen:** "ST: a FORTRAN IV subroutine for the solution of large systems of linear algebraic equations with real coefficients by use of sparse technique". Report No. NI-76-05. Institute for Numerical Analysis, Technical University of Denmark, Lyngby, Denmark, **1976**.

325. **Z. Zlatev and P. G. Thomsen:** "Application of backward differentiation methods to the finite element solution of time-dependent problems". Internat. J. Numer. Methods Engng., **14(1979)**, 1051-1061.

326. **Z. Zlatev and P. G. Thomsen:** "Automatic solution of differential equations based on the use of linear multistep methods". ACM Trans. Math. Software, **5(1979)**, 401-414.

327. **Z. Zlatev and P. G. Thomsen:** "Sparse matrices-efficient decomposition and applications". In: **"SPARSE MATRICES AND THEIR USES"** (I. S. Duff, ed.), pp. 367-375. Academic Press, London, **1981**.

328. **Z. Zlatev, Ph. Vu, J. Wasniewski and K. Schaumburg:** "Computations with symmetric, positive definite and band matrices on vector processors". Parallel Computing, **8(1988)**, 301-312.

329. **Z. Zlatev and J. Wasniewski:** "Package Y12M: solution of large and sparse systems of linear algebraic equations". Preprint Ser., No. 24. Mathematics Insitute, University of Copenhagen. Copenhagen, Denmark, **1978**.

330. **Z. Zlatev, J. Wasniewski and K. Schaumburg:** "A testing scheme for subroutines solving large linear systems". Comput. Chem., **5(1981)**, 91-100.

331. **Z. Zlatev, J. Wasniewski and K. Schaumburg:** "Y12M-solution of large and sparse systems of linear algebraic systems". Lecture Notes in Computer Science, No. 121. Springer, Berlin, **1981**.

332. Z. Zlatev, J. Wasniewski and K. Schaumburg: "Comparison of two al-
     gorithms for solving large linear systems". SIAM J. Sci. Statist.
     Comput., 3(1982), 486-501.

333. Z. Zlatev, J. Wasniewski and K. Schaumburg: "On some useful options in
     a code for solving stiff systems of linear ordinary differential
     equations". Report 1983-9. Regional Computing Center at the Univer-
     sity of Copenhagen, Copenhagen, Denmark, 1983.

334. Z. Zlatev, J. Wasniewski and K. Schaumburg: "Subroutine DENS1 for
     solving stiff systems of linear ordinary differential equations
     (basic algorithms, documentation, demonstration programs)". Report
     No. 83-7. Regional Computing Centre at the University of Copen-
     hagen (RECKU), Copenhagen, Denmark, 1983.

335. Z. Zlatev, J. Wasniewski and K. Schaumburg: "Subroutine SPAR1 for
     solving stiff systems of linear ordinary differential equations
     (basic algorithms, documentation, demonstration programs)". Report
     No. 85-6. Regional Computing Centre at the University of Copen-
     hagen (RECKU), Copenhagen, Denmark, 1985.

336. Z. Zlatev, J. Wasniewski and K. Schaumburg: "Exploiting the sparsity in
     the solution of linear ordinary differential equations". Comp.
     Math. Appl., 11(1985), 1069-1087.

337. Z. Zlatev, J. Wasniewski and K. Schaumburg: "Numerical treatment of
     models arising in nuclear magnetic resonance spectroscopy".
     Advances in Engineering Software, 8(1986), 223-233.

338. Z. Zlatev, J. Wasniewski and K. Schaumburg: "Condition number estimators
     in a sparse matrix software". SIAM J. Sci. Statist. Comput.,
     7(1986), 1175-1186.

339. Z. Zlatev, J. Wasniewski and K. Schaumburg: "Subroutine BAND1 for
     solving stiff systems of linear ordinary differential equations
     (basic algorithms, documentation, demonstration programs)". Tech-
     nical Note. Regional Computing Centre at the University of Copen-
     hagen (RECKU), Copenhagen, Denmark, 1988.

340. Z. Zlatev, J. Wasniewski and K. Schaumburg: "Running conjugate gradient
     algorithms on three vector machines". Supercomputer, 29(1989),
     31-41.

341. O. Østerby and Z. Zlatev: "Direct methods for sparse matrices". Lecture
     Notes in Computer Science, No. 157. Springer, Berlin, 1983.

AUTHOR INDEX

SUBJECT INDEX